Fluidization XVI

Guilin, China
26-31 May 2019

ISBN: 978-1-7138-0732-2

Printed from e-media with permission by:

Curran Associates, Inc.
57 Morehouse Lane
Red Hook, NY 12571

Some format issues inherent in the e-media version may also appear in this print version.

AIChE
120 Wall Street, FL 23
New York, NY 10005-4020

Phone: (800) 242-4363
Fax: (203) 775-5177

www.aiche.org

Additional copies of this publication are available from:

Curran Associates, Inc.
57 Morehouse Lane
Red Hook, NY 12571 USA
Phone: 845-758-0400
Fax: 845-758-2633
Email: curran@proceedings.com
Web: www.proceedings.com

TABLE OF CONTENTS

1A: FLUIDIZATION IN NATURE

1B: FUNDAMENTALS OF FLUIDIZATION

1C: FLUIDIZED BED APPLICATIONS

1D: MODELING AND SIMULATION

2A: FUNDAMENTALS OF FLUIDIZATION

3B: FLUIDIZED BED APPLICATIONS

3C: FLUIDIZED BED APPLICATIONS

3D: MODELING AND SIMULATION

PLENARY TALK & PANEL DISCUSSION

POSTER SESSION

5A: FUNDAMENTALS OF FLUIDIZATION

5B: FLUIDIZED BED APPLICATIONS

5C: MODELING AND SIMULATION

5D: MODELING AND SIMULATION

6A: FUNDAMENTALS OF FLUIDIZATION

6B: FLUIDIZED BED APPLICATIONS

6C: MODELING AND SIMULATION

6D: MODELING AND SIMULATION

7A: FUNDAMENTALS OF FLUIDIZATION

7B: FLUIDIZED BED APPLICATIONS

7C: MODELING AND SIMULATION

7D: MODELING AND SIMULATION

8A: FUNDAMENTALS OF FLUIDIZATION

8B: FLUIDIZED BED APPLICATIONS

8C: MODELING AND SIMULATION

8D: MODELING AND SIMULATION

PLENARY SESSION #3

PLENARY SESSION #4

2B: FUNDAMENTALS OF FLUIDIZATION

2C: FLUIDIZED BED APPLICATIONS

2D: MODELING AND SIMULATION

4A: FUNDAMENTALS OF FLUIDIZATION

4B: FLUIDIZED BED APPLICATIONS

4C: FLUIDIZED BED APPLICATIONS

4D: MODELING AND SIMULATION

Chaotic Vibration Characteristics of a Graphite Tube with an Internal Vapor-Liquid-Solid Flow Boiling

Conference: Fluidization

Year: 2019

Proceeding: Fluidization XVI

Group: General Paper Pool

Session: 1A: Fluidization in Nature

Time: Monday, May 27, 2019 - 2:31pm-2:43pm

Authors:

Xu, X., Tianjin University
Liu, M. Sr., State Key Laboratory of Chemical Engineering (Tianjin University)
An, M., Tianjin University
Ma, Y., Tianjin University

The vibration of a graphite tube in a fluidized bed evaporator is related to its safe operation and running. In this study, the chaotic vibration behavior of a graphite tube with internal vapor-liquid and vapor-liquid-solid flows were investigated by analyzing the vibration acceleration signals as well as pressure drop. The main results are as follows. After the wavelet decompositions of vibration acceleration signal and pressure drop signal and the reconstructions for the vibration acceleration signal, the characteristics of the middle and low frequency motions were displayed. Nonlinear vibration behavior of reconstructed vibration acceleration signal in the multi-phase boiling flow system was discussed at varied solid holdup and heating steam pressure. The values of time delay with minimum points in the correlation integral curves of the reconstructed signals are the same, and are little affected by the solid holdup and the heating steam pressure, which indicates that the influence of the solid holdup and the heating steam pressure can be negligible in the choice of time delay. The chaotic attractor of the vibration acceleration and the pressure drop signals is unstable. The volume of chaotic attractor in the space depends on the intensity of boiling. It increases gradually with the increase of solid holdup and heating steam pressure. Compared to the attractor of the pressure drop signal, the attractor of the vibration acceleration signal changes more significantly. In addition, the influence degree of solid particles on the volume of attractor is larger than that of the heating steam pressure, which demonstrates that the energy generated by the solid is large.

Cycle Times in Sawdust Conical Spouted Beds Equipped with Fountain Confiner and Draft Tube

Conference: Fluidization

Year: 2019

Proceeding: Fluidization XVI

Group: General Paper Pool

Session: 1A: Fluidization in Nature

Time: Monday, May 27, 2019 - 1:55pm-2:07pm

Authors:

Saldarriaga, J. F. Sr., Universidad de los Andes
Estiati, I., University of the Basque Country
Aguado, R. Sr., University of the Basque Country
Olazar, M. Sr., University of the Basque Country

A study has been conducted on the joint influence of the fountain confiner and draft tube on the solid circulation in a conical spouted bed made up of sawdust. Knowledge of the performance of these fine particles of low density and irregular texture is required in both physical chemical and processes, such as drying, pyrolysis, or combustion of these materials in conical spouted beds. The study has been carried out according to a factorial design of experiments. The cycle times follow different distributions depending on the configuration of the contactor-draft tube-fountain confiner. The spouted beds equipped with only fountain confiner are those with the more uniform distribution and lowest cycles times. Therefore, this device provides high stability and ensures high turbulence to the beds formed by fine and irregular particles, with a behavior similar to that of stable beds made up of coarse particles, in which no internal device is required.

Experimental Study on Flow Deposition Characteristics of Inorganic Salt Particles in a Transpiring Wall Reactor for Supercritical Water Oxidation

Conference: Fluidization

Year: 2019

Proceeding: Fluidization XVI

Group: General Paper Pool

Session: 1A: Fluidization in Nature

Time: Monday, May 27, 2019 - 3:19pm-3:31pm

Authors:

Xu, D., Xi'an Jiaotong University
Wang, Y., Xi'an Jiaotong University
Liang, Y., Xi'an Jiaotong University
Wei, N., Xi'an Jiaotong University
Guo, S., Xi'an Jiaotong University
Wang, S., Xi'an Jiaotong University

Supercritical water oxidation has a promising future in the treatment of highly toxic and refractory organic wastewater. However, harsh working environment hinders its commercial application due to reactor corrosion and salt deposition problems. Transpiring wall reactor has good performance in resistance to the two problems above because it can form a protective water film on its inner surface to dilute, dissolve and/or wash inorganic salts and corrosive species. In this work, the influences of operating parameters on the flow and deposition characteristics of type 1 salt, type 2 salt, and their mixture in a transpiring wall reactor were explored systematically under water film interference conditions. The results show that water film could significantly decrease the inorganic salt concentration near the porous wall and the salt deposition rate in the reactor. There is a negative linear relation between transpiration intensity and the concentration of type 1 salt, type 2 salt, or their mixture near the porous wall. The dilution multiple reduced with an increase in the initial salt concentration and in the dilution and solubility of water film. Runtime had a slight effect on salt distribution in the transpiring wall reactor, but water film could always exert protective action in the range of experimental measurement time. This information helps to theoretically guide the optimization of operating parameters of the transpiring wall reactor.

Fine Generation Ratio of Iron Ore at the Cyclone in a Gas-Solid Circulating Fluidized Bed

Conference: Fluidization

Year: 2019

Proceeding: Fluidization XVI

Group: General Paper Pool

Session: 1A: Fluidization in Nature

Time: Monday, May 27, 2019 - 2:19pm-2:31pm

Authors:

Sung, W. C., Sungkyunkwan University
Lee, D. H., Sungkyunkwan University
Kim, J. Y., University of British Columbia
Ko, C. K., POSCO

Fine generation ratio of iron ore at the cyclone was modeled with time. The variables were cyclone inlet velocity (20.6 - 30.9 m / s), solid loading (0.5-1) and elapsed time (0.5 - 13 hour) and experiments were conducted in a circulating fluidized bed (0.076 m ID x 3.7 m height) reactor. Fine generation ratio was divided into sifting effect and attrition ratio. The sifting effect was rapidly decreased with increasing cyclone inlet velocity and solid loading. The attrition ratio was positively correlated with cyclone inlet velocity and negative with solid loading as mentioned in the previous study. Since the sifting effect decreases with time, as the elapsed time increased, the contribution of sifting effect to fine generation was decreased. Thus, when the elapsed time was 0.5 h, contribution of sifting effect was 86-93 % and when elapsed time was 5 h, contribution of sifting effect was 57 %.

Improving Performances of Gas Cyclones for Hot Gas Cleaning

Conference: Fluidization

Year: 2019

Proceeding: Fluidization XVI

Group: General Paper Pool

Session: 1A: Fluidization in Nature

Time: Monday, May 27, 2019 - 2:07pm-2:19pm

Authors:

Sun, G., China University of Petroleum, Beijing

Improving performances of gas cyclones for hot gas cleaning

Guogang Sun *, Yuming Zhang, Jianfei Song

State Key Laboratory of Heavy Oil Processing, China University of Petroleum, Beijing, 102249, P. R. China;

Beijing Key Laboratory of Process Fluid Filtration and Separation, Beijing, 102249, P.R. China

The design of a gas cyclone operating at high temperature and high pressures is still a challenging task. This design is of interest in CFB coal generation, coal gasification, solid fossil fuel pyrolysis, waste incineration and cement industry, but also for petrochemical industries (e.g. FCCU). The development of new/advanced processes and more rigorous environmental protection requirements lead to a continuous interest in improving the performances of gas-solid cyclone separators. This presentation discusses major issues with respect to improving cyclone performances: 1) The prediction of the separation efficiency and pressure drop. There exist various models or procedures for the calculation of cyclones. For practical cyclone design a rough calculation of the separation efficiency and pressure drop is often sufficient. 2) The precise design and optimum design. Accurate operating Data information of the gas and particles is essential for the cyclone design. The maximum efficiency inlet gas velocity and maximum efficiency height should be determined beforehand, which are two constraints for the optimal design of cyclone separators. Some research has been devoted to developing the calculation method to aforesaid two maximum efficiency parameters. The optimized cyclone shows obvious improvement in performances. 3) Structure improvement and innovation. A series of structural measures have been developed to suppress short-circuit flow and re-entrainment of particles in cyclone separators. The performance of the new cyclone has been significant improved.

Key words: Cyclone separator, Hot gas cleaning, Design, Efficiency, Pressure drop

Keynote Talk: An Overview on Should-be-Challenged Issues Relevant to Fluidization in Nature

Conference: Fluidization

Year: 2019

Proceeding: Fluidization XVI

Group: General Paper Pool

Session: 1A: Fluidization in Nature

Time: Monday, May 27, 2019 - 1:30pm-1:55pm

Authors:

Horio, M., Tokyo University of Agriculture and Technology

A variety of fluidization phenomena taking place in nature are briefly reviewed to evaluate sufficiency/insufficiency of our knowledge on fluidization by covering avalanche of earth and snow, volcanic eruptions and plateau formation, potential fluidization on the Moon, hailstorm from cumulonimbus clouds and crowd dynamics. It is suggested that since our present datasets are insufficient to deal with the large-scale fluidization phenomena in nature, new challenges are needed to provide comprehensive fluidization phase diagrams for a wide range of particles and fluidizing conditions.

Radial Distribution of Particle Concentration in Downer

Conference: Fluidization

Year: 2019

Proceeding: Fluidization XVI

Group: General Paper Pool

Session: 1A: Fluidization in Nature

Time: Monday, May 27, 2019 - 2:43pm-2:55pm

Authors:

Zhao, Y., China University of Petroleum
Fan, Y., China University of Petroleum, Beijing

Fluid catalytic cracking(FCC) is one of the most critical oil secondary processing units in the refinery and it undoubtedly plays a vital role in China petroleum processing industry. More than 70% gasoline as well as about 30% diesel oil is produced via the FCC process. In a modern FCC unit, the catalytic cracking reaction mainly occurs in the riser reactor.However, the core-annular structure and gas-solid backmixing property of the riser reduced the conversion rate of heavy oil and product selectivity.Compared with the riser, the downer has the advantages of close to horizontal thrusting flow pattern, uniform gas-solid distribution, short gas-solid contact time, small gas-solid axial backmixing and narrow residence time distribution, etc. It has a wide application prospect in the rapid catalytic cracking process of heavy oil and rapid cracking of coal powder.

In downer reactor, the flow and distribution of oil & catalysts will influence the whole reaction directly. However, the actual condition is far from the ideal uniform contact of gas with solids phase.The particle concentration distribution of downer has obvious non-uniform structure. This non-uniform radial concentration distribution of particle/gas may result in an uneven contact of two phases and lead to undesirable effects on the cracking reaction.

In these years, there have been many theoretical explanations for the distribution of particle concentration in circulating fluidized bed.However, most of the theories are only qualitative descriptions without a quantitative analysis.At present, there is no ideal explanation for the cause of the radial distribution of particle concentration. Most studies focus on describing the phenomenon or quantifying the distribution of particle concentration based on experiment or numerical simulation.The current explanations include boundary layer theory, gas-solid phase drag theory, local pressure drop theory and turbulent diffusion theory.However, the main problem with these explanations is that they only stay at the qualitative description stage and fail to provide convincing quantitative analysis.

In this paper, the aerodynamic theory and field theory is â€œtransplantedâ€ into the two-phase flow. A lateral force, namely the Kutta-Joukowski force, is used to explain and quantify the formation of non-uniform structure in a downer.

Both the theoretical analysis and analyse others data were employed in this paper. The method to calculate local Kutta-Joukowski force is proposed via theoretical analysis. This force is the function of particle velocity curl (The particle velocity curl does not mean the actual rotation, but a function of velocity gradient), slip velocity and gas density.

Combined with that proposed in the published literatures, the distribution of Kutta-Joukowski lateral force in the downer is given. On the one hand, and effects of operating parameters on the Kutta-Joukowski lateral force distribution profile is analysed. The result shows Kutta-Joukowski lateral force is the main reason of the non-uniform structure occurence. The direction of the force is the lower part of the concentration point to the higher part of the concentration; while its valus were mainly determined by the particle physical property, the particle flux, the superficial gas velocity, the radial/axial position.The empirical correlation of the radial distribution of Kutta-Joukowski lateral force in the descending bed was obtained by regression of the experimental data.Finally,On the other hand, the particle concentration gradient force is introduced and quantified. This force balances the Kutta-Joukowski lateral force in the downer. The quantitave correlation of these two forces is established.The empirical correlation formula of concentration gradient force coefficient K is regression.

Key Words: Downer; non-uniform structure; Particle Velocity Curl; Slip Velocity; Kutta-Joukowski Lateral Force

Reactor Design of CanmetENERGY's Pilot-Scale Pressurized Chemical Looping Conversion

Conference: Fluidization

Year: 2019

Proceeding: Fluidization XVI

Group: General Paper Pool

Session: 1A: Fluidization in Nature

Time: Monday, May 27, 2019 - 3:07pm-3:19pm

Authors:

Lu, D., Natural Resources Canada

Canada is taking a leading role in conducting innovative research in carbon capture, utilization, and storage (CCUS) using a cost effective and efficient technique. Chemical looping combustion (CLC) is a newly emerging CCUS technology for fossil fuel conversion with the inherent advantage of economic carbon dioxide separation. In CLC, the oxygen for combustion is provided by a metal oxide-based oxygen carrier, which cycles between an air reactor and a fuel reactor. The oxygen carrier supplies the oxygen for combustion in the fuel reactor and then cycles into the air reactor where it is re-oxidized. The two reactors ensure that the reduction and oxidation reactions remain separate. Therefore, CLC fuel conversion is performed in the absence of air thereby producing a stream of flue gas concentrated in CO_2. CanmetENERGY aims to design and build a 600 kW_{th} pressurized-CLC (PCLC) pilot-plant to combust natural gas using ilmenite ore as the oxygen carrier. The pilot-plant includes feedstock supply systems (natural gas, air, oxygen carrier, loop seal gas, and water), two fluidized bed reactors (fuel reactor and air reactor), effluent processing, heat exchangers, and utility systems. The core of the pilot-plant is a dual fluidized bed system.

The reactors operate as gas-solid fluidized beds. To promote gas-solid mixing, and therefore increase the rates of reaction, these reactor sections should operate in the turbulent fluidized bed regime. CanmetENERGY has considered a configuration using a riser to transfer the oxygen carrier from the air reactor to the fuel reactor. This paper presents an analysis of hydrodynamics and heat transfer related to this configuration.

The air reactor riser must operate in the fast-fluidized regime to ensure the particles are carried over the top of the bed. The fluidization regime of the beds depends on the properties of the gas, the properties of the solid particles, and the inner diameter of the bed (i.e., the superficial gas velocity). The diameter was optimized to maintain the desired fluidization regime under the widest range of thermal input and pressure. Based on the results of the study, it was determined that the inner diameters of the air reactor bottom, air reactor riser, and fuel reactor should be between 203-305 mm, 51-76 mm and 102-127 mm, respectively.

The oxidation reactions are highly exothermic, whereas the reactions in the fuel reactor are net endothermic. Sensible heat from the hot solid particles entering the fuel reactor are used to drive the endothermic reactions. Therefore, it is necessary to minimize heat loss from each reactor in order to maintain the high temperature of the solids; this is particularly important for small pilot-plants such as the one under design here. A separate case study was completed to determine the optimal outer diameter of each reactor pressure vessel section based on the thickness of insulation required to limit heat loss to acceptable levels. From this study, it was determined that outer diameters of 508 mm, 203 mm, and 406 mm for the air reactor bottom, the air reactor riser and the fuel reactor, respectively, are sufficient to provide enough resistance to heat transfer without compromising cost effectiveness. From the results of this study, the reactor heights can be determined based on the residence time required in each reactor section to meet target fuel and oxygen carrier conversions.

Separation Characteristics of Feedstock at Different Regions in the Gas-Solid Dense-Phase Separation Fluidized Bed

Conference: Fluidization

Year: 2019

Proceeding: Fluidization XVI

Group: General Paper Pool

Session: 1A: Fluidization in Nature

Time: Monday, May 27, 2019 - 2:55pm-3:07pm

Authors:

Fu, Y., China University of Mining & Technology
Luo, Z., China University of Mining & Technology
Lv, B. Sr., China University of Mining & Technology
Qin, X., China University of Mining & Technology
Chen, C., China University of Mining & Technology
Xu, X., China University of Mining & Technology

Air dense medium fluidized bed technique has been proposed as an effective technique for dry coal beneficiation. However, different fluidization characteristics and flow conditions of medium solids in different regions are the key factors influencing the separation efficiency of coal. In this study, MATLAB software was employed for investigating the bed spatial structure characteristic under different bed heights and fluidization numbers. The force analysis of particles in radial and axia zones was compared. By dividing the bed regions, the separation effect of the materials in different regions was explored. The results show that the standard deviation and average density of bed density raised with static bed height increased, and the stability of the bed becomes worse. The minimum pressure gradients ratio at axial and radial in fluidized state is 10. A better separation efficiency was achieved under the upper and middle region feeding.

Effect of Temperature on Hydrodynamics of a Gas-Solid Fluidized Bed

Conference: Fluidization

Year: 2019

Proceeding: Fluidization XVI

Group: General Paper Pool

Session: 1B: Fundamentals of Fluidization

Time: Monday, May 27, 2019 - 1:42pm-1:54pm

Authors:

Huang, K., Dalian National Laboratory for Clean Energy and National Engineering Laboratory for MTO, Dalian Institute of Chemical Physics, Chinese Academy of Sciences
Meng, S., DICP
Ye, M., Dalian National Laboratory for Clean Energy and National Engineering Laboratory for MTO, Dalian Institute of Chemical Physics, Chinese Academy of Sciences
Liu, Z., Dalian National Laboratory for Clean Energy and National Engineering Laboratory for MTO, Dalian Institute of Chemical Physics, Chinese Academy of Sciences

Effect of temperature on hydrodynamics of a gas-solid fluidized bed

In this study, white alundum of diameter about 0.180 mm belonging to Geldart B classification powder was used to study the effect of temperature on hydrodynamics of fluidized bed. Fluidization tests were conducted at different temperatures (293-873 K) and different superficial gas velocities, covering fixed bed state, bubbling, and turbulent fluidization regime. Differential pressure transducer measuring the total bed pressure drop and the home-made high-temperature electrical capacitance tomography sensor measuring the cross section, 18 cm height above distribution plate, voidage of fluidized bed were used to characterize the influence of temperature. Experimental results indicated that not only gas properties had effect on performance of fluidized bed, but interparticle force induced by temperature had significant influence on the fluidized bed characteristics, alerting the fluidization behaviour of the bed from Geldart group B behaviour to Geldart group A behaviour. Pressure drop curves at 293 and 873 K were shown in Figure 1, pressure drop overshoot indicated the behaviour change. Additionally, home-made electrical capacitance tomography sensor was also used to measure the variation of solids concentration and bubble diameters induced by the temperature. Figure 2 shows the variation of standard deviation of average solids concentration at 473, 673 and 873 K. It can be seen that the standard deviation of average solids concentration were down then up at the same superficial velocity when the fluidized bed temperature increasing. It can be owing to the interparticle forces and gas properties induced by temperature. Consequently, the particle properties induced by temperature and the gas properties should be seriously considered when attempting to successfully design and operate gas-solids fluidized bed at high temperature. More researches are undergoing.

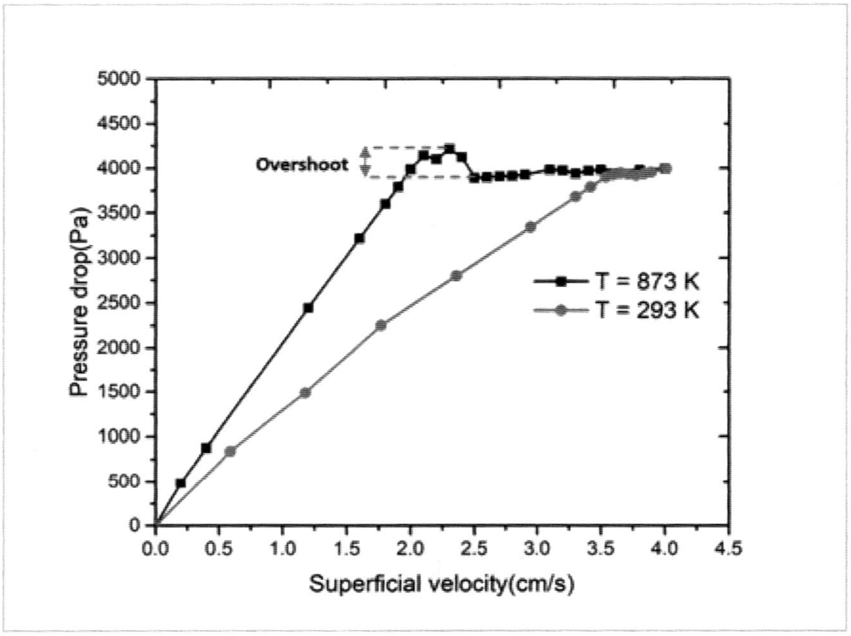

Figure 1 Pressure drop of fluidized bed at 293 and 873 K

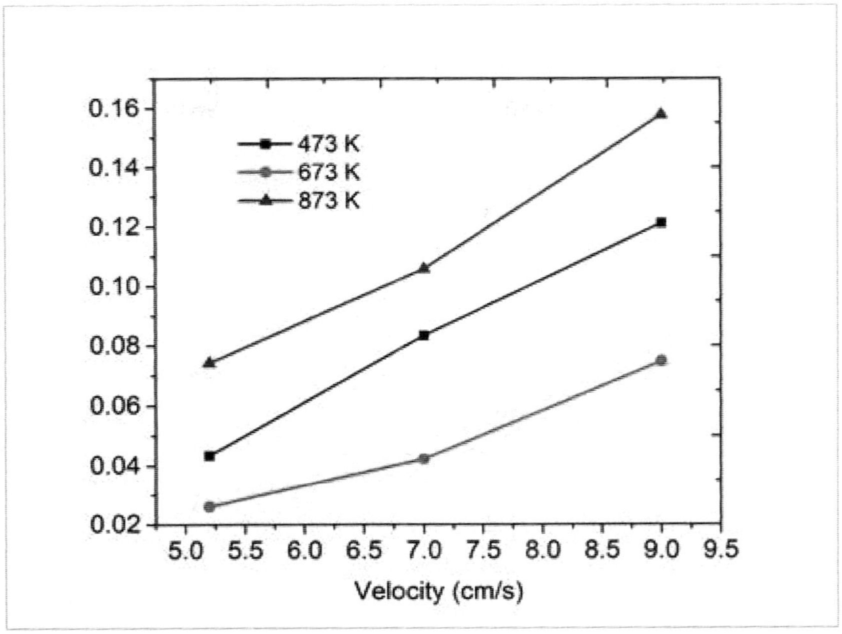

Figure 2 Standard deviation of average solids concentration at different temperature

Fluidization in Small-Scale Gas-Solid 3D-Printed Fluidized Beds

Conference: Fluidization

Year: 2019

Proceeding: Fluidization XVI

Group: General Paper Pool

Session: 1B: Fundamentals of Fluidization

Time: Monday, May 27, 2019 - 1:54pm-2:06pm

Authors:

Zivkovic, V., Newcastle University
Reay, D. A., Newcastle University
McDonough, J., Newcastle University
Law, R., Newcastle University

Additive manufacturing could be used to facilitate the rapid fabrication and testing of small-scale fluidized beds for use in screening applications, such as adsorbent screening for carbon capture. In this work experiments were performed in order to map the different flow regimes produced in small-scale (D_h = 3â€“15 mm) gas-solid fluidized beds that were fabricated using additive manufacturing using the stereolithography approach. Here, the effects of bed hydraulic diameter (D_h), static particle height (H_s), particle type/size (D_p) and bed aspect ratio were considered. Pressure drop data and high speed camera images were used to develop simple flow regime maps for these printed beds showing the operating windows for packed bed, bubbling, slugging and turbulence applicable to a wide range of bed size to particle diameter ratios (D_h/D_p = 20â€“200) and gas velocities (U_g = 1â€“400 mm/s). Fast Fourier transforms of the pressure drop signals were also used to study the evolution of bubbling/slugging behaviour as the gas velocity was increased by creating 2D colour maps of the frequency spectra. These allowed a new method to be proposed for the identification of slugging. It is concluded in this study that the rougher surfaces generated by additive manufacturing do not influence the fluidization characteristics nor modify the wall effects of small-scale beds. Macro-scale fluidization can also be achieved at smaller D_h/D_p ratios in these 3D printed beds compared to more conventional Plexiglas beds ($D_h/D_p = 75$ compared to $D_h/D_p = 300$).

Fluidized Bed Rheology: Direct Measurements of Apparent Viscosity through Stokesian and Bingham Approaches

Conference: Fluidization

Year: 2019

Proceeding: Fluidization XVI

Group: General Paper Pool

Session: 1B: Fundamentals of Fluidization

Time: Monday, May 27, 2019 - 2:18pm-2:30pm

Authors:

Schütz, D., Anton Paar
Aschl, T., Anton Paar Gmbh

Gas Fluidization as a process is an important part of many industrial processes, however compared to collapsed states the measurement of its intrinsic flow properties is in its infancy.

Our fluidized bed rheology setup consisting of a powder containing cylinder with a gas permeable bottom, through which a defined controlled gas flow is run into the powder sample, and a geometry placed in the powder. Attached to a rotational rheometer such a setup allows real rheological measurements over large ranges of shear rates and shear stresses on powder systems in a well-defined fluidization state. Being able to measure small torques in a reproducible manner enables a proper rheological study of fluidized beds similar to the rheology of complex fluids.

To obtain absolute viscosity values a linear velocity profile is required and slip needs to be prevented. Different concentric cylinders smooth, rough and profiled surfaces were tested. Viscosity curves of a glass bead sample in the un-fluidized and the fully fluidized state measured with two profiled cylinders with different sizes are in good agreement indicating that the no slip condition is met. Results of viscosity curves in different fluidization states are presented and are compared to the behavior of liquid suspensions. While exhibiting some unique effects, the shear rate dependent behavior of a gas fluidized bed can be for the most part be linked to that of particulate suspension.

Furthermore comparisons on multiple concentric cylinder geometries and a novel ball measurement system based upon stokesian surface viscosity that allows the measurement of more cohesive powders. We will illustrate the capabilities of this approach on multiple industrially relevant powders and applications and their applications for process design and control.

Methodology for the Estimation of Sphericity of Fiber and Hull Shaped Biomass Particles

Conference: Fluidization

Year: 2019

Proceeding: Fluidization XVI

Group: General Paper Pool

Session: 1B: Fundamentals of Fluidization

Time: Monday, May 27, 2019 - 2:42pm-2:54pm

Authors:

Guío-Pérez, D. C., Universidad Nacional de Colombia
Rincon-Prat, S. L., Universidad Nacional de Colombia
Moreno-Chaves, C. A., Universidad Nacional de Colombia

Methodology for the estimation of sphericity of fiber and hull shaped biomass particles

C.A. Moreno-Chaves; D.C. Guío-Pérez*; S.L. Rincón-Prat
Universidad Nacional de Colombia, Department of Mechanical and Mechatronics Engineering

Research group on Biomass and Optimization of Thermal Processes (BIOT)

Carrera 30 N° 45 A 03; Bogotá D.C - Colombia

**(0057 1) 3165000 Ext. 11207, dcguiop@unal.edu.co*

Abstract

Estimation of particles sphericity is a critical issue when aiming to model or predict the behavior of a fluidized bed. Accurate techniques for the determination of sphericity, such as micro-tomography and laser scanning are expensive, time-consuming, and therefore not available for routine measurements. Image analysis is in turn more accessible, but accurate only in cases where the particles have a high sphericity [1]. Natural materials such as residual biomass suppose a great challenge on this regard given the variety of shapes and the shape differences between size fractions [2,3,4]. The present work proposes a methodology for the estimation of sphericity of selected biomass particles (rice husk, oil palm fiber and oil palm rachis). Ideal models based on the approximate geometrical shape of particles (cylinders for fibers and flat flakes with constant width for hulls) were deduced, and different shape factors were calculated and tested on its ability to correlate experimental measurements with the ideal proposed shapes. For the selected materials, perimeter, projected area and length were measured through image analysis; additional width measurements were performed for hull-shaped particles. Focus variation microscopy was performed to measure the particles surface and volume, and to determine their actual sphericity. Experimental data were used to find the most suitable ideal shape and shape factor that describe the actual shape of the selected particles, these relations were then used to predict the sphericity. Results showed that diameter to length ratio could be used to predict the sphericity of fiber-like particles, while width to perimeter ratio and Sympatec sphericity parameter can be used to predict the sphericity of hull-like particles. The findings strengthen the potential of image analysis together with the ideal geometrical models for sphericity estimation, since they indicate that once an accurate model is adjusted to the particles shape, simple 2D image analysis can accurately be used to predict sphericity.

References:

[1] G. H. Bagheri, C. Bonadonna, I. Manzella, and P. Vonlanthen. On the characterization of size and shape of irregular particles. *Powder Technol.*, vol. 270, no. Part A, pp. 141â€“153, 2015.

[2] C. R. Cardoso, T. J. P. Oliveira, J. A. Santana Junior, and C. H. AtaÃde. Physical characterization of sweet sorghum bagasse, tobacco residue, soy hull and fiber sorghum bagasse particles: Density, particle size and shape distributions. *Powder Technol.*, vol. 245, pp. 105â €“114, 2013.

[3] M. Gil, E. Teruel, and I. Arauzo. Analysis of standard sieving method for milled biomass through image processing. Effects of particle shape and size for poplar and corn stover. *Fuel*, vol. 116, pp. 328â€“340, 2014.

[4] D. Barletta, R. J. Berry, S. H. Larsson, T. A. Lestander, M. Poletto, and Ã. RamÃrez-GÃ³mez. Assessment on bulk solids best practice techniques for flow characterization and storage / handling equipment design for biomass materials of different classes. *Fuel Process. Technol.*, vol. 138, pp. 540â€“554, 2015.

Particle Attrition in a Fluidised Bed

Conference: Fluidization

Year: 2019

Proceeding: Fluidization XVI

Group: General Paper Pool

Session: 1B: Fundamentals of Fluidization

Time: Monday, May 27, 2019 - 2:30pm-2:42pm

Authors:

Fulchini, F., University of Leeds
Ghadiri, M., University of Leeds
Borissova, A., University of Leeds
Yazdanpanah, M. M., TOTAL
Amblard, B., IFP Energies nouvelles
Bertholin, S., IFP Energies nouvelles
Forret, A., IFP Energies nouvelles

Particle attrition in fluidised beds and circulating fluidised beds causes operational problems as well as material losses. Prevailing mechanical, thermal and chemical stresses bring about a gradual degradation of the particles and formation of fines. The main mechanical sources of particle attrition in a fluidised bed are the gas jets, bubbling region of the bed and cyclones.

In this work a pilot scale fluidised bed is used to assess the particle attrition induced by a jet and the bubbling part bed of â€œGeldart group Bâ€ manganese dioxide particles, intended for use in the Chemical Looping Combustion. The effects of the jet gas velocity and orifice size on the rate of attrition are analysed. The study includes both ambient condition and operation at 700 Â°C.

The jet penetration height, for both cold and hot tests, is measured by monitoring the gas pressure using a pitot tube and compared with that obtained from an Eulerian-Eulerian CFD simulation. This is then used in the modelling of the effect of the jet velocity on the overall attrition rate. The model predictions of the rate of elutriated fines are compared with those of the existing models in literature.

Particle Cluster Acceleration and Stabilization in the Feeding Section of a Downer Unit

Conference: Fluidization

Year: 2019

Proceeding: Fluidization XVI

Group: General Paper Pool

Session: 1B: Fundamentals of Fluidization

Time: Monday, May 27, 2019 - 1:30pm-1:42pm

Authors:

Medina-Pedraza, C. A., Western University
de Lasa, H., Western University

Particle Cluster Acceleration and Stabilization in the Feeding Section of a Downer Unit

Cesar Medina Pedraza,
Hugo de Lasa (*)

(*) Corresponding Author

Chemical Reactor Engineering Centre, University of Western Ontario, Ontario, Canada

Particle cluster dynamics in the feeding/acceleration section of a downflow reactor is a topic of great importance for the implementation of downer units in catalytic chemical processes. To address this issue, particle cluster velocities, solids holdups and individual cluster sizes were measured in a cold-flow model unit with the help of the CREC-GS-Optiprobes. The experimental setup was equipped with 5.08" ID and 2 m height acrylic column. The feeding section contained a ring of 8 injectors angled at 45 degrees. The gas jets through these injectors impinged on a stream of particles fed from a particle separator cyclone.

To accomplish this, FCC catalyst particles with a mean particle diameter of 84.4 μm, a standard deviation of 32 μm, and particle density of 1722 kg/m3 were employed. More than 100 runs with measurements were effected at three different axial positions, 0.29 m, 0.49 m, and 1.89 m from the downer inlet, and at 23 positions along the diameter of the column. The CREC Optiprobes allowed measuring both particle cluster sizes and particle cluster velocities. With this end, six operating conditions were established with the superficial gas velocities varying in the 1 – 1.6 m/s range, and the solids mass flux ranging in the 30 – 50 kg/m²s.

Figure 1a reports a typical particle cluster size distribution performed at z= 0.29 m and z= 1.89 m. Individual particle clusters are viewed as a train of particles. This is the most stable particle cluster configuration as anticipated by Krol et al. (Krol, Pekediz, and de Lasa 2000). One can observe that at the 1.89 m level, there is a fully asymmetric distribution of particle clusters with a larger abundance of smaller clusters. One can notice as well, that at z = 0.29 m there is, in the particle cluster distribution, a larger fraction of bigger clusters containing 4 particles each on average. Thus, once particle clusters accelerate in the downer, they stabilize in size with a more abundant fraction of shorter clusters as compared to the particle clusters in the feeding section.

a)

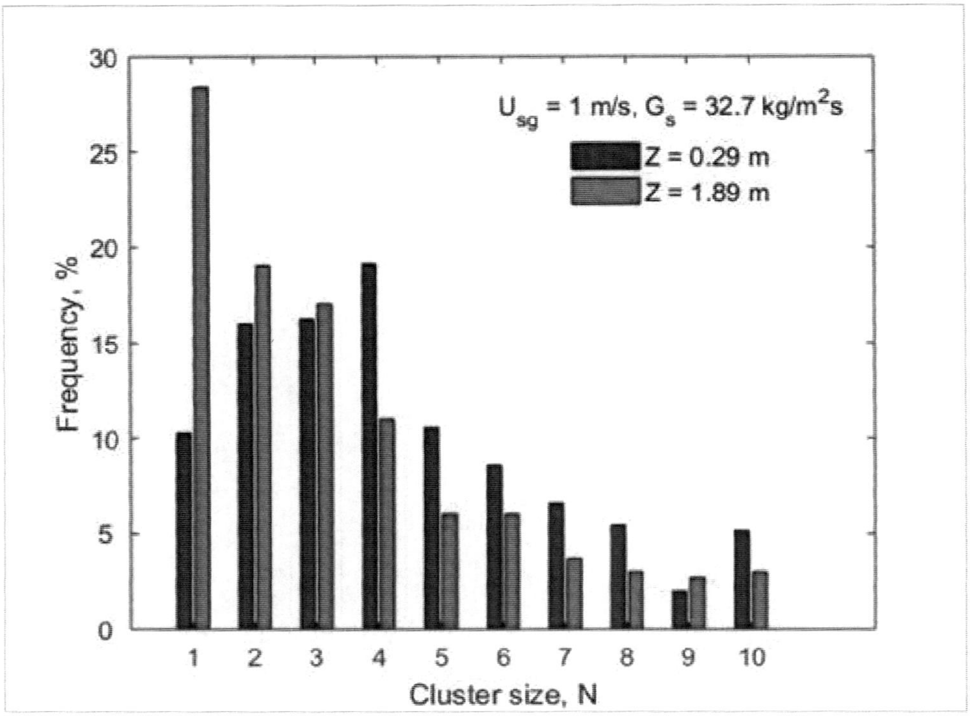

b)

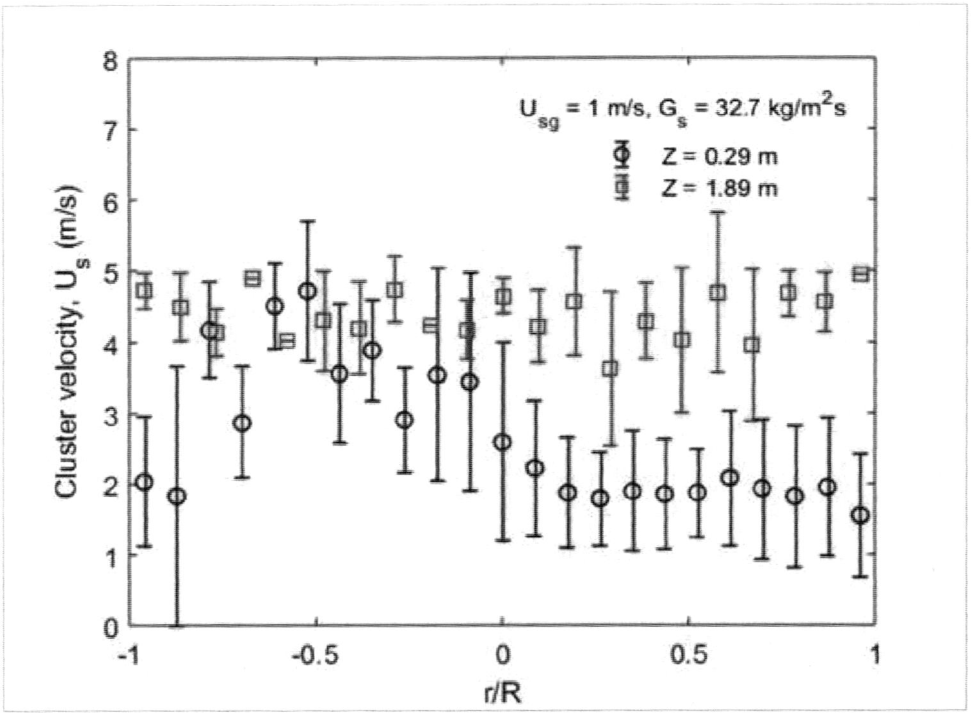

Figure 1.
a) Typical particle cluster size distribution, b) Radial cluster velocity profile, both in the acceleration and stabilized section of a downer unit at U_{sg} = 1 m/s, and G_s = 32.7 kg/m² s.

Figure 1b describes the particle cluster velocity distributions both at z = 0.29 m and z =1.89 m. This shows the progressive increase of particle cluster velocity when the particles move from the feeding flow section to the stabilized flow section.

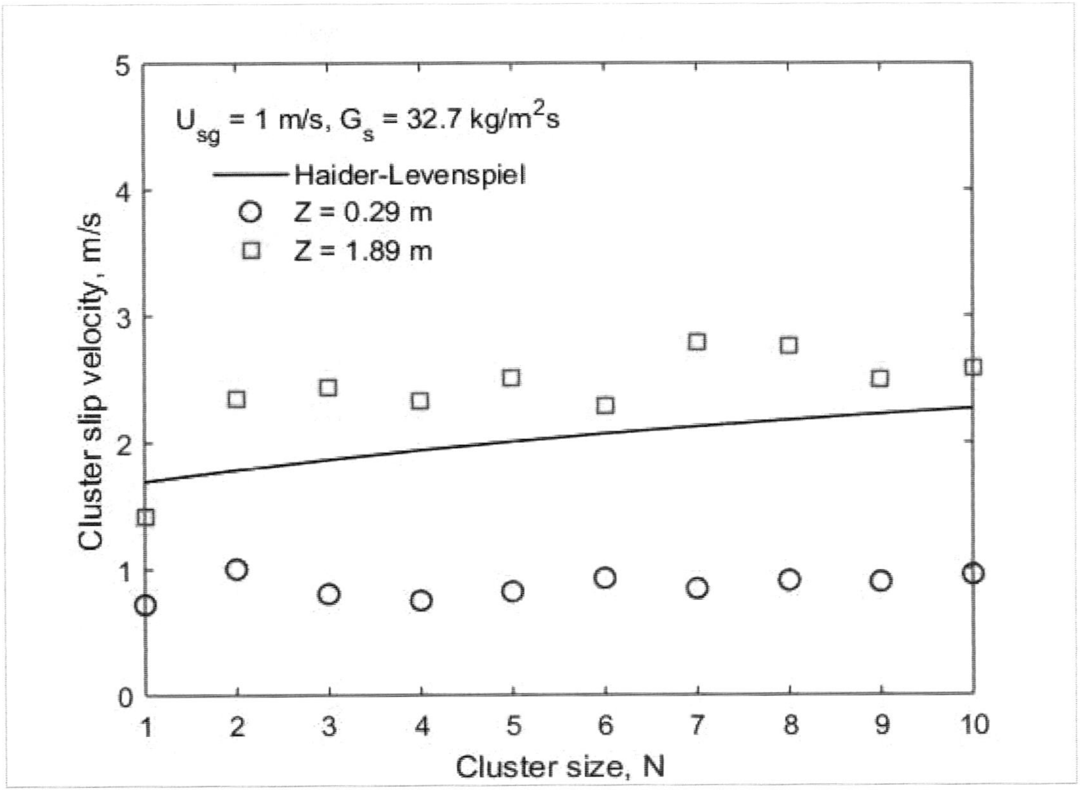

Figure 2.
Cluster slip velocity for different cluster sizes

One can see in Figure 2, that particle cluster velocities are closely correlated with terminal velocities, with a sphericity factor of 0.4. On the other hand, the particle cluster velocities in the feeding section are uncorrelated with particle cluster terminal velocity. Thus, particle velocity and particle size measurements are critical to comprehensively establish cluster dynamics in the downer feeding section. This study is the first attempt as far we are aware of, where individual particle cluster characteristics in the acceleration/feeding section are experimentally established.

References

Islam, Mohammad Ashraful, Stefan Krol, and Hugo I. de Lasa. 2010. "Slip Velocity in Downer Reactors: Drag Coefficient and the Influence of Operational Variables." Industrial and Engineering Chemistry Research 49 (15): 6735–44. doi:10.1021/ie901466p.

Krol, S., A. Pekediz, and H. de Lasa. 2000. "Particle Clustering in Down Flow Reactors."Powder Technology 108 (1): 6–20. doi:10.1016/S0032-5910(99)00196-5.

Lanza, A., M. A.
Islam, and H. de Lasa. 2012. “Particle Clusters and Drag Coefficients in Gas-Solid Downer Units.” Chemical Engineering Journal 200–202. Elsevier B.V.: 439–51. doi:10.1016/j.cej.2012.06.027.

Lanza, A., M. Islam,
and H. de Lasa. 2017. “Particle Cluster Sizing in Downer Units. Applicable Methodology across Downer Scale Units.” Powder Technology 316. Elsevier B.V.: 198–206. doi:10.1016/j.powtec.2016.11.035.

What Is Happening Above Your Fluidized Bed?

Conference: Fluidization

Year: 2019

Proceeding: Fluidization XVI

Group: General Paper Pool

Session: 1B: Fundamentals of Fluidization

Time: Monday, May 27, 2019 - 2:06pm-2:18pm

Authors:

Karri, S. B. R., Particulate Solid Research, Inc.
Cocco, R., Particulate Solid Research, Inc.
Issangya, A., Particulate Solid Research, Inc.
Freireich, B., Particulate Solid Research, Inc.

With a properly designed sparger or grid plate and internals, the maintenance cost in a fluidized bed can be easily managed. However, that is not the case for the equipment above the fluidized beds. All too often, the entrainment rates are not known and many entrainment calculations are lacking in providing an accurate representation of a commercial system. Furthermore, solids inlet feeds can also complicated entrainment rate predictions. This means that the design of the cyclone system and everything else in the freeboard may not be optimized for the actual conditions.

PSRI has been studying what is happening above the fluidized bed in an effort to reduce maintenance costs, increase reliability and increase operational times for commercial fluidized bed units. Three aspects need to be considered. First, fluidization quality is paramount in maintaining reliability. Gas bypassing or jet streaming, which commonly occurs in deep or dense fluidized beds or induced from solids feeds, can cause significant erosion of internals above the bed due to large particle-laden plumes ejecting into the freeboard region. In addition, gas bypassing leaves regions of the bed stagnant or very poorly fluidized. Cyclone diplges discharging into these regions can become flooded, resulting in significant catalyst losses. Fortunately, gas bypassing can be detected and mitigated. Second, particle fines (particles smaller than 44 Î¼m), may not increase entrainment rates. In fact, it may actually decrease the entrainment rate. PSRI, using high speed video with custom design borescopes, have observed and measure particle clusters, consisting mostly of fines, in and above a fluidized bed. Third, PSRI has been examining erosion in 20-inch diameter cyclones of various configurations.

Erosion was found to be depending on loading and inlet gas velocity but not in a linear fashion. In addition, the effect of vortex breakers in secondary cyclones was quantified over a range of operating conditions and designs.

Thus, ensuring good fluidization, understanding the actual entrainment rate and designing the cyclones for the actual conditions can increase the cycle time between maintenance cycles. Even existing units can be optimized by simply adding more fines or baffles in the bed to reduce gas bypassing and/or increase particle clustering in the freeboard. There are pitfalls, but fortunately there are options to mitigate those risks.

An Experimental Study of CO2 Separation Using a Bench-Scale Dual-Fluidized Bed Calcium Looping Process

Conference: Fluidization

Year: 2019

Proceeding: Fluidization XVI

Group: General Paper Pool

Session: 1C: Fluidized Bed Applications

Time: Monday, May 27, 2019 - 1:55pm-2:07pm

Authors:

Shimizu, T., Niigata University
Kato, H., Niigata University
Hara, O., Niigata University
Li, L., Niigata University

Ca-Looping (CaL) process that utilizes CaO as absorbent has been developed as a low energy-penalty CO_2 separation process. In this work, a bench-scale dual-fluidized bed solid circulating system which comprised a bubbling bed carbonator (bed height: 30 cm) and a fast fluidized bed regenerator (riser height 1.93 m) was operated. Calcined limestone (CaO) was employed as bed material. Simulated flue gas from air-blown combustor (CO_2 15%) was fed to the carbonator in which the temperature was maintained at 876 K. Most part of the fed CO_2 was captured by CaO in the carbonator. The produced $CaCO_3$ was transported to the regenerator and the decomposition of $CaCO_3$ to CaO and CO_2 was conducted at 1073 â€“ 1223 K. The produced CaO was transported to the carbonator again. Material balance of CO_2 was evaluated using two different methods, gas concentration measurement and flow rate measurement using a thermal mass flow meter. Material balance of CO_2 under steady state conditions nearly closed. Transient change in CO_2 transportation rate after starting CO_2 feed to the carbonator was predicted using a complete-mixing vessel model. The solid circulation rate was measured by injecting tracer particles (cold limestone) into the stand pipe and by measuring the delay of the temperature change between two points in the stand pipe. The experimental results of the transient CO_2 transportation rate agreed well with the predicted results. In addition to CO_2 transportation, coal feed to the regenerator was conducted. It was found that the emissions of NOx from the regenerator using CaO as bed material was much higher than that for the case of inert silica sand bed. The high NOx emissions from the CaO bed was qualitatively explained by the literature results about the catalytic activity of CaO to oxidize volatile-N compounds (HCN and NH_3) to form NOx.

Discharge Characteristics of Lognormal Particle Size Distributions from a Conical Hopper

Conference: Fluidization

Year: 2019

Proceeding: Fluidization XVI

Group: General Paper Pool

Session: 1C: Fluidized Bed Applications

Time: Monday, May 27, 2019 - 3:07pm-3:19pm

Authors:

Chew, J. W., Nanyang Technological University
Zhao, Y., Nanyang Technological U
Yang, S., Nanyang Technological U

This study employs the discrete element method (DEM) to investigate the impact of the widths of lognormal particle size distributions (PSDs) with the same mean particle diameter on hopper discharge behaviors, namely, discharge rate, particle velocities and size-segregation. Results reveal that (i) the hopper discharge rate decreases as PSD width increases; (ii) the mean discharge rates are constant with time, but the fluctuations increase as the PSD width increases; (iii) the overall size-segregation increases with PSD width; (iv) the overall mean particle diameters of the narrower PSDs do not exceed the initial mean of 5 mm, whereas that of wider ones do; (v) the relationship between PSD width and particle velocities is non-monotonic with no consistent trends; and (vi) no direct correlation exists between particle velocity and size-segregation. The results here provide valuable insights on the behavior of the prevalent polydisperse mixtures in hoppers.

Electrostatic Charging of Powders during Pneumatic Conveying

Conference: Fluidization

Year: 2019

Proceeding: Fluidization XVI

Group: General Paper Pool

Session: 1C: Fluidized Bed Applications

Time: Monday, May 27, 2019 - 2:31pm-2:43pm

Authors:

Taghavivand, M., University of Ottawa
Sowinski, A., University of Ottawa
Mehrani, P., University of Ottawa

Electrostatic charge generation which is mainly observed during particle handling, transport (e.g., pneumatic conveying) and processing (e.g., gas-solid fluidized beds), is an unavoidable and, in some cases, an undesired phenomenon that can result in operational challenges. Particle-wall adhesion, electrostatic discharge, particle agglomeration and segregation are problems associated with electrostatic charging that have been reported in gas-solid processes such as gas-solid fluidized beds [1]. As an example, electrostatic charging has been clearly observed in catalytic gas-solid fluidized bed reactors in polyethylene production plants. Adhesion of the charged polyethylene and catalyst particles to the reactor wall, leads to a problem known as â €œsheetingâ€ [2]. This is a major drawback since particle sheets can break off and block the reactor, causing production and economical losses. A potential source for the sheet formation is continuous collisions between the polyethylene and catalyst particles as well as collisions with the reactor wall. The electrostatic charge generated on the surface of these particles as a result of the collisions attracts a portion of them towards the reactor wall due to formed image and electrostatic forces and in turn forming a sheet. It is also anticipated that the catalyst particles, which are pneumatically conveyed into these reactors through narrow tubes, to be highly charged upon entering the reactor and therefore contribute significantly to the bed electrification and subsequently the sheeting formation. Thus, the aim of this work was to investigate electrostatic charging behaviour of various powders through their pneumatic conveying. The effects of conveying tube length and its pathway (addition of elbow and bend), conveying gas velocity, and solids loading ratio are also investigated. Two charge measurement techniques were employed in this work. A Faraday cage containing a filter bag was used to directly measure the powder electrostatic charge as they exited the conveying line. Current was also measured from the conveying line as this technique can be employed as an alternative charge measurement method where implementing the Faraday cage technique is not feasible in a commercial process.

References

[1] P. Mehrani, M. Murtomaa, D.J. Lacks, An overview of advances in understanding electrostatic charge buildup in gas-solid fluidized beds, J. Electrostat. 87 (2017) 64â€"78.

[2] B.D. Fulks, S.P. Sawin, C.D. Aikman, J.M. Jenkins, Process for reducing sheeting during polymerization of alpha-olefins, US 4532311, 1985.

Experimental Study of Gas-Solid Separation in Gas-Solid Concurrent Axial Moving Bed

Conference: Fluidization

Year: 2019

Proceeding: Fluidization XVI

Group: General Paper Pool

Session: 1C: Fluidized Bed Applications

Time: Monday, May 27, 2019 - 2:43pm-2:55pm

Authors:

Lv, H., China university of petroleum (Beijing)

In recent years, a great deal of dust from industrial exhaust emissions has seriously polluted the atmosphere. Particle bed dedusting technology is generally recognized as an effective high temperature gas dedusting technology, which has a wide range of applications and has a good prospect of development.

A new type of axial moving-bed gas-solid separation technology was proposed in this paper, which promote the fluency of granule flow by the method of gas-solid downstream and add multi-discharge cone, with using the spout-regeneration device to regenerate the captured granule. This technology can make the equipment has a low pressure drop and capture fine dust efficiently in case of fulfilling a certain gas residence time.

In this paper, we focus on gas-solid separation, flow characteristics and pressure drop of the gas-solid concurrent axial moving-bed. The appropriate operating conditions and improvement measures were determined by observing and analyzing the influence of operational control factors on the separation characteristics of the moving-bed. By investigating the distribution of internal pressure drop and pressure, moving velocity of the granule phase and gas residence time and so on, the internal gas-solid flow law is determined, which provides reference for structural optimization and industrial design of the equipment.

Fluidization Characterization of Coal Powder in a Novel Fluidized Bed with the Inner Cyclone

Conference: Fluidization

Year: 2019

Proceeding: Fluidization XVI

Group: General Paper Pool

Session: 1C: Fluidized Bed Applications

Time: Monday, May 27, 2019 - 2:07pm-2:19pm

Authors:

Song, W., University of Chinese Academy of Sciences
Na, Y., University of Chinese Academy of Sciences
Li, S., University of Chinese Academy of Sciences
Ouyang, Z., University of Chinese Academy of Sciences

A novel fluidized bed with the inner cyclone was developed to realize pulverized coal self-sustained preheating in this study. The temperature of the coal can be heated up to 850 °C-950 °C by the coal partial combustion in the novel inner fluidized bed combustor (IFBC). The hot pulverized coal is fed into the furnace of pulverized coal (PC) boiler. The integrated PC boiler combined IFBC will lead to the easer ignition and reduce the NOx emission for the low-rank coal. In order to investigate the operating performance and gas-solid flow characteristics in the IFBC , the cold experiments were carried out in a small-scale IFBC installation of 0.15m in diameter and 1m in height, continuous feeding using Geldart Group A particles. The influences of the superficial gas velocity on the pressure profile and particle balance were studied. The experimental results showed that the IFBC can be operated stably, and the hot coal output rate is steady when the superficial gas velocity is in the range of 1.5-2.25m/s, while the hot coal output is nearly equal to the cold coal input in the condition of continuous feeding. Based on the cold test, the hot inner fluidized bed combustor was designed which thermal input is 2MW. The hot experimental results showed that the temperatures along the riser are distributed uniformly that verifies the good fluidization in the IFBC.

Investigation of the Dynamic Behaviour of the Spray Granulation in Continuously Operated Horizontal Fluidised Beds

Conference: Fluidization

Year: 2019

Proceeding: Fluidization XVI

Group: General Paper Pool

Session: 1C: Fluidized Bed Applications

Time: Monday, May 27, 2019 - 3:19pm-3:31pm

Authors:

Heinrich, S., Hamburg University of Technology
Diez, E., Hamburg University of Technology

In continuous spray granulation processes very often horizontally constructed fluidised beds with rectangular cross sections are used, which are divided by a variable number of plates (weirs) into several chambers of different functionalities. Commonly an external product processing, consisting of pneumatic conveying, screening, grinding of the oversize granules and recycling of grinded oversize and undersize granular material into the fluidised bed granulator is applied. The resulting internal and external networks of solids process unit operations as well as gas, liquid and solid flows lead to a complex and dynamic process behaviour.

This work focuses on the dynamic behaviour of spray granulation processes in horizontal fluidised beds that affect the process conditions, as well as the product properties. Studying the dynamic effects of this process, on the one hand a holistic view on the macroscopic process is considered to characterize the process behaviour with respect to the stability and time constants of the rate processes. On the other hand the process is divided into local compartments to analyse several sub-processes and their effects within the granulation individually, such as the spraying process within the spray zone and the solids residence time and transport through the multi-staged process chamber under different weir configurations.

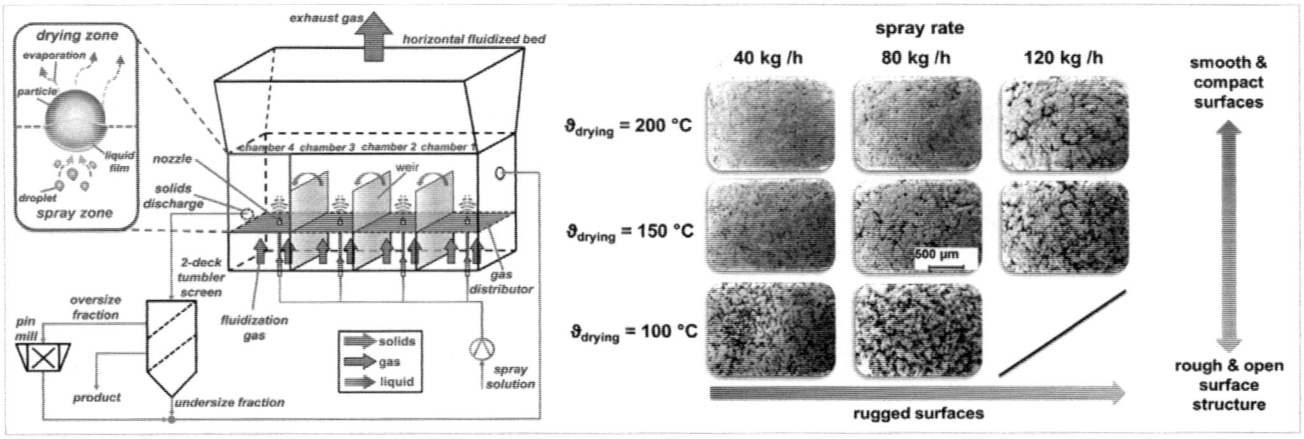

Figure: Simplified process flow diagram
with schematic representation of the internal zone formation and external process chain (left); Influence of the process conditions onto the morphological structure (SEM pictures) of the product granules (right).

The financial support of the German
Research Foundation ("Deutsche Forschungsgemeinschaft") within
the priority program SPP1679 is gratefully acknowledged.

Keynote Talk: Investigation of the Bubble Dynamics in a Semi-Cylindrical Gas-Solid Fluidized Bed

Conference: Fluidization

Year: 2019

Proceeding: Fluidization XVI

Group: General Paper Pool

Session: 1C: Fluidized Bed Applications

Time: Monday, May 27, 2019 - 1:30pm-1:55pm

Authors:

Mostoufi, N., Uniersity of Tehran
Okhovat-Alavian, M., Razi University
Behin, J., Razi University

Fluidized beds are used in various processes such as mixing, catalytic and non-catalytic reactions, drying, and adsorption since they provide uniform temperature distributions, high mass/heat transfer rates, and low pressure drops. Almost every aspect of fluidized beds operating in the bubbling regime, including gas interchange rate between phases, gas and particle residence times, reaction conversion, particle entrainment and elutriation, and mass and heat transfer are significantly controlled by the bubble characteristics of size, shape, and velocity. Accordingly, a great deal of effort in measuring and characterization of bubbles has been spent in the published fluidization literature. More researches have been dedicated to the development of various correlations for predicting bubble size and velocity. However, it is not easy to visualize motion of bubbles in three-dimensional fluidized beds. Although two-dimensional beds can be used for direct visual observation, their results are qualitative and have a limitation in applicability to three-dimensional fluidized beds. It has been shown that semi-cylindrical bed is a useful tool which allows direct observation of in-bed phenomena with minor wall interactions. Therefore, the semi-cylindrical fluidized bed is appropriate for direct visual observation of bubbles due to its flat wall. Moreover, the measurements would be more accurate and applicable to three-dimensional beds compared with two-dimensional beds.

In the present work, bubble dynamics was studied (i.e., bubble frequency, bubble size, bubble rise velocity, and gas hold up) in a gas-solid semi-cylindrical fluidized bed. A semi-cylindrical fluidized bed with a transparent flat wall, in order to observe bubbles visually, and a cylindrical fluidized bed were employed to investigate the effect of cross-section of the column on the bubble behavior. Both beds were made of Plexiglas with diameter of 14 cm filled with glass beads as the bed material. Air was entered into the column through a perforated plate distributor and its flow rate was controlled by a mass flow controller. All experiments were carried out in atmospheric pressure cold fluidized beds while the superficial gas velocity was varried in the range of 0.2-0.8 m/s. In order to investigate the effect of particle size on the bubble dynamics in both beds, experiments were conducted at three glass bead sizes 120, 290 and 450 Î¼m (Geldard B) at the same static bed height of 21 cm (L/D=1.5). In the semi-cylindrical bed, the bubble size and bubble rise velocity were directly measured at the bed height of 14 cm (L/D=1) by digital imaging technique, when using a camera in front of the flat wall. Bubble size and bubble rise velocity were measured by a reflection-type fiber optic probe in the cylindrical fluidized bed, located at 14 cm above the distributor. Gas hold up in both beds was measured by the same fiber optic probe.

It was shown that the semi-cylindrical fluidized bed is proper for direct visual observation of bubbles through its flat wall. The experimental results were compared with the existing correlations obtained for three-dimensional gas-solid fluidized beds. Results showed that increasing the particle size results in increase in the bubble chord length and bubble rise velocity in both beds. Besides, no effect of particle size on bubble chord length and bubble rise velocity was observed for different particle sizes when compared at the same excess gas velocity (U_G-U_{mf}). Correlation of Mori and Wen for bubble growth was found to be appropriate for both fluidized beds. Also, the expressions for bubble rise velocity under the present conditions developed for three-dimensional beds by Davidson and Harison, was found to be applicable to both beds. This finding indicates that in one hand the bubble size and bubble rise velocity in semi-cylindrical fluidized bed is in compliance with the cylindrical fluidized bed. On the other hand, measuring these parameters in the semi-cylindrical bed is much easier than in the cylindrical bed. This is ensuring that the semi-cylindrical fluidized bed is a very useful tool for being employed increasingly for cold model works in laboratories.

Positron Imaging of Segregation, Mixing and Transport of Plastics in Fluidised Beds: Toward a Circular Economy for Plastics

Conference: Fluidization

Year: 2019

Proceeding: Fluidization XVI

Group: General Paper Pool

Session: 1C: Fluidized Bed Applications

Time: Monday, May 27, 2019 - 2:19pm-2:31pm

Authors:

Windows-Yule, C., University of Birmingham

Plastic waste represents one of the foremost global challenges of our age. A promising technique for the creation of a true circular economy in plastics is the use of heated fluidised beds to pyrolise waste plastics, cracking long-chain molecules into shorter-chain hydrocarbons which are then fractionated and condensed into valuable `virgin' petrochemical feedstocks. While this methodology has been proven to be viable, and the fractionation processes are well-understood and refined, knowledge of the internal dynamics of the fluidised bed systems remains heavily limited. As such, current attempts at optimisation remain purely empirical, and reliant on measurements of basic, bulk properties.

In order to address this issue, we must gain an improved understanding of the fundamental mixing and transport behaviours of mixed sand-plastic systems, and how these behaviours may be optimised. To this end, we have performed detailed imaging of such systems using Positron Emission Particle Tracking, gaining novel insight into their internal dynamics and their variation with key system parameters. We demonstrate that the large disparities in both particle size and density in the systems of interest lead to behaviours significantly divergent from those of binary systems more commonly studied in academic research.

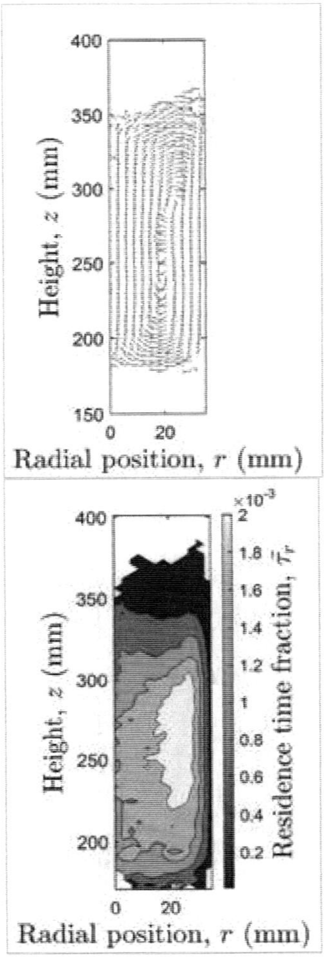
400
350
300
250
200
150
0
20
Height, z (mm)
Radial position, r (mm)
×10⁻³
2
1.8
1.6
1.4
1.2
1
0.8
0.6
0.4
0.2
Residence time fraction, τ̄_r

Study on Particle Concentration Distribution in a Dense-Phase Gas-Solid Separation Fluidized Bed

Conference: Fluidization

Year: 2019

Proceeding: Fluidization XVI

Group: General Paper Pool

Session: 1C: Fluidized Bed Applications

Time: Monday, May 27, 2019 - 2:55pm-3:07pm

Authors:

Duan, C., China University of Mining and Technology
Li, G. Sr., China University of Mining and Technology
Zhao, Y., China University of Mining and Technology
Dong, L., China University of Mining and Technology
Enhui, Z. Sr., China University of Mining and Technology

Dense-phase gas-solid fluidized bed is the most precise dry coal separation technology, which is of great significance in the field of coal preparation. However, due to the technical limitations, the research on the distribution of particle concentration in fluidized bed is still in the theoretical stage. In order to verify the existing theory about particle concentration distribution in gas-solid fluidized bed system, the accuracy of concentration distribution in dense gas-solid fluidized bed was tested by electrical capacitance tomography (ECT) method, and was compared with the theoretical model on particle concentration. By fitting the distribution rules of particle concentration, the existing model was modified, and a more suitable particle concentration distribution model of separation fluidized bed was obtained finally.

CFD-DEM Modeling of Bubble Movement in a Rectangular Fluidized Bed

Conference: Fluidization

Year: 2019

Proceeding: Fluidization XVI

Group: General Paper Pool

Session: 1D: Modeling and Simulation

Time: Monday, May 27, 2019 - 3:07pm-3:19pm

Authors:

Mostoufi, N., Uniersity of Tehran
Mostafaei, F., Uniersity of Tehran
Golshan, S., University of Tehran
Zarghami, R., University of Tehran
Sotudeh-Gharebagh, R., Uniersity of Tehran

In this research, CFD-DEM was implemented to study a single bubble formation, elevation and eruption in a rectangular fluidized bed. In order to validate the CFD-DEM model, experiments were conducted in a 1(H) × 0.2(W) × 0.02(D) m bed. Glass beads particles (d_p = 1300 μm, density = 2500 kg/m^3) were used to fill the bed at the static bed height of 0.22 m. Pressure probe was placed on the bed wall at the elevation of 0.025 m from the distributer to record the pressure fluctuations with the frequency of 400 Hz. Furthermore, a high-speed camera (240 fps) was installed in front of the bed (with the distance of 30 cm from the bed wall) to capture the bubble different states in the bed. While the bed was kept at the minimum fluidization (U_{mf} = 0.129), a pulse air flow ($U_{injection}$=0.85 m/s) was injected through embedded inlet (inlet diameter = 0.02m) for 0.23 s to form the single bubble. CFD-DEM model was conducted for the same bed and particle conditions. 463000 glass bead particles filled the rectangular bed. In order to validate the model, images from high-speed camera at different bubble states were compared with the images gained from CFD-DEM. For a more quantitative comparison, bubbles diameter was calculated from the images at different states, and plotted through time. The diagram revealed that CFD-DEM predicted the bubble diameter and rise time accurately. Estimated pressure fluctuations were showed that CFD-DEM was able to predict experimental pressure fluctuations.

It was observed that pressure fluctuations from the simulations had a maximum peak and a minimum. According to the images and pressure fluctuations time steps, when the bubble was introduced to the bed this maximum peak was observed. It was observed that after the start of the injection, the gas velocity started to increase near the injection tube. After 0.085 s the whole bed had a uniform velocity, which was in the order of magnitude of the injection velocity, and after this time the gas velocity decreased. Particles velocities distribution were also studied at the beginning of the injection step. Although the same pattern as the gas velocity was found for the particles, their maximum velocity reached smaller number. The particles hold-up was also drawn through time at the location where pressure signals were gained. The solid hold-up plot illustrated an increasing until 0.085 s, and after this time hold-up decreased. Regarding this information, it was concluded that after the beginning of the injection, the upward travelling compression wave increased the gas velocity through the whole bed. Then due to the friction force between gas and particles, particles were also accelerated and moved through the void spaces in the bed. By bubble growing on the distributer, the bed was expanded and hold-up decreased. Therefore, pressure decreased during bubble formation on the distributer. After the bubble detachment, injection was stopped and a minimum was observed on the pressure-time figure. Due to the stop of injection, compression wave stopped in the bed, and the mentioned minimum was observed. During bubble elevation in the bed, particles at the bottom part of the bubble formed the bubble wake. It was observed that particles move from the regions near the bottom of the bubble inside the bubble wake. Therefore, the hold up at the bottom of the bubble was less than the hold-up of the fluidization. When bubble passed the pressure probe and the distance between the bubble wake and the probe increased, particles moved back to the fluidization state. This phenomenon explained the pressure increase trend from the minimum to the steady part of the figure. Eruption of the bubble took place when it reached the surface of the bed. When the bubble erupted, a small increase was observed in the pressure. Eruption leaded to a relative weak compression wave (comparing to the compression wave caused by the bubble injection) which travelled downward through the bed. The increase at the end of the pressure signal was due to this downward compression wave.

The gas velocity profile inside the bubble was also studied during the formation, elevation, and eruption of the bubble using CFD-DEM. Results revealed that at the formation phase, gas velocity inside the bubble was uniform upward. During bubble rising in the bed, two vortices (one clockwise and one anticlockwise) were observed inside the bubble. When the bubble moved to the upper parts of the bed, these vortexes tend to separate. At this moment, particles start to fall into the bubble through the border of the vortexes. At the eruption point, the inside gas of the bubble traveled through the breakage area, and exited from the bubble from the breakage point. Particles velocity around the bubble at different phases were also illustrated using CFD-DEM results. It was shown that by starting the injection, the velocity vectors of all the particles were upward. Particles at the sides of the bubble in its growing phase, tend to travel downward and make the bubble wake. In the elevation phase same pattern of particles orientation as formation phase was observed. In this phase, particles at the bottom of the bubble had larger velocity than particles at the top of the bubble. Furthermore, there are few particles inside the bubble which their velocity vector is in accordance with the gas vortex inside the bubble.

CFD-PBM Simulation on Gas-Liquid Mass Transfer in a Gas-Liquid-Solid Circulating Fluidized Bed

Conference: Fluidization

Year: 2019

Proceeding: Fluidization XVI

Group: General Paper Pool

Session: 1D: Modeling and Simulation

Time: Monday, May 27, 2019 - 2:43pm-2:55pm

Authors:

Zhou, X., Tianjin University
Liu, M. Sr., State Key Laboratory of Chemical Engineering (Tianjin University)
Ma, Y., Tianjin University

The three-dimensional transient flow field and gas-liquid mass transfer behavior in a gas-liquid-solid circulating fluidized bed (CFB) riser were simulated using E-E-E model coupled with PBM model (population balance model) and gas-liquid mass transfer model. With PBM model, the bubble size distribution could be obtained through bubble coalescence and breakup, and the effects of the bubble size on the interphase forces were also considered. The gas-liquid interfacial area could be calculated from gas holdup and average bubble diameter. Mass transfer coefficient calculated from four classical mass transfer models, named as penetration model, surface renewal model, eddy cell model and rigid model, were compared and tested. Reasonable simulations of gas-liquid interfacial area and mass transfer coefficient were achieved under varied operating conditions, e.g. superficial gas velocity, superficial liquid velocity and solid holdup. The simulation results shows that CFD-PBM coupled model is an effective method for predicting the hydrodynamics and mass transfer behavior in a gas-liquid-solid CFB riser.

Circulating Fluidized Beds in Three Different Scales

Conference: Fluidization

Year: 2019

Proceeding: Fluidization XVI

Group: General Paper Pool

Session: 1D: Modeling and Simulation

Time: Monday, May 27, 2019 - 2:07pm-2:19pm

Authors:

Tu, Q., Institute of Engineering Thermophysics, Chinese Academy of Sciences
Wang, H., University of Chinese Academy of Sciences

Due to its excellent heat and mass transfer characteristics, circulating fluidized bed has gained increasing interest in industrial applications. Although it has been studied widely, most available data are obtained from the lab scale and pilot scale models due to the restriction and complexity of experimental data acquisition in large scale models. Furthermore, scale-up of fluidized beds is difficult because it is widely known that the hydrodynamics of small fluidized beds can be significantly different from the hydrodynamics of larger beds. So, the data from small scale cannot be used in the large scale directly.

With the development of computational algorithm and computational capacity, computational fluid dynamics (CFD) is becoming as a powerful and efficient tool for studying the flow behaviour of gas and solids in CFBs design, optimization and scale-up. In this study, computational particle fluid dynamics (CPFD) based on the multi-phase particle-in-cell (MP-PIC) methodology is used to model the solid circulation behaviour in the three full-loop CFB systems. In the three systems, they all have the rectangular cross section risers, but the first one is in the lab-scale with the dimension of 6.7 cm x 20.1 cm x 200 cm and one cyclone; the second one is in the pilot scale with the dimension of 42 cm x 92 cm x 580 cm and six cyclones; the third one is in the industrial scale with the dimension of 2823 cm x 922.5 cm x 5000 cm and three cyclones. Homogeneous drag model Wen-Yu or heterogeneous drag model energy minimization multi-scale (EMMS)/matrix drag models are employed in the simulation and the results will be compared with the available experiment data.

In the lab-scale CFB, the simulated pressure drop of the riser is 3500 Pa when the primary inlet air flow is 2.7 m/s and static bed height is 35 cm, and both the simulated and experimental results show that double bubbles appear in the cross section of the bottom region and the shape is irregular and stretched along the long side. In the pilot-scale CFB, the simulated pressure drop of the riser is 5976 Pa when the primary inlet air flow is 4 m/s and bed material is 200 kg, and Wen-Yu drag model based simulation shows the drag force between the gas and the solid particles is over-predicted while EMMS drag model under-predicts the force. In the industrial scale CFB, the simulated pressure drop of the riser is 7400 Pa when the primary inlet flow is 180 kg/s and static bed height is 100 cm, and half of the pressure drop happens in the bottom of the riser.

CPFD Simulation of a Full-Scale Calciner Operating with Refuse Derived Fuel

Conference: Fluidization

Year: 2019

Proceeding: Fluidization XVI

Group: General Paper Pool

Session: 1D: Modeling and Simulation

Time: Monday, May 27, 2019 - 2:31pm-2:43pm

Authors:

Nakhaei, M., Technical University of Denmark
Wu, H., Technical University of Denmark
Grévain, D., FLSmidth A/S
Glarborg, P., Denmark Technical University
Jensen, L. S., FLSmidth A/S
Dam-Johansen, K., Technical University of Denmark (DTU)

In this work, a full-scale cement calciner firing refuse derived fuel (RDF) has been simulated using computational particle fluid dynamics (CPFD). To accurately predict the aerodynamics and conversion of RDF particles, a series of sub-models have been developed. A physical characterization framework has been established to quantify the size and shape distribution of the RDF particles. This framework is based on aerodynamic separation in a wind sieve followed by mass measurement and 2D photographing of individual particles. A proper drag model based on the terminal velocity of particles is adopted. Besides, a reliable iso-thermal model for conversion of plastic particles in suspension condition is developed based on the experiments in a single particle combustor (SPC).

The CPFD simulation results are compared with detailed temperature and gas composition measurements in the full-scale calciner. It has been observed that the overall trends of all measured (i.e., gas temperature and O_2 and CO_2 concentrations) and estimated (i.e., calcination degree and fuel burnout) parameters are well-predicted along the calciner. The burnout degree of fuel particles in the calciner is well-predicted as compared to the measurements. The fuel burnout mainly takes place in the calciner lower and upper vessels and is nearly completed when the particles exit the calciner. The calcination degree, on the other hand, is underpredicted; and in general, the non-uniformity of cross-sectional profiles is overpredicted.

Investigation on Bubble Characteristics in a Gas Fluidized Bed: Evaluation of Different Drag Models

Conference: Fluidization

Year: 2019

Proceeding: Fluidization XVI

Group: General Paper Pool

Session: 1D: Modeling and Simulation

Time: Monday, May 27, 2019 - 2:55pm-3:07pm

Authors:

Zhao, J., Harbin Institute of Technology
Liu, G., Harbin Institute of Technology

In this work, a 3-D calculation model coupled with different drag models was established to investigate the bubble behaviors in the gas-solid fluidized bed, and the numerical simulation was carried out based on the TFM method. The first aim of this study is to investigate the effect of particle size and gas velocity on bubble size and bubble rise velocity. Then the simulation results calculated by different drag models were compared with the value of van Lareâ€™s experiment and Davidsonâ€™s correlations. By the comparison with experiment and theory, the most accurate drag model for predicting bubble behaviors in bubbling bed will be acquired.

Keynote Talk: Massively Parallel Numerical Simulation of Hydrodynamics and Transfers in a Polydispersed Reactive Gas-Particle Fluidized Bed at Industrial Scale with a Very Fine Mesh, over One Billion of Cells

Conference: Fluidization

Year: 2019

Proceeding: Fluidization XVI

Group: General Paper Pool

Session: 1D: Modeling and Simulation

Time: Monday, May 27, 2019 - 1:30pm-1:55pm

Authors:

Neau, H., CNRS
Fede, P., Université Paul Sabatier
Ansart, R., Université de Toulouse, CNRS-Toulouse
Simonin, O., Université de Toulouse, CNRS-Toulouse
Renon, N., Université Paul Sabatier
Barbaresco, P., Université Paul Sabatier
Baudry, C., EDF R&D
Mérigoux, N., EDF R&D

Dense particle-laden reactive flows are encountered in a wide range of industrial applications such as coal/biomass combustion (CO_2 capture / chemical looping), catalytic polymerization, uranium fluorination or medicine drugs production. The development of their numerical modeling with particular focus on the hydrodynamics of fluidized beds has been a very active topic of research over the last three decades. From a physical point of view, the key challenges are about accounting for complex phenomenon occurring in reactive multi-scale flows: turbulence-particle fluctuating kinetic energy, fluid-particles interactions, interfacial transfers of mass, momentum and energy as well as chemical reactionsâ€¦ On top of these modelling aspects, this work primarily focuses on numerical challenges such as numerical schemes, mesh refinement, partitioning, highly parallel computing (HPC), data and post-processing.

Igci et al. [1] pointed out that the numerical simulation of fluidized beds is very expensive (CPU time), both at industrial and academic scales, in particular due to the 3D unsteady solid structures of small size (clusters). Predicting accurately the macroscopic flow properties such as bed expansion or solid flow rate in these systems requires a detailed description of these small-scale structures. To accurately describe and predict the formation of these small clusters, a very fine mesh having million/billion of cells is required. Therefore, massively parallel computations are required and are accessible thanks to recent progresses in supercomputer capabilities. However, this induces complex issues in terms of geometry and mesh generation, data management (I/O and storage), computation time and HPC.

The purpose of the work is the numerical simulation of an industrial fluidized bed whose dimensions are 30 m height and 5 m in diameter. The reactor contains 100 tonnes of large particles (1600 µm) and 4 injections (located at 0.6 m height) of very fine particles (80 µm) which are subjected to an exothermic reaction. Therefore, separate enthalpy transport equations are computed for gas and both particle species with local reactive source terms, heat transfers between gas and particles and dispersion by fluctuant velocity. To sum up the overall modeling, simulations are including particle drag, turbulent and kinetic model with four-way coupling, for gas and particles, respectively. Finally, a polydispersed collision model describes large and fine particle-particle interactions (Batrak et al. [2]).

Nowadays it is possible to perform realistic 3D simulations of industrial geometries using an unsteady Eulerian multi-fluid approach for turbulent polydisperse reactive particle mixtures (Hamidouche [3], Fede et al. [4]). This approach is implemented in the massively parallelized code NEPTUNE_CFD. This software is a multiphase flow solver developed in the framework of the NEPTUNE project, financially supported by CEA, EDF, IRSN and Framatome. The main numerical characteristics of this software are (i) unstructured meshes with all types of cell, (ii) non-matching meshes and/or rotating meshes, (iii) â€œcell-centerâ€ type finite-volume method, (iv) calculation of co-localized gradients with reconstruction methods and (v) distributed-memory parallelism by domain splitting (mesh partitioning and MPI parallelization). The algorithm core is based on an elliptic fractional step method using iterative linear solvers or direct matrix inversion (MÃ©chitoua et al. [5]). The algorithm accounts for density variation as a function of pressure and enthalpy during each time step. The pressure is resolved by a parallel multigrid solver. NEPTUNE_CFD scalability has already been proven on this kind of configuration up to 2,560 cores (Hamidouche et al. [3]). Previous runs have been performed using up to 4,000 cores with linear speed-up and perfect efficiency as long as each core handles at least 25,000 cells.

The key novelties of this work are a highly refined meshing and the use of the whole new CALMIP supercomputer, Olympe, to perform a reactive polydispersed simulation. The spatial discretization consisted in an unstructured mesh of 1,002,355,456 hexahedron cells. The characteristic volume of these cells is about $1.25*10^{-7}$ m^3 that must be compared to a total reactor volume of 600 m^3.

This massively parallel unsteady simulation of a 3D industrial reactive polydispersed fluidized bed with such a refined mesh and using more than 13,000 cores is a worldwide premiere. NEPTUNE_CFD ran on the 362 nodes of Olympe, i.e. 13,032 cores (SkyLake 6140) and relied on Intel compiler and IntelMPI library. The numerical simulation used 5 millions of CPU hours over 15 days (elapsed time) and 120 TB of generated raw data. The physical time simulated is about 15.5s. A special attention was given on massively parallel computation performances of NEPTUNE_CFD in terms of profiling and code efficiency, obtained speed-up, MPI communications and mesh partitioning.

The use of the latest generation of supercomputers coupled with the massively parallel CFD software NEPTUNE_CFD allows performing computations that could not be considered hitherto. Obtained results have a yet unmatched accuracy which will allow for fine understanding of complex flows encountered at industrial scale. These results can be considered as reference results, converged and mesh independent for an industrial geometry. They will be used to develop and evaluate sub-grid modeling approaches to enable low cost simulations while accounting for small scale phenomenon. This is the first time that highly resolved simulations can be used to develop such models at an industrial scale.

REFERENCES

[1] Igci, Y. & Sundaresan, S., Verification of filtered two-fluid models for gas-particle flows in risers
AIChE Journal, Wiley Subscription Services, Inc., A Wiley Company, 2011 , 57 , 2691-2707.

[2] Batrak O., Patino G., Simonin O., Flour I., Le Guevel T. & Perez E., Unlike particles size collision model in 3D unsteady polydispersed simulation of circulating fluidised bed, Circulating Fluidized Bed Technology VIII, Kefa Cen Ed., International Academic Publishers / Beijing Word Publishing Corporation, pp.370â€“378, 2005.

[3] Hamidouche Z., Masi E., Fede P., Ansart R., Neau H., Hemati M. & Simonin O., Numerical Simulation of Multiphase Reactive Flows, in: Advances in Chemical Engineering, volume 52, 2018, pp. 51â€“124. doi:10.1016/bs. ache.2018.01.003.

[4] Fede P., Simonin O. & Ingram, A., 3D numerical simulation of a lab-scale pressurized dense fluidized bed focussing on the effect of the particle-particle restitution coefficient and particle--wall boundary conditions, Chemical Engineering Science, 2016, 142, 215-235.

[5] MÃ©chitoua N., Boucker M., LaviÃ©ville J., HÃ©rard J.M., Pigny S., & Serre G., An unstructured finite volume solver for two-phase water/vapour flows modeling based on elliptic oriented fractional step method. NURETH 10, Séoul, Corée du Sud, 2003.

Modelling of a Combined Biomass Clc Combustion and Renewable-Energy-Based Methane Production System for CO2 Utilization

Conference: Fluidization

Year: 2019

Proceeding: Fluidization XVI

Group: General Paper Pool

Session: 1D: Modeling and Simulation

Time: Monday, May 27, 2019 - 3:19pm-3:31pm

Authors:

Bareschino, P., Università degli Studi del Sannio
Salatino, P., Università degli Studi di Napoli Federico II
Chirone, R., Consiglio Nazionale delle Ricerche
Coppola, A., University of Naples Federico II
Mancusi, E., Università degli Studi del Sannio
Pepe, F., Università degli Studi del Sannio
Solimene, R., Consiglio Nazionale delle Ricerche
Urciuolo, M., Istituto di Ricerche sulla Combustione

In this work, an innovative process layout to promote the integration among the chemical looping combustion of biomass, solar hydrogen, and carbon methanation is proposed and numerically investigated.

The core of the layout consists of a multiple interconnected fluidized bed system (MFB) equipped with a two-stage fuel reactor (t-FR), a riser used as Air Reactor (AR), a cyclone, a L-valve return leg, and a loop-seal. The two-stage FR is based on the concept presented in a previous work [1], where two bubbling beds were placed in series with respect to both gaseous and solids streams in order to overcome the limitations of a single-stage fuel reactor (poor char conversion, slip of unburnt volatiles, extensive elutriation of char fines). In the first stage, where fuel and fresh oxygen carrier (OC) from the AR enter, the solid carbon combustion took place. Volatile combustion occurs mainly in the second stage, where the partly reduced OC is present. An internal riser connects the two stages, thus allowing the solids to move from the first reactor to the other. The t-FR showed the best performances in terms of combustion efficiency, volatile matter and char conversion, carbon-to-CO_2 conversion efficiency and loss of elutriated carbon for all the operating conditions investigated in [1]. Solids from the second stage go through the loop-seal into the AR where the oxidation capability of the OC is restored. A cyclone collects the regenerated OC that is sent through the L-valve to the first stage of the t-FR. At the exit of the second stage, water and fines were separated from flue gas. This latter is partly conveyed to a methanation unit in order to react with a hydrogen stream coming from an-electrolysis cells (EC) array, while the remnant is recycled to FR.

A coupled hydrodynamic and reactive model of the multiple interconnected fluidized bed system, developed on the basis of a previous work [2], is applied in order to evaluate solid circulation rate, solid bed levels in different parts of the multiple interconnected fluidized bed system, flue gas composition and flow rate, and power production. The hydrodynamics of the proposed MFB system is dynamically modelled by considering the different parts as separate blocks mutually interconnected. The performance of the system has been evaluated by considering chemical and physical properties of olive wood and of CuO supported on zirconia as fuel and oxygen carrier, respectively. Olive wood was selected as representative of residual biomasses diffused into Mediterranean area. CuO has been selected on the basis of the favourable thermodynamic equilibrium of oxygen release in the temperature range of interest for the chemical looping combustion. The utilization of ZrO_2 as support, instead, ensures good chemical and physical stability. The mass fraction of the active phase in the oxygen carrier has been set at 50%. A complex kinetic scheme comprising both gasâ€"solid heterogeneous reactions taking place in the fluidized beds - involving both carbon (direct oxidation, Boudouard, and water gas reactions) and oxygen carrier (oxidation of copper and cuprous oxide, reduction of cupric and cuprous oxides) - and gas-phase homogeneous reactions occurring in the freeboard of fuel reactors (oxidation of volatiles and hydrogen) has been considered, extending the reaction scheme previously [1] developed.

The methanation unit has been modelled developing a thermodynamic calculation method based on minimization of the free Gibbs energy. Such a model does not need the application of a detailed reactions stoichiometry or kinetics model, only requiring that the system have enough time to reach equilibrium. At steady state, it was verified that the performances of the Gibbsâ €™ model match those obtained from a detailed kinetic model. This approach ensures the evaluation of methane yield without the need of hydrodynamic and kinetic modelling of the methanation unit itself, thus greatly reducing the computational effort. In fact, several reactor concepts have been proposed and investigated in literature [3], ranging from multiple adiabatic layers fixed beds to fluidized bed reactors, in order to contain the temperature rise associated to the highly exothermic Sabatier reaction and to avoid catalyst sintering while approaching the best thermodynamic condition for the process. The performance of the system has been evaluated by considering that the CO/CO_2 stream coming from the t-FR reacts with a H_2 pure stream coming from the EC array over NiO supported on alumina catalyst.

The number of ECs arranged in the array was evaluated by considering that a constant hydrogen production able to convert the whole CO/CO_2 stream produced by the CLC process should be attained. Moreover, by considering that only energy coming from renewable sources (such as photovoltaic panels or wind turbines) was fed to the EC array, it was assessed the capability of the proposed process to be used as an energy storage system.

Under the operating conditions that enable steady state operation of the proposed process, solid circulation rate, flue gas composition, thermal power output, methane yield and mass flow rate, and electrolysis cells number, along with the number of wind turbines and/or the photovoltaic panels area needed, were evaluated.

[1] Coppola A., Solimene R., Bareschino P., Salatino P., â€œMathematical modeling of a two-stage fuel reactor for chemical looping combustion with oxygen uncoupling of solid fuelsâ€, *Appl. Energy*, (157) 449-461, 2015.

[2] Diglio G., Bareschino P., Solimene R., Mancusi E., Pepe F., Salatino P., â€œNumerical simulation of hydrogen production by chemical looping reforming in a dual fluidized bed reactorâ€, Pow. Tech., (316) 614-627, 2017.

[3] Rei-Yu Chein , Ching-Tsung Yu, Chi-Chang Wang, â€œNumerical simulation on the effect of operating conditions and syngas compositions for synthetic natural gas production via methanation reactionâ€, Fuel, (185) 394-409 (2016).

Numerical and Experimental Study of Electrostatic Charge in Gas-Solid Fluidized Bed

Conference: Fluidization

Year: 2019

Proceeding: Fluidization XVI

Group: General Paper Pool

Session: 1D: Modeling and Simulation

Time: Monday, May 27, 2019 - 1:55pm-2:07pm

Authors:

Nasro-Allah, Y., Université de Toulouse, CNRS-Toulouse
Montilla, C., Laboratoire de Génie Chimique, Université de Toulouse
Ansart, R., Université de Toulouse, CNRS-Toulouse
Fox, R. O., Iowa State University
Simonin, O., Université de Toulouse, CNRS-Toulouse

Numerical and experimental study of electrostatic charge in gas–solid fluidized beds

Youssef Nasro-Allah[1] , Carlos Montilla[1] , Renaud Ansart[1] , R. O. Fox[2] and O. Simonin[3]
[1] Laboratoire de Génie Chimique, Université de Toulouse, CNRS–Toulouse, France

[2] Department of Chemical and Biological Engineering, Iowa State University, Ames, IA, USA

[3] Institut de Mécanique des Fluides de Toulouse, Université de Toulouse, CNRS–Toulouse, France

youssef.nasroallah@inp-toulouse.fr
Keywords: fluidized beds, tribocharging, fluid–particle flow, CFD simulation

Abstract

Electrostatic charges have been one of the major issues in gas–solid fluidized beds, including wall fouling due to particle accumulation on the walls, defluidization, and security issues such as sparks and dust explosions. At a molecular scale, the particle–wall or particle–particle contact generates electron transfer, inducing a charge on each particle. As a result, the surrounding gas carries an electromagnetic field which results in an electromagnetic force known as the Lorentz force. The aim of this work is to model this force and take its contribution into account in the particle-phase momentum balance. In this work, a tribocharging model, coupled with the multi-fluid model, is proposed in an Eulerian approach to model the phenomenon. The numerical simulations are carried with NEPTUNE_CFD to test the model. An experimental setup is build to develop an accurate database for comparison with simulations, and to calibrate the model parameters.

Introduction

In polymerization reactors, electrostatic charge generation can cause the formation of larger granules, as well as reactor wall fouling, which is the formation of layers of the particles on the reactor wall. Hendrickson (2006) reviewed electrostatic effects in polymerization fluidized-bed reactors and the causes of reactor fouling, explained the charge distribution in fluidized beds, and compared the effect of electrostatic forces with other forces such as drag and gravity on particle entrainment. Sowinski et al. (2012) studied the effect of particle size of a polyethylene resin received directly from industrial reactors on electrostatic charge generation and reactor wall fouling using a Faraday cup. Regarding the effect of relative humidity on electrostatic charge generation, Fotovat et al. (2016) performed experiments at different values of relative humidity using a mixture of glass beads as coarse particles and fine particles of different materials. These experiments aimed to study the relationship between fluidized bed electrostatics and entrainment.

On the numerical side, many models have been proposed to describe the phenomenon. Rokkam et al. (2010) developed an electrostatic model based on basic laws describing electromagnetic phenomena using an Eulerian approach. For tribocharging phenomenon,Lindell et al. (1993) analyzed the interaction of two dielectric spheres to find their reaction to external sources based on a static theory image. On the other hand, some authors proposed models based on the surface state theory (Ali et al. 1998; Laurentie et al. 2013). Kolehmainen et al. (2016) built a Lagrangian model on the same works and tested it on a miniaturized fluidized bed in a DEM simulation. More recently,Kolehmainen et al. (2018) proposed the same model in an Eulerian approach.

Numerical Methods

Building on the well-known Maxwell equations for an electromagnetic field, a Poisson equation is solved for the electric potential:

â^‡ â‹…(Îµmâ^‡ Ï†) = ---Î²=1--pÎ²--pÎ² Îµo
" src="https://www.aiche.org/sites/default/files/aiche-proceedings/conferences/..." class="documentimage"> (1)

where α_g is the gas volume fraction, α_{p_β} is the volume fraction on the βth solid phase, N is the number of solid phases in the mixture, ε_m and ε_o are the dielectric permittivity of the mixture and the vacuum, respectively, and φ is the electric potential. The walls are assumed to be grounded ($\varphi = 0$). Note that the effect of magnetic field is neglected here.

The gradient of the potential results in an electric field E:

" src="https://www.aiche.org/sites/default/files/aiche-proceedings/conferences/..." class="documentimage"> (2)

This electric field induces an electrostatic force F_{qp_β}:

" src="https://www.aiche.org/sites/default/files/aiche-proceedings/conferences/..." class="documentimage"> (3)

Numerical simulations are carried with NEPTUNE_CFD, an unstructured code using unsteady Eulerian multi-fluid approach for dilute and dense particle-laden reactive flows (Hamidouche et al. 2018). The coupling between the electrostatic model and the multi-fluid was verified by simulating experimental results of Sowinski et al. (2010). The weakness of this model is that charges are fixed on particles, which does not take into account the tribocharging phenomenon.

The triboelectric charging mechanism assumes that the charge carried by the solid phase obeys to the following transport equation:

Î±pÏp--p-+ --- Î±pÏpUp,iÏ‡cp = â^‚t â^‚xi (c) -â^‚- Î±pÏpD Ï‡â^‚Ï‡p- + Ï^p,q â^‚xi â^‚xi " src="https://www.aiche.org/sites/default/files/aiche-proceedings/conferences/..." class="documentimage"> (4)

where $\chi_p{}^c$= mpp--" align="middle" src="https://www.aiche.org/sites/default/files/aiche-proceedings/conferences/..." class="documentimage"> is the average charge per mass unit. α_p and ρ_p are, respectively, the volume fraction and the density of the pth phase. $U_{p,i}$ is the ith component of the p^{th}phase velocity. D_χ is the diffusion coefficient. $\psi_{p,q}$ is a source term representing the charge generation due to collisions between particles of different solid phases. The transfer between particles inside the same solid phase is included in the diffusion coefficient. The boundary conditions are expressed by a charge flux from the wall towards the particles. This flux represents the charge generation due to the particles-wall collision.

Experimental setup

The experiments are performed in a laboratory scale facility of 0.1 m inner diameter and 1 m height Plexiglass column. The electrostatic charges are measured by means of two Faraday cups connected to an electrometer. The distributor plate was welded in a valve attached to an actuator to allow particles to drop into the Faraday cup. Entrainment flux is captured via a box at the end of the cyclone. The box contains a Faraday cup placed on a balance to measure the evolution of charge and entrainment flux simultaneously. This facility is made of the same material (Plexiglass) starting from the distributor to the end of the cyclone in order to have the same effect of walls on particles. The effect of humidity is studied by means of a humidity generator. This generator ensure a constant relative humidity in a range up to 40% without heating the system. It is connected to the compressed air supplier.

The experimental work can be divided in two main parts. The first one is the study of a bubbling fluidized bed. The net charge of the particles is measured after each experiment. This aims to compare to literature findings and to estimate the saturation charge of the particles. The second part is the study of entrainment flux. The entrained mass and charge evolution are measured at the end of the cyclone. This allows to study the effect of electrostatic charges on entrainment flux. Both parts are performed at different relative humidity to highlight its effect. These experiments provide an accurate database to adjust the tribocharging model parameters depending on relative humidity and particles properties (size, material).

References

Hendrickson, G. Electrostatics and gas phase fluidized bed polymerization reactor wall sheeting. Chemical Engineering Science, Vol 61(4), 1041–1064 (2006)

Sowinski, A., Mayne, A. & Mehrani P. Effect of fluidizing particle size on electrostatic charge generation and reactor wall fouling in gas-solid fluidized beds. Chemical Engineering Science, Vol 71, 552–563 (2012)

Fotovat, F., Alsmari, T. A., Grace, J. R. & Bi X. T. The relationship between fluidized bed electrostatics and entrainment. Powder Technology, Vol 316, 157–165 (2017)

Rokkam, R. G. & Fox, R. O. & Muhle, M. E. Computational fluid dynamics and electrostatic modeling of polymerization fluidized-bed reactors. Powder Technology, Vol 203(2), 109–124 (2010)

Lindell, I. V., Sten, J. C.-E. & Nikoskinen, K. I. Electrostatic image method for the interaction of two dielectric spheres. Radio Science, Vol 28(3), 319–329 (1993)

Ali, F. S., Ali, M. A., Castle, G. S. P. & Inculet, I. I. Charge exchange model of a disperse system of spherical powder particles. In The 1998 IEEE Industry Applications Conference, 1998. Thirty-Third IAS Annual Meeting, Vol 3, 1884–1891 (1998)

Laurentie, J. C., Traoré, P. & Dascalescu, L. Discrete element modeling of triboelectric charging of insulating materials in vibrated granular beds. Journal of Electrostatics, Vol 71, 951–957 (2013)

Kolehmainen, J., Ozel, A., Boyce, C. M. & Sundaresan,S. Triboelectric charging of monodisperse particles in fluidized beds. AIChE Journal, Vol 63, 1872–1891 (2016)

Kolehmainen, J., Ozel, A. & Sundaresan, S. Eulerian modelling of gas-solid flows with triboelectric charging. Journal of Fluid Mechanics, Vol 848, 340–369 (2018)

Hamidouche, Z., Masi, E., Fede, P., Ansart, R., Neau, H., Hemati, M. & Simonin, O. Numerical simulation of multiphase reactive flows. Advances in Chemical Engineering, Vol 52, 51–124 (2018)

Sowinski, A., Miller, L. & Mehrani, P. Investigation of electrostatic charge distribution in gas-solid fluidized beds. Chemical Engineering Science, Vol 65(9), 2771–2781 (2010)

Numerical Investigation of Electrostatic Effect on Particle Dispersion in Turbulent Pipe Flows

Conference: Fluidization

Year: 2019

Proceeding: Fluidization XVI

Group: General Paper Pool

Session: 1D: Modeling and Simulation

Time: Monday, May 27, 2019 - 2:19pm-2:31pm

Authors:

Yao, J., China University of Petroleum-Beijing
Zhao, Y., China University of Petroleum

Numerical Investigation of Electrostatic Effect on Particle Dispersion in Turbulent Pipe Flows

Yanlin Zhao, Jinzhui Li, Jun Yao

Beijing Key Laboratory of Process Fluid Filtration and Separation, College of Mechanical and Transportation Engineering, China University of Petroleum-Beijing, 18 Fuxue Road, Changping, Beijing, 102249, Peopleâ€™s Republic of China

* Corresponding author, Email address: yaojun@cup.edu.cn (J.Yao)

Keywords: Electrostatics, Numerical simulation, Particle dispersion, Turbulence

Abstract

The electrostatic effect on particle dispersion in turbulent pipe flows with an additional electrostatic field generated by particle-particle and particle-wall collisions is investigated by RANS approach, combined with a Lagrangian tracking technique of one-way coupling. Particle motion equation is governed by drag force, gravity, lift force and electrostatic force. A wide range of particle size (50~ 2800*Î¼m*) at three high Reynolds number cases (Re=31k, 39k, 58k) are considered. Results obtained for the single-phase flow show good agreement with the available experimental and numerical data. Three particle-laden flow patterns can be found within the Reynolds number range studied, which agrees well with previous experimental observation. It is found that particles tend to move toward the pipe walls under the effect of electrostatic force and some of them stick to the wall in the end. In addition, the electrostatic force is observed to promote particle dispersion first and then inhibit it independent of Reynolds number and particle size. A dynamic analysis demonstrates that particle dispersion in near-wall regions is dominated by the electrostatic force rather than other forces because its magnitude is at least two orders larger than others.

A Machine Learning Approach for Electrical Capacitance Tomography Measurement of Gas-Solid Fluidized Beds

Conference: Fluidization

Year: 2019

Proceeding: Fluidization XVI

Group: General Paper Pool

Session: 2A: Fundamentals of Fluidization

Time: Monday, May 27, 2019 - 5:01pm-5:13pm

Authors:

Guo, Q., Dalian Institute of Chemical Physics, Chinese Academy of Sciences
Ye, M., Dalian Institute of Chemical Physics, Chinese Academy of Sciences
Yang, W., The University of Manchester
Liu, Z., Dalian Institute of Chemical Physics, Chinese Academy of Sciences

A Machine Learning
Approach for Electrical Capacitance Tomography Measurement of Gas-Solid Fluidized Beds

Gas-solid
fluidized beds are commonly used in industry, such as coal conversion, power generation, pharmaceutical granulation, and polymerization. Accurate measurement of the hydrodynamic characteristics is of paramount importance to the design, control, and optimization of fluidized beds in these processes. In the past decades, both intrusive and non-intrusive measurement techniques have been developed. Normally, intrusive methods provide only single point measurements and tend to disrupt the hydrodynamics in the region of the vicinity of the probe. Non-intrusive methods such as process tomography, meanwhile, can be used to visualize the entire flow field without causing any disturbance to the flow. Compared to other process tomographic methods, electrical capacitance tomography (ECT) is the most mature and ideal for measurement of gas-solid fluidized beds because of the advantages of no radiation, high temporal resolution, robustness, withstanding high temperature

and high pressure, and low cost.

In ECT measurement, a set of electrodes is mounted around the periphery of the fluidized bed under investigation. Because the gas and fluidized particles have different permittivity, once the distribution and/or concentration of the fluidized particles vary, the inter-electrode capacitance will change accordingly. These changes in capacitance are measured by the sensing electronics and further used to reconstruct an image, in which each pixel is assigned a gray level to represent the material distribution, i.e., the solid concentration, by a specific image reconstruction algorithm. By post-processing the obtained image, some key hydrodynamic parameters, such as the overall solid concentration, number of bubbles, bubble position, and bubble size can then be estimated. Therefore, the key hydrodynamic parameters are obtained in an indirect manner.

However, there are two main difficulties associated with ECT image reconstruction. First, it is severely under-determined due to the number of independent capacitance measurements is much less than the number of pixels in an image. Second, the characteristics of soft-field sensing make the reconstructed images sensitive to noise in raw capacitance measurements. Many image reconstruction algorithms have been proposed so far, and the Landweber iteration algorithm is the most popular for generating high quality images in most cases. However, the iterative process is time-consuming and hence not suitable for on-line measurement. In addition, it is laborious to carry out post-processing of the reconstructed images to obtain key hydrodynamic parameters since additional computing time is needed. Therefore, it is essential to develop an alternative approach for on-line monitoring of key hydrodynamic parameters for flow regime identification, fluidization quality characterization, feedback control, and fault detection of gas-solid fluidized beds by use of ECT.

Note that although the two-phase flow in gas-solid fluidized beds exhibits very complex chaotic behavior, many researches indicated recurrence of similar flow patterns. In this regard, most of the expensive calculations needed for the material distribution reconstruction and post-processing are repetitive and can be possibly avoided via a universal data-driven key parameter prediction recipe.

The aim of this work is to apply machine learning to ECT without image reconstruction for on-line measurement of key hydrodynamic parameters in real gas-solid fluidized

beds. To this end, a machine learning approach was proposed. Figure 1 shows the flowchart of the proposed approach. At first, u_g
linear-increasing strategy was used to perform high-throughput experiments to collect a large amount of training samples that traverse many possible flow patterns in a short time period. Then at the second step, the key hydrodynamic parameters in the training samples were calculated by post-processing the material distribution reconstructed by the Landweber iteration algorithm off-line. Next, supervised machine learning was used to train a map from the measured capacitance to the calculated key hydrodynamic parameters. With the trained model, the key hydrodynamic parameters can be monitored on-line. The proposed machine learning approach can be used to predict any key hydrodynamic parameters in gas-solid fluidized beds, such as the overall solid concentration, number of bubbles, bubble position, and bubble size using the same procedure with little modification.

Ã’Ã‘Ã‰ÃºÂ³É¼Â«Â¸ß¿Ã‰ÃÅ¶ÈµÃ„ËµÃƒÃ·"
src="https://www.aiche.org/sites/default/files/aiche-proceedings/conferences/..."
height="208" class="documentimage">

Figure 1. Flowchart
of the proposed machine learning
approach for the measurement of gas-solid fluidized beds using ECT.

Heat Transfer Study in Corrugated Wall Bubbling Fluidized BED Reactor: Experiments and CFD Simulations

Conference: Fluidization

Year: 2019

Proceeding: Fluidization XVI

Group: General Paper Pool

Session: 2A: Fundamentals of Fluidization

Time: Monday, May 27, 2019 - 5:13pm-5:25pm

Authors:

Wardag, D. A. N. K., Pakistan Institute of Engineering and Applied Sciences
Larachi, D. F., Laval University

Gasification technology must ideally be bestowed with the following traits: auto/allothermal process, non-diluted biosyngas abolishing downstream N_2 separation or upstream O_2 enrichment, thermal coupling via micro-segmentation between endothermic and exothermic steps to improve heat exchange and enhance thermal efficiency, and high yield and heating value of biosyngas. These attributes represent the foremost challenges, next-generation biomass steam gasifiers must cope with. With the endeavor of approaching such ideal configuration, Iliuta et al. proposed a reactor concept of allothermal cyclic multicompartment bubbling fluidized beds. Gas solid fluidized bed reactors would be used to obtain the enhanced heat and mass transport and conversion performances as compared to packed beds. In the current work, corrugated walls were installed in narrow gas-solid bubbling fluidized bed (CWBFB) enclosures to decrease minimum bubbling velocity, reduce bubble sizes, improve gas distribution, offer stable operation and minimize the particles carryover or loss. Thorough analyses of wall-to-bed heat transfer coefficient in flat- (FWBFB) and corrugated- (CWBFB) wall bubbling fluidized beds were performed for a variety of wall declinations and operating conditions covering a range of corrugation angles, inter-wall clearances (*C*), initial rest bed heights (*Hi*) and ratios of gas superficial velocity to minimum bubbling velocity, *Ug*/*U*mb (1-1.55). Fast response self-adhesive heat flux probes and thermocouples were employed to simultaneously measure the wall-to-bed heat flux, surface temperature and bed temperature. These instruments were used to measure the heat transfer coefficient (HTC) at different (18) axial and lateral locations. For a given set of parameters, significant increase in HTC was observed at lower gas flow rate in CWBFB as compared to FWBFB. It showed that CWBFB inventory required lesser U_{mb} (gas flow rate) and offered more economical gas solid fluidization phenomena as compared to FWBFB. Full 3-D transient Euler-Euler CFD simulations employing kinetic theory of granular flow were also carried out which corroborated the experimental findings.

Hydrodynamics and Heat Transfer during Fluidization of Fine Particles

Conference: Fluidization

Year: 2019

Proceeding: Fluidization XVI

Group: General Paper Pool

Session: 2A: Fundamentals of Fluidization

Time: Monday, May 27, 2019 - 4:25pm-4:37pm

Authors:

Garcia Triñanes, P., School of Engineering; Fluid, Heat and Reaction Engineering Group (FHRENG)

Recently, new approaches were taken towards the use of dense upward-flowing suspensions as heat transfer fluids and in thermal energy storage systems. Gas/solid two-phase flows are employed as the heat-transfer fluid with the objective of maximizing the energy capture, and enhance the power conversion efficiency using fluidized-bed technology for heat exchange. In most dense gas-solid fluidization systems, particle circulation (e.g., that induced by the bubbles) is the primary cause of particle convective heat transfer. This work concerns the measurement and analysis of the heat transfer coefficient in dense suspensions of fine particles and the dependence on the hydrodynamics of the system. Results show that the heat transfer coefficient first increases as the gas flowrate increases (because the rate of particle renewal at the wall increases to a maximum), after which it decreases (because the density of particles at the wall decreases and this effect becomes dominant). Further experiments establish also the influence of temperature on the flowability and mechanical properties of the suspensions. This was achieved with a designed heat transfer probe that allows the measurement of the heat-transfer coefficient and using a quartz fluidized bed oven-enclosed. A detailed review of heat-transfer mechanisms will also point out the heat transfer mechanisms in this distinctive kind of suspensions.

References:

Botterill, J.S.M., 1975. Fluid-Bed Heat Transfer. Academic Press, London.
Mickley, H.S., Fairbanks, D.F., 1955. Mechanism of heat transfer to fluidized beds. AIChE J. 1 (3), 374â€“384. Baskakov, A.P., Berg, B.V., 1966. Heat transfer between a

fluidized bed and an immersed cylinder. J. Eng. Phys.
10, 434.
Thring, R.H., 1977. Fluidized bed combustion for the
stirling engine. Int. J. Heat Mass Transfer 20, 911â€"918. Ozkaynak, T.F., Chen, J.C., 1980.
Emulsion phase residence time and its use in heat transfer models in
fluidized beds. AIChE J. 26, 544â€"550.
Wong, Y.S., Seville, J.P.K., 2006. Single-particle motion
and heat transfer in fluidized beds. AIChE J. 52, 4099â€"
4109.
GarcÃa-TriÃ±anes, P., Seville, J.P.K., BoissiÃ¨re, B., Ansart, R.,
Leadbeater, T.W., Parker, D. J., 2016a. Hydrodynamics and particle motion in upward flowing dense particle suspensions: application in solar receivers. Chem. Eng. Sci. 146, 346â€"356.
Benoit, H., Ansart, R., Neau, H. , GarcÃa-TriÃ±anes, P., Flamant, G. and Simonin, O. (2018), Threeâ€dimensional numerical simulation of upflow bubbling fluidized bed in opaque tube under high flux solar heating. AIChE J, 64: 3857 - 3867
GarcÃa-TriÃ±anes P., Seville J.P.K., Ansart R, Benoit H, Leadbeater T.W., D.J. Parker D.J., (2018) Particle motion and heat transfer in an upward-flowing dense particle suspension: Application in solar receivers, Chemical Engineering Science, Volume 177, 2018, 313- 322

Investigation the Strong Swirling Gas-Solids Flow in Cyclone Separators Based in a Circulating Fluidized Bed Based on Process Tomography and CFD Simulation

Conference: Fluidization

Year: 2019

Proceeding: Fluidization XVI

Group: General Paper Pool

Session: 2A: Fundamentals of Fluidization

Time: Monday, May 27, 2019 - 5:25pm-5:37pm

Authors:

Wang, H., University of Chinese Academy of Sciences

Cyclone separator plays an important role on the separating of fine ash from flue gas in a circulating fluidized bed (CFB) system. The gas-solids flow is strong swirling turbulent and the flow characteristics, i.e. flow velocity profiles both for tangential and axial velocity vector, will affect the fine particle separating efficiency. Especial, with the scaling up of a CFB system from small scale to industrial scale, the cyclone diameter increases and the separating efficiency decreases. In addition, with the increase of thermal capacitive, the number of cyclone also increases and the mal-distribution of solids flux among different cyclones becomes serious. This mal-distribution of solids flux will affect the heat and mass transfer balance inside the CFB riser and lead to low combustion efficiency. Therefore, it is necessary to investigate the flows in cyclones and to understand the flow characteristics change with the increase of cyclone dimensions and cyclone number in CFB system. In this research, details computation fluid dynamics (CFD) based on different turbulent models, i.e. standard k-eps two equations models, RNG k-eps model and Reynold Stress Turbulent model, will be given to investigate the swirling flow fields in cyclones. To account for the effect of the stochastic characteristic of the instantaneous gas velocity on the particles, the improved Lagrangian stochastic model based on the Reynolds stress model for gas phase was adopted, which attributed to the successful prediction of the turbulence anisotropy and particle crossing-trajectories effects. Meanwhile, electrical capacitance tomography (ECT) will be applied and conical shape ECT sensors will be designed. Solids concentration will be given based on ECT measurement results. Different arrangement of cyclones in a CFB system are investigated and details CFD simulation results will be given to reveal the dimensions effect on the gas-solids flow characteristics. Optimized design based on CFD simulation and ECT measurement will be given in the end of this research.

Keynote Talk: Visualisation of Pharmaceutical Fluidised Beds By Electrical Capacitance Tomography

Conference: Fluidization

Year: 2019

Proceeding: Fluidization XVI

Group: General Paper Pool

Session: 2A: Fundamentals of Fluidization

Time: Monday, May 27, 2019 - 4:00pm-4:25pm

Authors:

Wang, H., University of Chinese Academy of Sciences
Yang, W., The University of Manchester
Ye, J., Institute of Engineering Thermophysics, Chinese Academy of Sciences

Gas-solids fluidised beds are commonly used in the pharmaceutical industry and other industries. However, the understanding of the fluidisation processes is still limited because the measurement methods and tools are limited, which can provide both real-time monitoring and online measurement. Therefore, it is difficult to optimise the operation of fluidised beds. The consequence is low efficiency, more energy consumed, more pollution to the environment and more importantly it is difficult to guarantee the product quality. Electrical capacitance tomography (ECT) has been developed as a visualisation and measurement tool for deal with complicated industrial processes, such as fluidised beds. In recent years, ECT has been used to investigate granulation, drying and coating processes in pharmaceutical fluidised beds. For measuring drying and granulation processes, a novel online calibration method was proposed and implemented to deal with continuous change in moisture content and hence the continuous change in equivalent permittivity. This method provides a means of continuous measurement of the solids' moisture level and its distribution, in addition to gas/solids distribution. Experiments were carried out using different scales of fluidised beds, lab-scale, pilot-scale and production-scale, and good results were achieved. For example, the online measurements of moisture level by ECT are in agreement with those measured offline by a moisture meter. For monitoring the coating process in a lab-scale Wurster fluidised bed, a unique ECT sensor is designed with three sets of electrode, one set on internal wall of drying bowl, one set on outside wall of Wurster tube and one set on internal wall of Wurster tube. By this arrangement, all regions of the fluidised bed can be monitored. Because the coating process involves in gas/solids/liquid three phases, it is more difficult to interpret the measured capacitance data and more difficult to implement image reconstruction than a normal fluidised bed containing gas and solids only. Based on the work done so far, it is concluded that ECT is very useful for monitoring and measurement of pharmaceutical fluidised beds. The remaining challenge is how to deal with engineering and pharmaceutical regulation issues.

Particle Residence Time Measurement in Dual Fluidized Bed Pyrolysis Unit

Conference: Fluidization

Year: 2019

Proceeding: Fluidization XVI

Group: General Paper Pool

Session: 2A: Fundamentals of Fluidization

Time: Monday, May 27, 2019 - 4:37pm-4:49pm

Authors:

Pfeifer, C., University of Natural Resources and Life Sciences
Frohner, A., University of Natural Resources and Life Sciences, Vienna
Tondl, G., University of Natural Resources and Life Sciences, Vienna

A measurement system for particle tracking is designed and implemented in a scaled dual fluidized bed cold-flow model. A riser operating as combustion section, is connected to a bubbling fluidized bed operating as pyrolysis section. The riser provides the energy for the pyrolysis section transported by the hot bed material circulating in the system. This system can be applied to produce a high calorific value fuel gas out of waste plastic material. The majority of the feedstock material is converted to gaseous as well as liquid products which can be directly utilized e.g. in burners for rotary kilns (cement industry, waste incineration, â€¦). Solid residues such as char can be utilized as energy carrier and ash as well as metal fraction is sorted out of the system.

The particle tracking system consists of two inductors, which are placed upstream and downstream of the pyrolysis section. Metal particles with a mass fraction of 1 percent of the bed material are used as tracking particles. The particles are released on the upstream side of the pyrolysis zone and increase the inductance when passing the inductors. The inductance change is measured with two measurement bridges. Residence time distribution and circulation rates are analyzed for different operating points. Moreover, the behavior of the dual fluidized bed systems for different geometries of the pyrolysis section is determined. The measurements are set into reference to calculation of mass and energy balances of the conversion of feedstock to syngas respectively thermal energy. Optimal process operation window is determined as basis for the design of a hot pilot scale unit.

The Effect of Kaolin As Bed Material in an Oxy-CFB Combustor

Conference: Fluidization

Year: 2019

Proceeding: Fluidization XVI

Group: General Paper Pool

Session: 2A: Fundamentals of Fluidization

Time: Monday, May 27, 2019 - 4:49pm-5:01pm

Authors:

Nguyen, H. K., Korea Institute of Energy Research
Song, B., Kunsan National University
Moon, J., Korea Institute of Industrial Technology (KITECH)
Lee, J. G., Korea Institute of Energy Research
Mun, T., Korea Institute of Energy Research
Bae, D., Korea Institute of Energy Research
Jo, S., Korea Institute of Energy Research
Park, S., Korea Institute of Eerngy Research
Oh, G., Korea Institute of Energy Research
Seo, M., Korea Institute of Energy Research
Ra, H., Korea Institute of Energy Research
Yoon, S., Korea Institute of Energy Research
Yoon, S., Korea Institute of Energy Research

Kaolin ($Al_2Si_2O_5(OH)_4$) has been considered as an effective addition for bed material in a circulating fluidized bed combustion (CFBC) system due to its advantages to reduce agglomeration, fouling and deposit problems. In the present paper, the impact of kaolin addition as bed material during the oxy-combustion in a 0.1 MWth oxy-CFB system with sub-bituminous coal and wood pellet were investigated through the combustion characteristics. The above aspects are operating temperature and pressure profile, flue gas composition including CO_2, O_2, SO_2, NO, CO and the sampled ash properties by analyzing XRF, SEM-EDS, unburned carbon (combustion efficiency) and their particle size distribution. The results revealed that kaolin in steady of silica sand has contributed to smooth fluidization and could keep the oxy-fuel combustion stable with the exit gas of having a high concentration of CO_2 (above 90 vol.% on dry basis) and low emission of CO, SO2, and NOx. The attrition index (AI) value of raw silica sand and fly ash sampled with experimentally added kaolin were 0.46 and 0.58%, respectively. No agglomeration was found in the bed for the most operating period. Consequently, it can be said that the feasibility of a kaolin material as substitutive silica sand is reasonable.

Plenary Talk: Fluidized Bed Mineral Roasting: From Fundamental to Application

Conference: Fluidization

Year: 2019

Proceeding: Fluidization XVI

Group: General Paper Pool

Session: Plenary Session #1

Time: Monday, May 27, 2019 - 9:10am-9:50am

Authors:

Zhu, Q., Institute of Process Engineering, Chinese Academy of Sciences

China has large reserves of low grade minerals, which are difficult to be beneficiated by conventional processes, and are currently not utilized. To utilize these minerals, it is therefore often needed to transform ores from difficult-processed phases and structures to more easily processed phases and structures, normally through chemical processes like roasting via oxidation, reduction, etc. While roasting can be processed via reactors like shaft furnaces, rotary kilns, fluidized beds, etc., fluidized bed (FB) reactors have been proven to be the most efficient reactors for mineral roasting. In the present presentation, fluidized bed roasting practices of our group will be introduced.

Over the past decade, several fundamental aspects related to FB mineral roasting have been explored, like the influence of roasting conditions on the phase and structure change, process intensification via FB configuration and internals, adjustment of means residence time (MRT) for the coarse and fine particles, optimization of FB scale-up through CFD simulations, etc., have been systematically investigated. Moreover, our practices on engineering implementation of fluidized bed mineral roasting will also be introduced, including oxidative-reductive roasting of Panzhihua ilmenite for the production of synthetic rutile, magnetizing roasting of low grade iron ores for iron ore beneficiation, direct reduction of iron ore concentrate for the development of non-blast furnace ironmaking technology, calcination of aluminum hydroxide to produce 50-70% a-Al_2O_3, reduction of low grade pyrolusite for $MnSO_4$production, etc. Several pilot and commercial scale plants have been established with cooperation of various companies, e.g. 200 kt/a reduction plant has been successfully put in operation for over 3 years for the reduction of low grade pyrolusite in Yunn province, China.

Plenary Talk: Recent Advances in the Multi-Scale Simulation of Mass, Momentum and Heat Transfer in Dense Gas-Particle Flows

Conference: Fluidization

Year: 2019

Proceeding: Fluidization XVI

Group: General Paper Pool

Session: Plenary Session #1

Time: Monday, May 27, 2019 - 8:30am-9:10am

Authors:

Kuipers, H. J. A. M., Eindhoven University of Technology

Dense gas-particle flows involving mass, momentum and heat transfer are frequently encountered in a variety of large scale industrial processes such as coating, granulation, polymerization and the production of base chemicals and synthetic fuels. In such flows the interactions between the dispersed particles and the continuous gas or liquid phase as well as the complex interactions among these particles govern the prevailing phenomena, i.e. the formation and evolution of heterogeneous flow structures. These structures have a significant impact on the quality of the contact between the phases and consequently strongly affect the overall performance of the process. Substantial additional complexities arise from the presence of non-spherical particles and particles which are partially wet due to the injection of a liquid.

Due to the inherent complexity of dense gas-particle flows a multi-scale modeling approach is used in which both fluid-particle and particle-particle interactions can be properly accounted for. The idea is essentially that fundamental models accounting for the relevant details of both fluid-particle (DNS) and particle-particle (DEM) interaction, are used to quantify the complex phase interactions in terms of closures. These closures are required in continuum models (i.e. the two-fluid model (TFM) or multi-fluid model (MFM)) based on the Kinetic Theory of Granular Flow (KTGF) suited to compute the flow structures on a much larger (industrial) scale. In this presentation recent advances in the multi-scale modelling of dense gas-particle flows will be highlighted with emphasis on coupled mass, momentum and heat transfer. In addition, areas which need substantial further attention will be briefly discussed including the challenges for non-invasive monitoring of such complex flows.

Tailor-Made Particles by Fluidized and Spouted Bed Spray Granulation: Opportunities and Recent Advancements

Conference: Fluidization

Year: 2019

Proceeding: Fluidization XVI

Group: General Paper Pool

Session: Plenary Session #1

Time: Monday, May 27, 2019 - 9:50am-10:30am

Authors:

Heinrich, S., Hamburg University of Technology
Pietsch, S., Hamburg University of Technology
Kieckhefen, P., Hamburg University of Technology
Diez, E., Hamburg University of Technology

Spray granulation and coating are important process steps in the production of many high-value products. Besides the common fluidized beds, spouted beds can be used for gas-solid processes with liquid injection as the mushroom-shaped spout pattern offers advantages in terms of heat, mass and momentum transfer. In both apparatuses, spray nozzles inject a solution, suspension or melt into the bed of fluidized particles. The solidification of the liquid layer results in an onion-like growth of the granules. The main challenges are the formation of a homogeneous layer without cracks on one single particle and the homogeneous coating of the entire particle bed in one batch. In-line coating layer measurements are advantageous compared to off-line methods because real-time data are available and the process is not disturbed. Two different approaches for in-line monitoring of the coating progress in a laboratory spouted bed are presented: digital image analysis and optical coherence tomography (OCT). While the digital image analysis approach is able to detect qualitatively the coating homogeneity, the OCT method allows the quantification of layer thicknesses.

For industrial applications, the operation of spray granulation and coating processes is preferred on larger scales and in continuous mode of operation. In continuously operated spray granulation processes, horizontally constructed fluidized beds with rectangular cross sections are commonly used, which are divided by a variable number of plates (weirs) into several chambers for different functionalities (e.g. for granulation, coating, drying, cooling). Commonly, an external product processing, consisting of pneumatic conveying, screening, grinding of the oversize granules and recycling of grinded oversize and undersize particles into the fluidized bed granulator is applied. These can be modeled by multi-dimensional Population Balance Models (PBM) on the process unit scale. The resulting internal and external networks of solids process unit operations as well as gas, liquid and solid flows lead to a complex and dynamic process behavior, which can be captured in flowsheet simulations by combining the individual PBMâ €™s. By adjustment of the process parameters the granule properties can be adjusted and optimized for example regarding the flowability or the morphology of the product. The spray rate and the drying temperature were found to have a major effect on the surface structure of granules.

Nowadays, computational methods offer great opportunities for understanding the details of processes on micro and macro scale. For spray granulation and coating processes, the CFD-DEM method is state of the art to simulate the interaction between fluid, gas and droplets. This method is very computationally expensive as both particle-particle and particle-fluid interactions are resolved. To cover the diversity of time scales present in industrial or even pilot scale plants, the recurrence CFD method (rCFD) offers remedy: by capturing the recurrent flow patterns present in the phase dynamics and extrapolating them in a physically meaningful manner, the computational effort is minimized and the numerically cheaper transport processes can be tracked on much larger time scales. The methodâ€™s applicability was validated by showing the accurate reproduction of instantaneous and average bed distributions and particle residence times in a spray zone. The speedup of rCFD over CFD-DEM for pure dynamics was 2100x for the laboratory spouted bed apparatus.

Plenary Talk: Fluidization and Fluid Particle Systems Research Contribution in Creating A Pathway to Sustainable Society

Conference: Fluidization

Year: 2019

Proceeding: Fluidization XVI

Group: General Paper Pool

Session: Plenary Session #2

Time: Monday, May 27, 2019 - 10:50am-11:30am

Authors:

Arastoopour, H., Illinois Institute of Technology

Securing future supplies of energy and water in view of continuous growth in the world population requires the creation of a pathway to sustainable energy and water that is both multifaceted and complex. An effective pathway aims to improve the reliability, security and affordability of energy and water by utilizing a least-cost strategy to reduce the negative impact of energy consumption on climate change; securing water availability; emphasizing the gradual decarbonization of the global energy system; increasing energy efficiency and conservation; and preserving natural resources and the environment. This presentation highlights the potential contribution and impact of Fluidization and fluid particle systems research to the successful creation of a pathway to sustainable society and economy.

In this presentation the role of fluidization and fluid particle computational fluid dynamics (CFD) research in carbon capture and sequestration processes, particle pulverization as a recycling tool for polymeric and elastomeric materials, production of natural gas from unconsolidated hydrate reservoirs, fluid particle heat transfer medium for concentrated solar power (CSP) systems and effect of rain droplets on wind turbines performance will be discussed.

Plenary Talk: Innovations in Fluidization, Inspired by Nature

Conference: Fluidization

Year: 2019

Proceeding: Fluidization XVI

Group: General Paper Pool

Session: Plenary Session #2

Time: Monday, May 27, 2019 - 11:30am-12:10pm

Authors:

Coppens, M. O., University College London

The non-uniform hydrodynamics of fluidized beds complicate their operation and scale-up. Over the past two decades, we have developed approaches that seek to structure bubbling fluidized bed hydrodynamics by drawing lessons from natural systems. A first example is a fractal injector, inspired by the structure of lungs and trees, to distribute fluids uniformly, promote local mixing, and preserve hydrodynamics over different length scales. A second example is pulsation of the gas flow, inspired by wave and wind-induced regular patterns on dunes and beaches, to organize a chaotic bubble flow into a regular, triangular tessellation of rising bubbles (deep, quasi-2D bed) or a square pattern (shallow, 3D bed) with controllable wavelength and bubble size, over a range of frequencies.

Simulations provide insights into the mesoscopic physics that induce the observed patterns by dynamic self-organization. Wave patterns in shallow beds display similarities with those observed in vibrated granular layers, although fluid-solid interactions lead to much richer behavior and qualitative differences as well, which are especially noted in the bubble patterns formed in deeper, quasi-2D beds. Common two-fluid models are unable to reproduce pattern propagation, because granular friction results in transitions in collective particle behavior between solid-like and fluid-like, during each pulsation period. On the other hand, discrete element model simulations, coupled with computational fluid dynamics, offer insights that do not only allow us to employ gas oscillation as a way to structure fluidized beds, but also to take pattern formation as a robust fingerprint to test computational models, which can be used to model and design fluidized beds more generally. Probing the rich behavior of pulsating fluidized beds helps us to augment our understanding of fluidization fundamentals. In this context, we are currently investigating annular beds, allowing us to study transitions in hydrodynamics between quasi-2D to 3D systems, which is essential in the application of this fascinating nature-inspired phenomenon to fluidized bed operation and design.

Accelerated Carbonation of Mineral Wastes in Fluidized Beds

Conference: Fluidization

Year: 2019

Proceeding: Fluidization XVI

Group: General Paper Pool

Session: 3A: Fundamentals of Fluidization

Time: Tuesday, May 28, 2019 - 9:07am-9:19am

Authors:

Garcia Triñanes, P., School of Engineering; Fluid, Heat and Reaction Engineering Group (FHRENG)
Hills, C., University of Greenwich

Minerals with high content of silicate and rich in calcium are known to readily react with carbon dioxide (CO2). The reaction products can cause rapid hardening and derives in the production of aggregate materials which can then be used as construction materials and other valuable purposes. This paper reviews recent developments in this emerging technology and provides an insight in the use of fluidized beds as suited reactors providing good mixing and adapted space-time. The effects of the accelerated carbonation reaction on different solid particles and possible strategies to achieve adequate fluidization such as regulating phase ratios, gas-solid contact and critical film layer are considered here.

References:

Scuzzarella, A, Fernandez Bertos, M, Simons, S J, Hills, Colin and Carey, Paula (2006) *Expansion of cohesive gas fluidized binary solid systems.*Powder Technology, 163 (1-2). pp. 18-22
Maries, Alan and Hills, Colin (2015) *Ultra-rapid hardening of cement by accelerated carbonation â€“ Past, present and future* In: 5th International Conference on Accelerated Carbonation for Environmental and Material Engineering 2015 (ACEME). AIChE, New York
Garcia-Trinanes, P., Zigan, S., Necas J. (2016) Assisted fluidization of cohesive powders. I: 22nd International Congress of Chemical and Process Engineering, CHISA 2016 and 19th Conference on Process Integration, Modelling and Optimisation for Energy Saving and Pollution Reduction, PRES 2016

An Experimental Study of Bubble Size Distribution in a Two-Dimensional Gas-Solid Tapered Fluidized Bed

Conference: Fluidization

Year: 2019

Proceeding: Fluidization XVI

Group: General Paper Pool

Session: 3A: Fundamentals of Fluidization

Time: Tuesday, May 28, 2019 - 9:43am-9:55am

Authors:

Sahoo, L., Indian Institute of Technology Madras
Sarkar, S., Indian Institute of Technology Madras

Bubbles are very common in gas-solid fluidized bed reactors. They form the core of fluidized bed hydrodynamics, which affects the heat transfer, mass transfer and reaction kinetics as well as mixing and segregation behaviour of fluidized beds. The bubble behaviour of columnar fluidized beds is well studied and reported in the literature. However, there is a lack of study of bubble behaviour in tapered/conical fluidized beds, which is significantly different from the columnar fluidized beds. To study bubble behaviour, experiments were carried out using Geldart group B glass beads in a laboratory-scale pseudo-two-dimensional tapered fluidized bed with air as the fluidizing medium using high speed imaging technique. The images that were captured using high speed camera, were further processed using digital image analysis technique to quantify the bubble phase from the emulsion phase. The bubbles were classified into three groups: small (0-3 cm), medium (3-6 cm) and large (>6 cm), based on their diameters. The effects of tapered angle and superficial air velocity on the bubble size and itsâ€™ distribution were studied along the vertical and lateral directions. It is found to be affected by these parameters significantly. Further, coalescence and break-up of bubbles that were observed in the bed is also found to affect the bubble size and itsâ€™ distribution.

Characterization of Dynamics of Binary Gas-Solid Flow of Particles with Different Density and Shape Using Digital Image Analysis

Conference: Fluidization

Year: 2019

Proceeding: Fluidization XVI

Group: General Paper Pool

Session: 3A: Fundamentals of Fluidization

Time: Tuesday, May 28, 2019 - 8:55am-9:07am

Authors:

Parvathaneni, S., Indian Institute of Technology Delhi
Buwa, V., Indian Institute of Technology Delhi

Introduction

Owing to their excellent mixing characteristics, gas-solid fluidized beds are used in several industrial processes like coal and, biomass gasification and combustion. These industrial processes involve mixtures of particles with different size, density and, shape (eg. coal and inerts, coal and ash, biomass and inert used in gasification process). The difference in the physical properties (size, density, and shape) of particles leads to a non-uniform mixing and segregation phenomena at a particular range of superficial gas velocity and bed composition. As the local fluidization behaviour governs the mixing and segregation phenomena, mass and heat transfer mechanisms, and eventually the performance of the fluidized bed reactor, it is important to characterize the dynamics of binary gas-solid flow. Therefore, time-and space-resolved phase volume fraction measurements are important to comprehend the dynamic characteristics of flow, which can elucidate the effects arising due to the mixing of particles that differ in size, density, and shape.

The mixing and segregation phenomena of binary gas-solid flow was investigated by several researchers and the main focus of their work was to predict the bed dispersed height, minimum fluidization velocity (U_{mf}), and flow regimes. Goldschmidt et al. (2003) and Olaofe et al. (2013) quantified the dynamics of mixing and segregation phenomena of binary system differing in the size through percentage of segregation and average bed height of solid phases with respect to time. However, the dynamics of mixing and segregation of binary system differing in density and shape is not reported in the open literature. In addition to the measurements of time-evolution of percentage of segregation and average bed height of solid phases, it is also important to measure the bubbling characteristics (bubble size distribution and bubble rise velocity) because of its significance in influencing and understanding the flow behaviour. These detailed measurements are required in understanding and characterizing the dynamics of fluidization behaviour of binary system differing in density and shape and to develop and to rigorously validate the Eulerian CFD models to simulate the binary gas-solid flow.

The present work is carried in a pseudo 3-D (rectangular) fluidized bed because of its advantage of visualizing the flow behaviour. The objectives are (i) to measure the local individual solid volume fraction (Î±$_s$) of two binary systems, one differing only in density (of same size) and another one only in shape (of same density) from the grey scale images obtained by Digital Image Analysis technique, (ii) to characterize the dynamics of mixing and segregation behaviour using the local Î±$_s$ measurements and (iii) to develop a methodology to measure the bubble size distribution and bubble rise velocities for a wide range of bed compositions and superficial gas velocities.

Experimental Set-up and Methodology

The experiments were performed in a rectangular column with height (H) of 1 m, width of 0.2 (m) and depth of 0.02 (m) with an initial static bed height of 25 cm. Air was used as the fluidizing medium and glass beads (GB) (d_p = 2.26 mm; Ï$_p$ = 2500 kg/m^3), mustard seeds (MS) (d_p = 2.18 mm; Ï$_p$ = 720 kg/m^3), cylindrical glass particles (CG) (d_p = 2 mm; L = 10 mm; Ï$_p$ = 2500 kg/m^3) were used as the solid phases. The experimental images were recorded with a highâ€“speed monochromatic camera at 640 Ã— 800 pixel^2 resolution and at a rate of 400 fps. The background was illuminated with a LED light and two solid phases were colored black and white to create a difference in the color intensities of the three phases. As a part of post-processing, the grey-scale images obtained are converted into RGB images and the local area fraction of three phases (two solids and one gas) was acquired using the image processing toolbox in MATLAB. The dynamics of fluidization of binary system-1 (GB: MS) and binary system-2 (CG: GB) in the weight ratio of 70%:30%, 50%:50%, and 30%:70% solids was investigated for superficial gas velocities (U_G) in the range 1.1-2 U_{mf} of dense phase where segregation and mixing of the particles was visually observed. Further details of the experimental set-up, methodology and the calibration procedure to identify all three phases (two solids and gas) from grey scale image will be presented in the full manuscript.

Results and Discussion

The time-evolution of percentage of segregation and average bed heights of individual solid phases are calculated from the RGB image and calculation criteria based on Goldschmidt et al. (2003) was used. For binary system-1 (GB: MS = 70:30 wt. %) at a superficial gas velocity of 1.73 m/s (1.1 U_{mf} of GB), it was observed that the time-evolution of average bed height of GB reached 15 cm from an initial static bed height of 25 cm after 25s, which elucidates that the lower portion of the bed is segregated and dominated by GB particles. This is due to the operating U_g is just above the U_{mf} of dense phase i.e., GB. The time-evolution of percentage segregation calculated was found to reach 60 % after 25 s and thereafter further change was no observed in the segregation pattern of the solid phases. The calculated quantities showing the final time required to attain the segregation and average dispersed bed heights reached by individual solid phases are in good agreement with the visual observations.

For binary system-2 with particles differing in shape, for a bed composition of CG: GB = 70:30 wt. %, segregation was not observed at lower U_G due to the channelling as the bed is dominated by CG particles. Therefore, experiments were performed at a higher U_G of 2.78 m/s where mixing was observed. In addition, from the information of the bubble area fraction and displacement of bubble with respect to time and axial location, bubble size distribution and bubble rise velocities are calculated and compared with binary system-1 at U_G = 2.78 m/s. The detailed measurements and analysis of the effect of U_G, and bed composition on dynamics of mixing and segregation, bubble size distribution and bubble rise velocity of the two binary systems will be presented in the full manuscript.

Conclusions

Local solid phase area fraction measured using high-speed imaging was used to characterize the segregation of binary gas-solid flow at lower gas velocities and fluidization behaviour at higher gas velocities. The effects due to the difference in solid phase physical properties (density and shape) and mixture compositions on dynamics of mixing and segregation of binary gas-solid flow are investigated for a wide range of superficial gas velocities. The dynamics of mixing and segregation were quantified in terms of time-evolution of percentage of segregation and averag bed heights of individual phases. In addition to this, to compare the fluidization behaviour of binary gas-solid flow differing in density and shape, bubble size distribution and bubble rise velocities are measured at higher gas velocities. Detailed analysis of the aforementioned measurements will be presented in the full manuscript. The present work will be useful to understand the differences in dynamic behaviour of binary gas-solid fluidization caused due to difference in density and shape of particles. Further, the comprehensive set of experimental data presented will be used to validate the Eulerian CFD models.

Acknowledgments

One of the authors (Sirisha Parvathaneni) gratefully acknowledges the NETRA of NTPC Limite for providing the research fellowship.

References

Goldschmidt, M.J.V., Link, J.M., Mellema, S. and Kuipers, J.A.M. Power Technology, Vol. 138, 135 - 159 (2003).

Olaofe, O.O., Buist, K.A., Deen, N.G., van der Hoef, M.A. and Kuipers, J.A.M. Power Technology, Vol. 246, 695 - 706 (2013).

Effects of Granular Temperature on Inter-Phase Heat Transfer in Gas-Solid Flows

Conference: Fluidization

Year: 2019

Proceeding: Fluidization XVI

Group: General Paper Pool

Session: 3A: Fundamentals of Fluidization

Time: Tuesday, May 28, 2019 - 9:55am-10:07am

Authors:

Huang, Z. Q., Xi'an Jiaotong University
Zhou, Q., Xi'an Jiaotong University

Direct numerical simulations with the immersed boundary-lattice Boltzmann method are used to investigate the effect of granular temperature on the heat transfer. The numerical method is validated first by comparing the simulation results for the heat conduction around a stationary particle immersed in stationary fluid with the analytical solution, then by comparing the results for forced convection around a single particle and around random arrays of spherical particles with empirical correlations published in literatures. To mimic the granular temperature of a typical gas-solid system, random velocities obeying the isotropic Maxwellian distribution are imposed to all particles in the computational domain. The simulation results show that the effect of granular temperature on the heat transfer is significant especially when the solid volume fraction and the particle Reynolds number are small. Based on the present results, a new term reflecting the effects of granular temperature and is added to the â€œstaticâ€ Nusselt number correlation to formulate a new heat transfer correlation.

Forces on Inserted Objects in Fluidized

Conference: Fluidization

Year: 2019

Proceeding: Fluidization XVI

Group: General Paper Pool

Session: 3A: Fundamentals of Fluidization

Time: Tuesday, May 28, 2019 - 9:31am-9:43am

Authors:

Zhang, Y., China University of Petroleum,Beijing
Liu, D., China University Of Petroleum ,Beijing

Various objects are often inserted in fluidized beds for different purposes. For example, immersed tube bundles are used fluidized beds for heat transfer and baffles made of multiple inclined slats are widely employed in industrial fluidized bed reactors to split bubbles, suppress gas/solids back-mixing for higher conversion and better product selectivity. As well, probes are usually inserted in beds to monitor operating parameters. However, all these immersed internals are subject to forces during the ddifferent operation states in fluidized bed reactors. Erosion and excessive vibration caused by these forces may lead to internal failure, especially in industrial fluidized beds with hostile environment and needing several years of continuous operation. Therefore, the information of forces on internals under different operation states is very helpful for their reasonable strength design and long-period reliability.

In this study, forces on a horizontal slat inserted in the bed were summarized during different operating states, i.e. defluidized, start-up and steady fluidized states. During the starting-up of a fluidized bed, the bed will experience the above three different operating states and force acting on immersed objects are also different. The size of the slat was 300 mm (long) x 50 mm (wide)x 4 mm (thick). The forces on the slat were measured by adhering strain gauges on its surface. All experiments were carried out in a cold fluidized bed with a square cross-section of 300 mm×300 mm. The bed materials were fine FCC particles (Group A) and coarse silica sand particles (Group B). Other main influencing parameters considered in this study are gas velocity, installation level, inclination angle etc.

Typical transient stress signals on the slat measured under the three different operation states of FCC and silica sand beds are measured by the test slat. In defluidized state, the slat was subject to a downward force due to the weight of the upper un-fluidized particles. When the fluidized air was supplied, i.e. during start-up state, the slat encountered a very high upward force impulse. Then, when the beds were completely transformed into a steady fluidized state, the slat suffered continuous fluctuating forces induced by the dynamic gas-solids behaviors in beds. The magnitudes of the fluctuating force signals are several times smaller than that measured during start-up state, especially in FCC bed. The results indicate that, for the three different operation states, start-up process is the most dangerous condition for internals immersed in fluidized bed reactors. The designer should be paid more attention to this special operating condition. If the baffles immersed fluidized bed reactors without consideration of the special condition in design, the high transient load may lead to serious damage of immersed internals and their supports. On the other hand, if the situation happened in experimental unit, it also may result in damage of inserted expensive measurement instruments.

Homogeneous, High Packing Fraction Fluidization of Coarse Particles Using a Combination of Gas Flow and Vibration

Conference: Fluidization

Year: 2019

Proceeding: Fluidization XVI

Group: General Paper Pool

Session: 3A: Fundamentals of Fluidization

Time: Tuesday, May 28, 2019 - 9:19am-9:31am

Authors:

Boyce, C. M., Columbia University
McLaren, C., ETH Zurich
Kovar, T. M., Columbia University
Penn, A., ETH Zurich
Müller, C. R., ETH Zurich

When subject to both upward gas flow and vertical vibration, Geldart Group B and D particles are shown to fluidize at gas flow rates below that needed without vibration in a state with high particle packing fraction and no gas bubbles. We investigate the effects of particle size, superficial gas velocity and vibration strength on the minimum fluidization and minimum bubbling velocities in 3D and pseudo-2D beds. Further, we use particle tracking velocimetry in a pseudo-2D bed to measure the particle velocity distribution and understand how it compares to molecular and granular gases. We use theoretical analysis to uncover the physics underlying this non-bubbling fluidized state and test the capabilities of computational models of fluidization to reproduce the phenomenon. Finally, we investigate the capability of this homogeneously fluidized state to form structured flow patterns.

Keynote Talk: Fluidization of Graphene Nanoplatelets: From Microstructure to Hydrodynamics

Conference: Fluidization

Year: 2019

Proceeding: Fluidization XVI

Group: General Paper Pool

Session: 3A: Fundamentals of Fluidization

Time: Tuesday, May 28, 2019 - 8:30am-8:55am

Authors:

van Ommen, J. R., Delft University of Technology
La Zara, D., Delft University of Technology
Grillo, F., ETH Zurich

Graphene is a material consisting of a single two-dimensional layer of sp2 hybridized carbon atoms arranged in a honeycomb lattice. Since its discovery in 2004, graphene has been attracting considerable attention due to its exceptional physical, mechanical, thermal, electronic and optical properties. In fact, it has an extremely high surface area (theoretically 2630 m^2/g) and it is the lightest (with a mass of 0.77 mg for 1 m^2), but at the same time the strongest material known (with an ultimate tensile strength of 130 GPa and a tensile stiffness of â^¼1 TPa) [1]. Moreover, it is the best conductor of heat at ambient conditions (â^¼5,000 W m^{-1} K^{-1}) and electricity (up to 200,000 cm^2 V^{-1} s^{-1}), and it is optically transparent (with a light absorption of 2.3% of white light) [2-4]. One of the most common methods to synthesize graphene is liquid-phase exfoliation, where graphite is milled into a powder whose particles are separated into tiny flakes (also called platelets) by mechanical forces in liquid. These flakes which consist of a few layers of graphene are then sorted based on their size and thickness, i.e., number of layers. Yet, controlling the number of layers in graphene manufacturing remains challenging and often the final product consists of a powder containing multilayer (i.e., â‰¤ 10 layers) flakes. Still, the properties of multilayer graphene powders are relevant to practical applications such as electric batteries and field emitters.

The number of layers of graphene determines its microstructure which affects the powder bulk behaviour (e.g., flowability, dispersibility, fluidizability and agglomerability). One of the most interesting applications of graphene is its use in composite materials and, for example, graphene powders are emerging in the field of catalysis as excellent support materials. An attractive route for the functionalization of graphene with metal and metal-oxide nanoparticle catalysts is atomic layer deposition carried out in a fluidized bed reactor. In order to achieve an uniform deposition of nanoparticles for high-performance catalysts, ensuring a good fluidization behaviour of graphene powders is crucial. Therefore, understanding how graphene powders fluidize depending on their microstructure is vital.

In this work, we study for the first time the fluidization behavior of three kinds of graphene powders: graphene nanoplatelets of 150 m^2/g (GNP150), graphene nanoplatelet aggregates of 300 m^2/g (GNPA300) and graphene nanoplatelet aggregates of 750 m^2/g (GNPA750). Graphene powders fall under group C of Geldartâ€™s classification, i.e., cohesive powders. Therefore, their fluidization is expected to be prevented by the strong interparticle forces, which exceed the drag force induced by the gas flow. However, the graphene platelets do not fluidize individually, instead they form into agglomerates, due to the aforementioned cohesive forces, as earlier observed for semi-spherical nanoparticles [5, 6]. As a result, the fluidization behavior depends on the properties of the agglomerates. We correlate the morphology of the graphene nanoplatelets at nano-, micro- and macro-scales to their hydrodynamic behaviour.

GNP150 at the nanoscale consist of agglomerates of mostly 2D flakes (6-8 nm thick and 15 Î¼m wide). The latter, in virtue of being relatively thick, tend to be rigid, or in other words they are not prone to fold and wrinkle. As such, they truly enjoy a 2D geometry. Rigid 2D-like objects pack poorly compared to 3D-like flexible objects. As a result of that, agglomerates consisting of 2D thick nanoplatelets are loosely packed, hence GNP150 have a low bulk density. At the macroscale, GNP150 form mm-sized spherical agglomerates, whose size distribution depends on the history of the powder. Upon vibration-assisted fluidization, the bed of particles segregates into mm-sized agglomerates at the bottom, increasingly smaller agglomerates along the bed height, and fines above the splash zone. No fluidization is observed without the use of vibration at any of the tested superficial velocities (i.e., up to 16.85 cm/s). The use of vibration results into bubbling fluidization at relatively low superficial velocities (1.35 cm/s).

GNPA300 and GNPA750 at the nanoscale are not found in 2D form. In fact, higher-surface-area graphene powder translates into thinner nanoplatelets that are more prone to fold and wrinkle. As a result of that, the nanoplatelets are closely packed, and thus have an higher bulk density. At the macroscale, the powders look very fine and the size of the agglomerates is hardly discernable by naked eye. For GNPA300, fluidization occurs only when assisted via vibration. Fluidization occurs at higher gas velocities than in the previous case, i.e., 2.7 cm/s. Once fluidization occurs, the bed height remains approximately constant with increasing gas velocity. High superficial velocities result in slugging behaviour and significant elutriation. Instead, for GNPA750, fluidization does occur even without the use of assisting methods upon breaking of channels. In fact, increasingly thinner nanoplatelets translate into more cohesive agglomerates and higher agglomerate density. Even though cohesiveness deteriorates fluidization quality, higher densities promote fluidization. In light of this, the fact that GNPA750 fluidizes even without vibration is no surprise, as we expect the increased cohesiveness between nanoplatelets to result into highly packed and thus dense agglomerates that act as micron-sized Geldart-B-like powders.

This work demonstrates the large influence of the structure at the nm- to mm-scale on the large-scale fluidization behaviour. Such insight is essential when scaling up processes for gas-phase production of nanostructured materials [7].

References

1. Lee, C., X. Wei, J.W. Kysar, and J. Hone, *Measurement of the Elastic Properties and Intrinsic Strength of Monolayer Graphene*.Science, 2008. **321**(5887): p. 385-388.
2. Balandin, A.A., S. Ghosh, W. Bao, I. Calizo, D. Teweldebrhan, F. Miao, and C.N. Lau, *Superior Thermal Conductivity of Single-Layer Graphene*. Nano Letters, 2008. **8**(3): p. 902-907.
3. Bolotin, K.I., K.J. Sikes, Z. Jiang, M. Klima, G. Fudenberg, J. Hone, P. Kim, and H.L. Stormer, *Ultrahigh electron mobility in suspended graphene*. Solid State Communications, 2008. **146**(9): p. 351-355.
4. Nair, R.R., P. Blake, A.N. Grigorenko, K.S. Novoselov, T.J. Booth, T. Stauber, N.M.R. Peres, and A.K. Geim, *Fine Structure Constant Defines Visual Transparency of Graphene*. Science, 2008.**320**(5881): p. 1308-1308.
5. van Ommen, J.R., J.M. Valverde, and R. Pfeffer, *Fluidization of nanopowders: a review*. Journal of Nanoparticle Research, 2012.**14**(3): p. 1-29.
6. Yao, W., G. Guangsheng, W. Fei, and W. Jun, *Fluidization and agglomerate structure of SiO2 nanoparticles*. Powder Technology, 2002. **124**(1): p. 152-159.
7. Salameh, S., J. Gómez-Hernández, A. Goulas, H. Van Bui, and J.R. van Ommen, *Advances in scalable gas-phase manufacturing and processing of nanostructured solids: A review*. Particuology, 2017.**30**: p. 15-39.

Study on the Influence of Different Carrier Gases on the Fluidization Properties of Glass Bead and FCC Powders

Conference: Fluidization

Year: 2019

Proceeding: Fluidization XVI

Group: General Paper Pool

Session: 3A: Fundamentals of Fluidization

Time: Tuesday, May 28, 2019 - 10:07am-10:19am

Authors:

Jin, Y., Key Laboratory of Coal Gasification and Energy Chemical Engineering of Ministry of Education, Shanghai Engineering Research Center of Coal Gasification, East China University of Science and Technology

In this paper, non-adsorbed glass bead and highly absorptive FCC were used as experimental media, respectively, using N2 and CO2 as carrier gases to study the powder's own physical properties and the difference of carrier gas properties for powder aeration and fluidization. The FT4 powder Rheometer was used to test the fluidization state under the the minimum fluidization velocity (U_{mf}) and the uniform expansion fluidization phase. The influence of carrier gas properties on the fluidity of aerated/fluidized powders was analyzed. At the same time, the difference between the inflated state and the fluidization of the bed in the fluidized state was combined with the bed stress analysis. The stress distribution was used to reveal the mechanism of the influence of different carrier gases on the fluidization properties of the powder.

Analysis of L-Valve Performance Feeding Spent Coffee Ground Powders into a Circulating Fluidized Bed

Conference: Fluidization

Year: 2019

Proceeding: Fluidization XVI

Group: General Paper Pool

Session: 3B: Fluidized Bed Applications

Time: Tuesday, May 28, 2019 - 8:42am-8:54am

Authors:

Sousa, L. M., Federal University of São Carlos
Ferreira, M. D. C., Federal University of São Carlos

Circulating Fluidized Beds (CFB) are widely used in industrial processes for combustion and gasification due to the high solids throughputs, temperature uniformity and effective gas-solid contact. A common challenge in operating CFBs is to ensure a stable and efficient feeding of the particulate material into the reactor. Solids feeding can be particularly limiting when handling biomass residues. Characteristics such as wide size distribution, variable moisture content, irregular shape and low bulk density may cause poor flowability, therefore the powders are likely to block into feeders and hopper discharge. Using non-mechanical L-valves as feeders to CFB reactors has already proven to be feasible and attractive when processing regular particles, but a quite limited number of studies were conducted to test such feeders with biomasses. In this study, our aim is to evaluate the performance of a non-mechanical L-valve to feed Spent Coffee Ground (SCG) powders into a cold-flow CFB with an overall height of 2 m and an internal diameter of 0.021 m. The L-valve is made of acrylic, with an internal diameter of 0.021 m and a length of 0.22 m. The static pressures throughout the CFB loop (silo, standpipe, L-valve, riser, and cyclone) and the solids mass fluxes were monitored using a data acquisition system from National InstrumentsÂ®, model NI-9205, and a digital balance coupled to a solidsâ€™ sampler. The performance of the L-valve was assessed through measurements of pressure loss in the valve (Î”P), solids mass flux (G_S) and by filming the gas-solid flow through the acrylic L-valve. The SCG samples were obtained brewing an ordinary Brazilian coffee, and the moisture content of the residue was reduced by keeping the samples in an oven for 24h at 105Â±2Â°C. Two samples of SCGs with different particle-size distributions were prepared: sample A was composed of 1.5 kg of SCG particles retained between the sieveâ€™s apertures of 500 and 300 Î¼m and sample B was composed of 1.35 kg of sample A and 0.15 kg of particles retained between the sieveâ€™s apertures of 300 and 150 Î¼m. An additional sample with moisture content equal to 30% in wet basis, denominated sample C, was prepared by mixing sample A with water in glass flasks, which were sealed and stored at 4Â°C for 60 h. To cover a wide range of operating conditions with samples A, B, and C, the L-valve performance was evaluated under four aeration velocities (U=0.34, 0.68, 1.02, 1.36 m/s) and three air velocities in the riser (Q=8.8, 10, 11.2 m/s). The assays were carried out in triplicate. The results showed that the aeration velocity overwhelmingly drives the values of Gs. Depending on the aeration velocity, three operating regions could be identified. Under a low aeration rate (Uâ‰ˆ0.34 m/s), the packed-bed resistance in the standpipe controlled the transport of solids. A high-pressure drop in the valve and a low solid mass flux were observed in this condition. Under intermediate aeration rates (0.34<U<1.02 m/s) the solids weight in the standpipe was in sync with the buoyancy force generated by aeration and the highest Gs values were observed in this aeration range. Finally, under high aeration rates (U>1.02 m/s), the solids transport was controlled by the gas phase. Turbulent vortices were observed in the base of the standpipe and a diluted solids transport was achieved in the L-valve, resulting in a low-pressure drop. Besides, we observed that the air velocity in the riser had a small effect on the L-valve performance. Depending on the aeration velocity, Gs varied between 1.7 and 3.8 kg/m^2s when operating with sample A. With sample B, which has 10% of fines mixed to sample A, the mass flux yielded by the valve was limited to only 1.7 kg/m^2s and a stable operation was achieved only for a narrow range of aeration velocities. The presence of fine particles in the sample contributes to increasing the specific surface area and contact among particles. Both the inter-particle attractive forces and mechanical interlocking are enhanced, thus limiting the valve operating range. Using moist samples, on the other hand, had a positive effect on Gs, with values from 2.8 to 24.4 kg/m^2s. This behavior can be attributed to a lubricant effect of the water under this saturation level. To burn SCG powders effectively in the soluble coffee industry furnaces, the recommended moisture should vary from 25 and 50%. According to the literature, the lower limit is set to avoid spontaneous combustion and the upper one to preserve burning efficiency. Our results showed that the L-valve did not perform well for samples containing particles of sizes under 300 μm at a mass fraction of 10%. However, it provided stable feeding in a wide range of solid mass fluxes (2.8<Gs<24.4 kg/m^2s) for SCG powders with a mean size of 400 μm and for moisture content of 30%. Besides, the solid mass fluxes can be easily varied by adjusting the aeration velocity in the L-valve.

Chemical Looping Combustion in a Pressurized Internally Circulating Fluidized-Bed Reactor Operated with a CaMnO3-5-Based Oxygen Carrier

Conference: Fluidization

Year: 2019

Proceeding: Fluidization XVI

Group: General Paper Pool

Session: 3B: Fluidized Bed Applications

Time: Tuesday, May 28, 2019 - 9:18am-9:30am

Authors:

Osman, M., Norwegian University of Science and Technology
Zaabout, A., SINTEF Industry
Cloete, S., SINTEF Industry
Amini, S., SINTEF Industry

Chemical looping combustion in a pressurized internally circulating fluidized-bed reactor operated with a $CaMnO_{3-\delta}$-based oxygen carrier

Mogahid Osman[1], Abdelghafour Zaabout[2], Schalk Cloete[2], Shahriar Amini[1,2*]

[1]Norwegian University of Science and Technology, [2]SINTEF Industry

Trondheim, Norway

Abstract

Chemical looping combustion, CLC, is one of the most promising technologies for power production with inherent CO_2 capture at minimal energy penalty. When using gaseous fuels, the full potential of this technology can only be achieved if operated at high pressure to maximize efficiency through a combined power cycle. However, a lack of pressurized chemical looping is observed, owing to the many challenges facing the pressurized operation of the conventional dual fluidized-bed reactor.

This study provides the prospects and opportunities that exist in application of the internally circulating reactor (ICR) to pressurized CLC. A novel reactor design, the ICR, was proposed incorporating many of the operational capabilities of the circulating fluidized-bed, while eliminating the complex separation system (i.e., cyclone and loop seals). The unit consists of a single reactor that combines two sections, (i.e., the fuel and air sections) with internal physical separations. Specifically, the concept aims to simplify the design, ease the solid circulation, operate at high pressure in a single pressurized vessel and eventually bring the chemical looping technology a step closer to commercialization [1,2]. The ICR unit has been constructed and commissioned as shown in Figure (1). The reactor was placed in a cylindrical shell to accommodate a maximum pressure of 12 bar. Cold flow experiments mapped out the operating conditions required for achieving steady solid circulation and minimum gas leakage between the two sections. The unit was recently successfully operated for chemical looping reforming under atmospheric pressure [3].

The primary aim of the current study is to demonstrate autothermal operation of the CLC process in the ICR under pressurized conditions using a $CaMnO_{3-\delta}$ based oxygen carrier, identified as one of the best candidates for chemical looping combustion. Tests will be completed at different temperatures, pressures and air/fuel ratios to gain a complete understanding of the ICR behaviour. The reactor performance will be quantified in terms of the fuel and air conversion achieved, solid circulation rate as well as the degree of undesired gas leakage between the two reactor sections. Gas leaking with the circulating solids through the interconnecting ports will decrease CO_2 separation efficiency, but previous work has shown that CO_2 capture efficiency and purity of ~95% can be achieved when reactor operation is optimized [2].

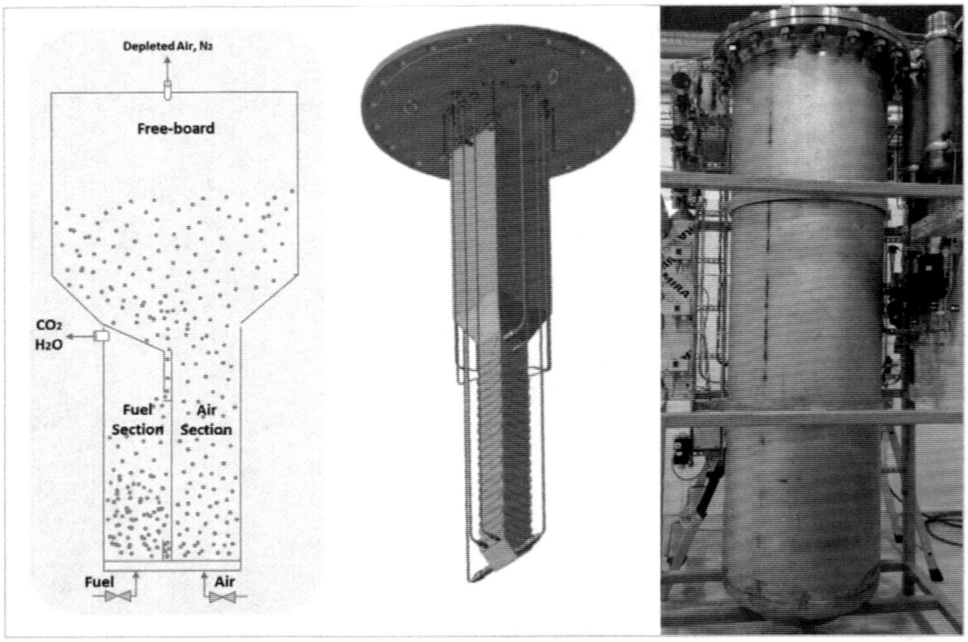

Figure 1. Simplified scheme of the ICR
design, CAD drawing of the ICR unit, and the ICR unit under operation inside
the shell. .

References

[1] Zaabout, S. Cloete, S. Amini,
Chemical Engineering & Technology, 39 (2016) 1413-1424.

[2] S. Cloete, A. Zaabout, S. Amini, Energy Procedia, 114 (2017) 446-457.

[3] M. Osman, A. Zaabout, S.
Cloete, S. Amini, Chemical Engineering Journal, 2018, doi.org/10.1016/j.cej.2018.10.013.

Defluidization Study on Industrial Reactive Powders at High Temperature

Conference: Fluidization

Year: 2019

Proceeding: Fluidization XVI

Group: General Paper Pool

Session: 3B: Fluidized Bed Applications

Time: Tuesday, May 28, 2019 - 8:54am-9:06am

Authors:

Macri', D., University College London
Sutcliffe, S., Venator UK
Lettieri, P., University College London - Torrington Place

Fluidized bed reactors are used in a wide range of industrial chemical processes and they are particularly useful in high temperature systems thanks to their ability to provide high heat transfer rates and rapid solids mixing, which lead to isothermal and more controllable conditions. Their applications spread the environmental, chemical, energy and process industries. For example, they are a key technology in the petroleum industry, catalytic chemicals synthesis processes, combustion and gasification of solid fuels, metals productions and many more.

Operations below incipient fluidizing conditions in combustion-type operations can lead to particle agglomeration and sintering which change the characteristics of the bed, leading to possible bed defluidization, unscheduled process downtime and additional costs. However, despite the countless researches that have been done so far, the details of the physical processes under such conditions are still not well understood, disclosing unrevealed features and challenging the researchers worldwide.

The objective of this work is to assess the effect of operative conditions, such as high temperature, air flowrate and particles size, on the flow properties and fluidization behaviour of multi-components systems of carbon coke and titanium ores. To this end, a simple approach for the early detection of aggregation and defluidization in the regime below and close to incipient fluidization - when the carbon combustion initiates the defluidization - is followed. The approach is based on the simultaneous measurements of local temperatures, the pressure drop across the particle bed. The mechanical and chemical characterization of the aggregated formed are performed offline and presented in details.

Experimental Characterization of Liquid Film Behavior during Droplets-Polyethylene Particles Collision Behavior

Conference: Fluidization

Year: 2019

Proceeding: Fluidization XVI

Group: General Paper Pool

Session: 3B: Fluidized Bed Applications

Time: Tuesday, May 28, 2019 - 8:30am-8:42am

Authors:

Ren, X., Zhejiang University
Sun, J., Zhejiang University
Wang, J., Zhejiang University
Yang, Y., Zhejiang University

Experimental Characterization of Liquid Film Behavior during Droplets-Polyethylene Particles Collision Behavior

Xiang Ren, Jingyuan Sun, Jingdai Wang, Yongrong Yang

State Key Laboratory of Chemical Engineering, College of Chemical and Biological Engineering, Zhejiang University, Hangzhou, 310027, China

The collision of droplets with solid particles is a common phenomenon in nature and industrial applications. For example, in the ethylene polymerization fluidized bed process under the condensed-mode operation, a large amount of condensate is sprayed into the fluidized bed, and the atomized droplets collide with high-temperature polyethylene particles, forming liquid films to wrap the particles or traversing the particles from the outside. The droplet-particle interaction significantly affects the heat and mass transfer characteristics, and even provides a basis for the formation of the liquid bridge and particle agglomeration. Therefore, it is essential to understand the complex hydrodynamic and thermal behaviors and the underlying mechanisms of droplet-polyethylene particle collisions.

In our study, the collision behaviors of water and n-hexane droplets with polyethylene particles at different temperatures have been studied. By controlling the falling velocity of the droplets to obtain different Weber numbers, high-speed camera and image analysis methods were used to investigate the liquid film thickness of the north pole point of the particle, the central angle of the liquid film extending farthest point, the dynamic contact angle, and the arc length. Characterize liquid film behavior.

At low Weber numbers, the water droplets collide with the polyethylene particles and stay dynamically on the particle surfaces. Under the atmospheric temperature, the underdamped oscillation of kinetic energy is gradually attenuated. With the increase of particle temperature, the underdamped oscillation amplitude is first attenuated and then enhanced. The horizontal axis of the underdamped oscillation is linearly correlated with temperature. In the case of high Weber number, the droplets of water collide to form liquid film-coated particles, and the liquid film drops slightly after rebounding to the particles. The film thickness of the regurgitation is correlated with time.

At high Weber numbers, the n-hexane droplets exhibit the collision behaviors as the water droplets at low Weber numbers, and the liquid film-coated particles are formed after the collision. Unlike water, no regurgitation phenomenon appears for the n-hexane liquid films at the at the atmospheric temperature.

Key words: droplets-particles collision, liquid film, high-speed camera , fluidized bed reactor

Experimental Study on Separation Characteristics of a Gas-Liquid Cyclone Separator in WGS

Conference: Fluidization

Year: 2019

Proceeding: Fluidization XVI

Group: General Paper Pool

Session: 3B: Fluidized Bed Applications

Time: Tuesday, May 28, 2019 - 9:30am-9:42am

Authors:

Zhou, W., China University of Petroleum-Beijing
Wang, K., China University of Petroleum-Beijing
Lu, C., China University of Petroleum-Beijing
E, C., China University of Petroleum-Beijing

Wet Gas Scrubbing (WGS) flue gas
desulfurization scrubber is an important equipment for Fluid Catalytic Cracking (FCC) to wet desulfurization. It is mainly used to separate and purify flue gas into the atmosphere and lye droplets which can be collected and recycled after separation. Sinopec Daqing Company found that there was serious rain and ice formation around the scrubbing tower, which affected the stable operation of the device due to the low gas-liquid separation efficiency of the scrubbing tower in actual operation. In order to improve the separation efficiency of the scrubber, it is necessary to understand the internal separation rule. In this experiment (Fig.1), a large-scale cold gas-liquid cyclone separator was established by the industrial device based on the similar amplification principle. The pressure and velocity field information inside the device was measured by pressure sensors and five-hole probes.

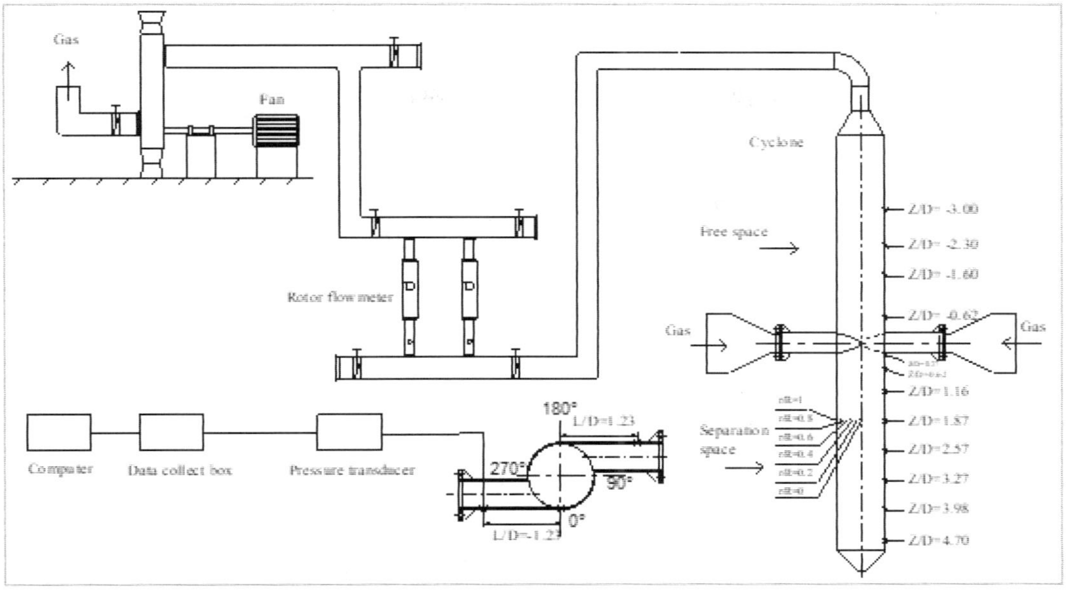

Fig.1 Schematic
diagram of device structure and measuring point

Then we analyzed separation rule of
the inside of the device. The results are as follows.

1. The pressure drop of the gas
liquid separator has good predictable performance and the total pressure drop is very small less than 30 Pa and the Eu
number ($Eu = \frac{2\Delta P}{\rho u^2}$) keeps at 4.7×10^{-4} at
different inlet gas velocities in the range of 3.7 to 9.8 m/s.

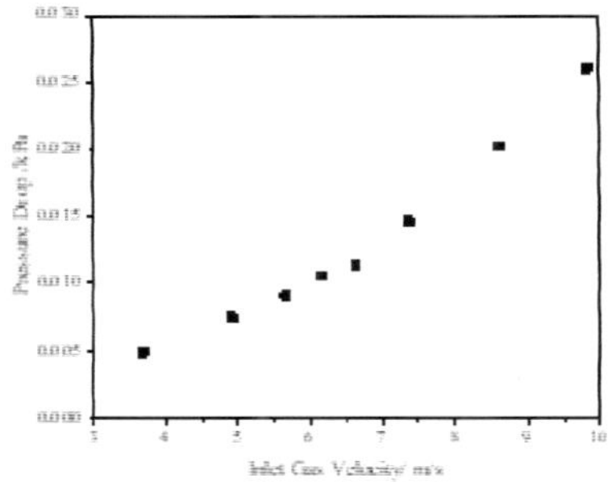

Fig.2 The
variation of pressure drop with inlet gas velocity

2. Through the analysis of mean,
standard deviation of pressure and velocity field at different measuring points of the gas-liquid separator, we found that the tangential velocity peaks at the region of r/R=0.12, the boundary of inner and outer swirling flow in the separation space is in the region of r/R=0.15-0.2 (Defined r/R=1 as the wall, r/R=0 as the center), and the boundary of upward flow and downward flow at r/R=0.57. The two boundary radial positions vary little in the whole separation space. In the free space, all flow is upward, and the variation of tangential velocity along the radial direction is the same as that in the stable separation space. In the whole space, the static pressure value is the lowest at the center and the maximum at the wall. The standard deviation at the center is the maximum and the value at the side wall is the minimum. A small amount of outer swirling gas along the separation space will enter the inner swirling flow, and the amount of gas increased gradually with the increase of the axial altitude. Most of the outer swirling flow reverses to the inner swirling occurs in the region of Z/D = 3.98 ~ 4.70 (Z- distance from the center of the feed pipes; D- inner diameter of the cylinder).

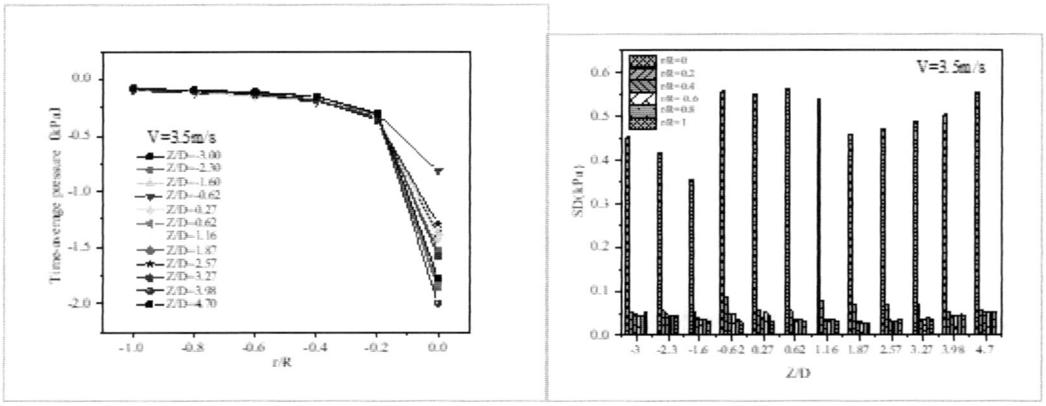

Fig.3 Time-average and standard deviations pressure at different tappings

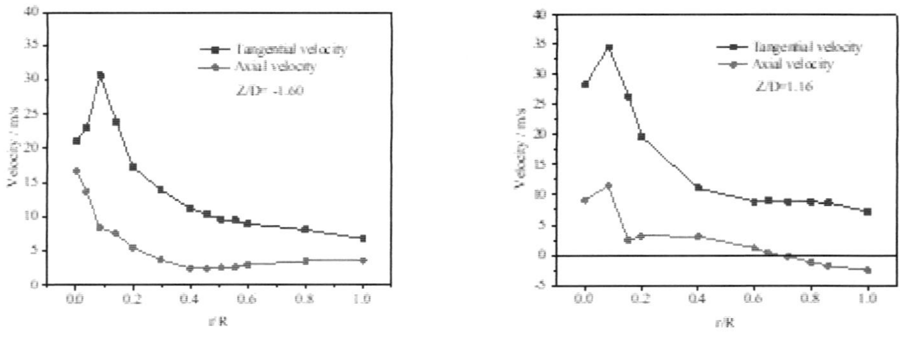

Fig.4 Tangential
and axial velocity at different tappings

3. Through the power spectrum
analysis of the pressure in different positions of the experimental device, the pressure fluctuation caused by the unstable intake air will be transmitted in the whole device. The high-frequency signal can be easily eliminated in the air flow movement but the low-frequency signal can not be easily eliminated which caused the high-amplitude clutter signal peaks in the center. In the separation space, the inner and outer swirling frequencies of 155 Hz and 45 Hz are found at the center, and the inner and outer swirling frequencies of 145 Hz and 35 Hz are found at the wall. It shows that the frequency and amplitude has certain transferability along the radial direction, and the frequency attenuates less from the center to the side wall, and the amplitude attenuates more. Because of the intersection of outer and inner swirling, the unique frequency peaks of 88 Hz was found at Z/D= -2.30, 3.98 and 4.70 sections in the free space and separation space.

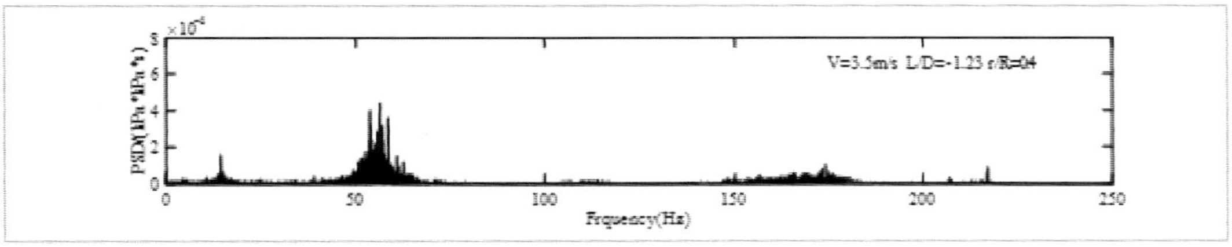

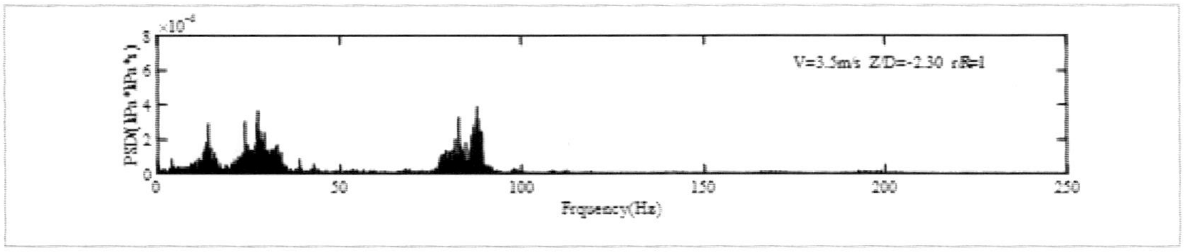

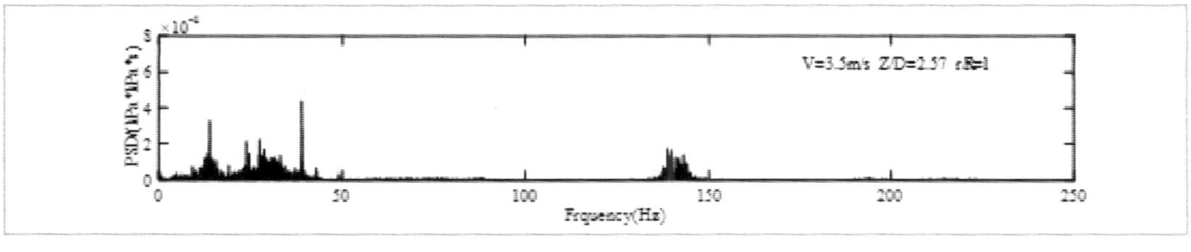

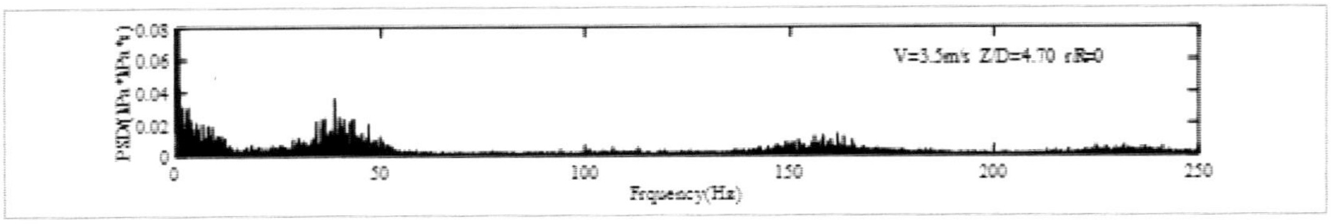

Fig.5 Power spectral density at different tappings

4. When there is a certain height of initial liquid level in the separation space, the liquid level will rotate under the influence of rotating airflow. The maximum distance and the minimum distance of the liquid level will increase with the increase of cross-sectional gas velocity (V). When the cross-sectional Reynolds number Re_c ($Re_c = \frac{DV\rho}{\mu}$) is greater than 4.69×10^{4} , the rotating distance will remain unchanged. Compared to the condition of no level, the dimensionless static pressure defined as pressure value of the measuring point divide value of the inlet and the standard deviation of static pressure decrease greatly at the center in the whole space, but the change is not obvious at the wall. The change of pressure in free space has nothing to do with the initial liquid height. In the separation space, it is found that the dimensionless static pressure and the static pressure standard deviation is similar in the whole section when the liquid level DH/D (DH- distance from the initial liquid level to the center of the feed pipes) is constant measured at Z/D=0.62. With the increase of the initial liquid level, the dimensionless static pressure does not change, but the static pressure standard deviation increases first then decreases and approach the peak when the initial liquid level is DH/D=3.01. In the separation space, when measured at the same distance above the initial liquid level, it is found that there is little difference in dimensionless static pressure and standard deviation of the whole section, that is, dimensionless static pressure and standard deviation in the center and the wall keeps the same change because of the existence of liquid level.

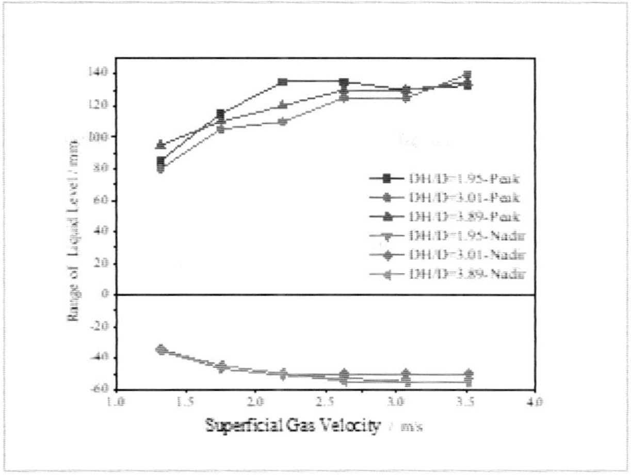

Fig.6
The distance of the liquid level in different cross section gas velocity.

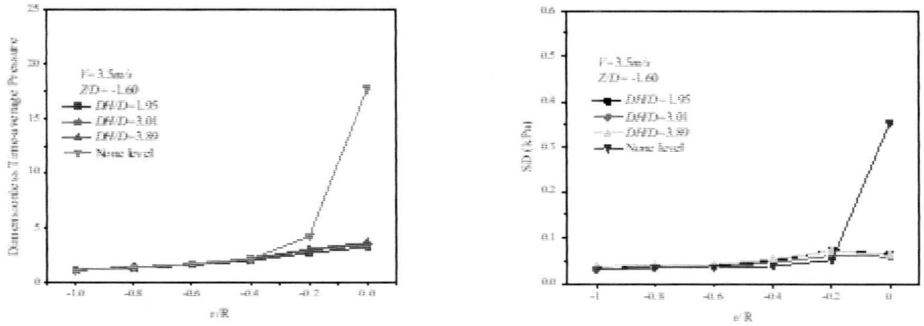

Fig.7 The dimensionless mean pressure and standard deviation in free space

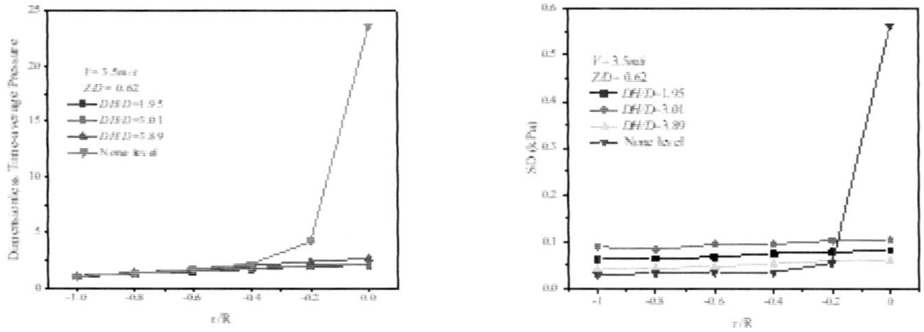

Fig.8 The dimensionless mean pressure and standard deviation in separation space

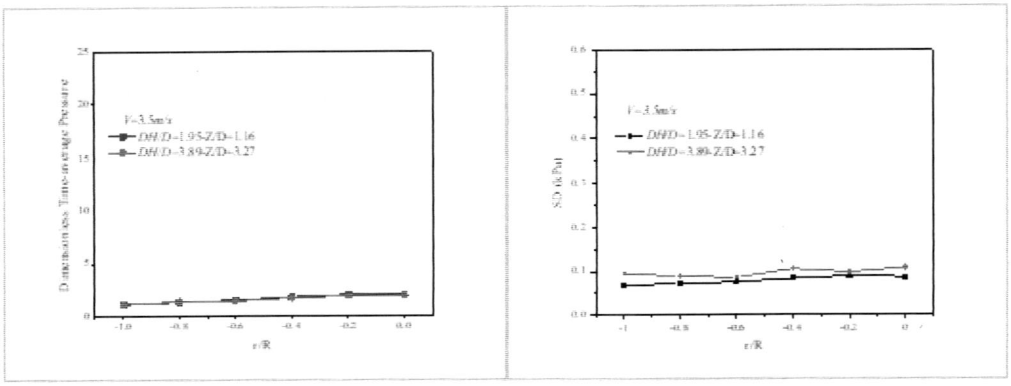

Fig.9 The dimensionless mean pressure and standard deviation at the same distance above the initial liquid level

5. The separation efficiency

increased from 82% to 93% with the increase of liquid concentration at 30 g/m3-60 g/m3 while the cross-section gas velocity remained unchanged. And the separation efficiency increased with the increase of cross-section velocity. If the water evaporation of import and export is not considered, the separation efficiency deviation is 5%-7%. It is also found that the mass flow rate of the separated water in the free space is approximately the same, that is, the separated water in the free space may be independent of the concentration of the liquid.

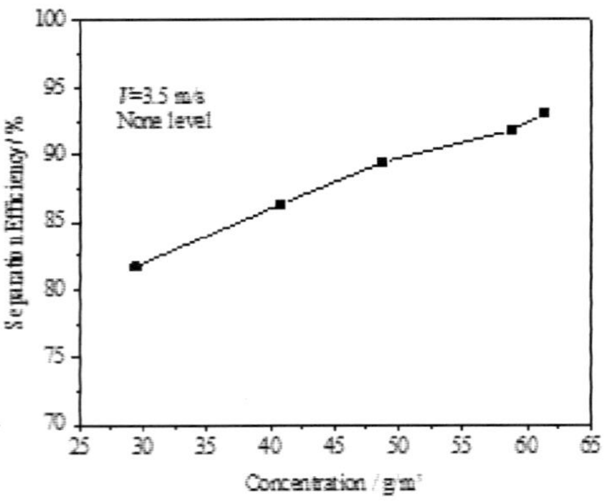

Fig.10
Relationship between separation efficiency and liquid concentration

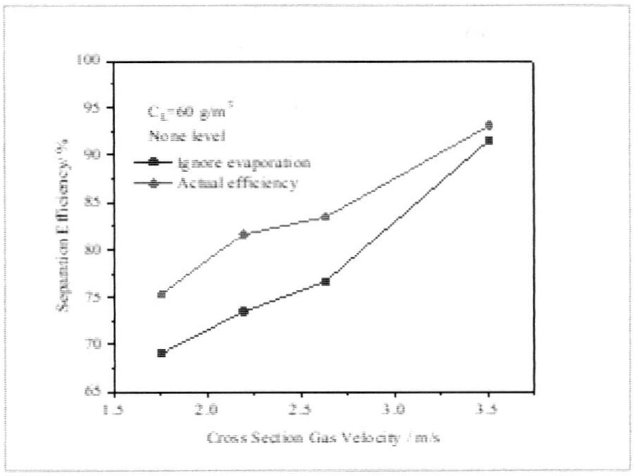

Fig.11
Relationship between separation efficiency and cross-section gas velocity

According
to the existing analysis, it has great potential structure to improve separation efficiency of the gas liquid cyclone separator.

High Efficient Hydrogen Production through Novel Gas Switching Water Splitting (GSWS) and Reforming (GSR) Using High Iron Content Oxygen Carrier

Conference: Fluidization

Year: 2019

Proceeding: Fluidization XVI

Group: General Paper Pool

Session: 3B: Fluidized Bed Applications

Time: Tuesday, May 28, 2019 - 9:06am-9:18am

Authors:

Ugwu, A., Norwegian University of Science and Technology
Zaabout, A., SINTEF Industry
Donat, F., Laboratory of Energy Science and Engineering, ETH Zürich
Müller, C. R., ETH Zurich
Albertsen, K., Euro SupportAdvanced Materials B.V
Cloete, S., SINTEF Industry
Amini, S., SINTEF Industry

This study is primarily motivated by the scale-up challenges of pressurized chemical looping applications. One promising alternative reactor concept, among several proposed in the last decade focusing on configurations with no external solid circulation, is the â€œGas Switching Technology (GST)â€, which has been proposed for power [1] and hydrogen production with integrated CO_2 capture (Gas Switching Combustion â€œGSCâ€ and Gas Switching Reforming â€œGSRâ€ respectively [2, 3]). Unlike the conventional chemical looping, this novel technology utilizes a single fluidized bed reactor and avoids the circulation of solid oxygen carrier by alternating the feeds of the oxidizing and reducing gases to complete the different redox reactions involved in the process. With this arrangement, scale-up challenges can be greatly reduced. Experimental demonstration campaigns have proven ease of operation of this concept under atmospheric and pressurized conditions [2-4]. To capitalize on this success, this study extends the GST to the water splitting, based on the steam-iron process, for pure hydrogen production. The suitability of iron-based oxygen carrier to steam methane reforming in the GSR is also further explored in this study.

Gas Switching Water Splitting (GSWS) is a three-step (steam, air and fuel steps) process that utilizes the different oxidation states of iron-based oxygen carrier (Fe_2O_3, Fe_3O_4 and FeO/Fe) to efficiently produce hydrogen from natural gas with integrated CO_2 capture in a single process [5]. The redox cycle starts with the steam step where hydrogen is produced by the steam oxidation of FeO (or Fe) to Fe_3O_4. Subsequently, the air step follows where Fe_3O_4 is fully oxidized by air to Fe_2O_3 in an exothermic reaction that generates the heat needed for the next reduction step using CH_4. At this step, the oxygen carrier is regenerated bringing Fe_2O_3 to FeO (or Fe) with inherent CO_2 capture.

As for the Gas Switching Reforming (GSR) with iron based oxygen carrier, it has also been designed as a three-step process (reduction, reforming and oxidation step) operating in a cyclic mode by alternating fuel, a fuel-steam mixture and air feeds to the reactor. The redox cycle starts with the fuel step where dry CH_4 is fed to reduce the iron based oxygen carrier with significant yield of pure CO_2 ready for storage. The reforming step follows the reduction step, where steam is co-fed with CH_4 to produce syngas. The air step follows to re-oxidize the oxygen carrier and provide the heat required for the consecutive reduction and reforming reactions.

However, gas mixing between the stages takes place when switching between them, thereby reducing the CO_2 capture efficiency, CO_2 purity and H_2 purity. It is therefore important that the stages are run longer to minimize the extent of the mixing. To tackle this issue, oxygen carrier of high active Fe_2O_3 content (70wt%) has been used in this study. The oxygen carrier was produced through spray-drying and calcined at 1300°C. For Gas Switching Water Splitting (GSWS), . GSR substantially increases the CO_2 capture efficiency relative to GSWS, making gas mixing between stages less important. In addition, thermogravimetric analysis (TGA) has shown that the oxygen carrier demonstrates high cyclic stability under relevant reaction conditions.

The experiments were conducted under the GSWS and GSR conditions at atmospheric pressure and elevated temperature from 700°C to 900°C using a 1 kW fluidized bed reactor. Process performance at different conditions was quantified through online gas composition and temperature measurements. The results from the GSWS experiments have shown a low degree of fuel conversion in the reduction step. In addition, the oxygen carrier started to agglomerate after about 34% of reduction (to FeO), thus not allowing the process to utilize even half of the oxygen carrier capacity thereby hampering the gas separation performance of the GSWS process. However, better performance was achieved with GSR, with no particle agglomeration, as the small reduction degree of the oxygen carrier was enough to initiate syngas production when steam was co-fed with methane. As a result, the degree of carbon deposition was also less for GSR than GSWS. In summary, results show that the proposed oxygen carrier is more suitable for reforming (GSR) than water splitting (GSWS) using the novel gas switching reactor concept.

References

[1] A. Zaabout, S. Cloete, S. T. Johansen, M. v. S. Annaland, F. Gallucci, and S. Amini, "Experimental Demonstration of a Novel Gas Switching Combustion Reactor for Power Production with Integrated CO2 Capture," *Industrial & Engineering Chemistry Research,* vol. 52, no. 39, pp. 14241-14250, Oct 2 2013.

[2] S. A. Wassie, F. Gallucci, A. Zaabout, S. Cloete, S. Amini, and M. van Sint Annaland, "Hydrogen production with integrated CO2 capture in a novel gas switching reforming reactor: Proof-of-concept," *International Journal of Hydrogen Energy,* vol. 42, no. 21, pp. 14367-14379, 2017/05/25/ 2017.

[3] S. A. Wassie *et al.*, "Hydrogen production with integrated CO2 capture in a membrane assisted gas switching reforming reactor: Proof-of-Concept," *International Journal of Hydrogen Energy,* vol. 43, no. 12, pp. 6177-6190, 2018/03/22/ 2018.

[4] A. Zaabout, S. Cloete, and S. Amini, "Autothermal operation of a pressurized Gas Switching Combustion with ilmenite ore," *International Journal of Greenhouse Gas Control,* vol. 63, pp. 175-183, 2017/08/01/ 2017.

[5] D. J. C. T. Sanfilippo, "One-step hydrogen through water splitting with intrinsic CO2 capture in chemical looping," vol. 272, pp. 58-68, 2016.

Hydrodynamics of a Few Coarse Particles in a Fluidized Bed of Fine Particles

Conference: Fluidization

Year: 2019

Proceeding: Fluidization XVI

Group: General Paper Pool

Session: 3B: Fluidized Bed Applications

Time: Tuesday, May 28, 2019 - 9:54am-10:06am

Authors:

Roy, S., Indian Institute of Technology Delhi
Roy, S., Indian Institute of Technology Delhi
Pant, H. J., Bhabha Atomic Research Centre

Fluidized bed reactors have been extensively employed for coal gasification. Typically, in Winkler type gasifier, feed coal (~ 1-3 mm) is fed in a bed of fine sand (~ 0.1- 0.5 mm) particles and fluidized with the aid of steam and air. However, more often than not, unburnt coal particles escape the bed, which is subsequently recovered in cyclone and recirculated back. It can also be lost along with ash recovery from the bottom discharge. Both these aspects lead to loss of feed and eventual lower carbon conversion. This further leads to frequent plant shut down and a drop in reactor performance. Until now, such issues are largely addressed based on variation in dynamic bed pressure drop and heuristic experience of operators. However, to develop deeper insights into the complex hydrodynamic behavior of coal particles in the bed, one needs to employ a particle tracking study, wherein, a typical path of the fresh feed are traced. This should be performed taking into account for various fractions of coarser coal and finer sand particles in the bed.

With this broad motivation, the present study thus employs Radioactive Particle Tracking (RPT) studies in which the coarser glass beads (2 mm) are tracked in a matrix of fine (~0.5 mm) glass beads in a binary fluidized bed (0.10 cm in ID). This in effect mimics an actual bed of fluidization of coal in the sand in a laboratory scale facility. Although several techniques have been implemented in the past, based on the principle of tracing the path of a single radiotracer particle, RPT remains a popular velocimetry technique [1, 2]. In order to effect such a study, it was planned to conduct the measurements in a mixture of particles of known distributions wherein a marker ^{46}Sc tracer was used as a solid tracer to track the coarser fraction of glass beads (2 mm). For the range of superficial velocities studied (2.97- 4.24 m/s), and bed composition (10- 50 weight % of 2mm glass beads) it was found that the bed composition did have a strong influence on the occurrences of solids in the bed. Further, the instantaneous tracer locations showed that it ensured an unbiased scanning of the entire test section. Also, the mean velocities revealed two distinctive zones â€“ a high-velocity region around core while relatively lower velocity towards the wall. Detailed analysis of the results will be discussed in the final presentation.

References

[1] Upadhyay, R. K., & Roy, S. (2010). Investigation of hydrodynamics of binary fluidized beds via radioactive particle tracking and dualâ€source densitometry. *Can. J. Chem. Eng.* 88(4), 601-610.

[2] Roy, S. (2017). Radiotracer and particle tracking methods, modeling and scaleâ€up. *AIChE J.* 63(1), 314-326.

Investigation on the Mass Transfer Rates in Fluidized Bed Membrane Reactors

Conference: Fluidization

Year: 2019

Proceeding: Fluidization XVI

Group: General Paper Pool

Session: 3B: Fluidized Bed Applications

Time: Tuesday, May 28, 2019 - 9:42am-9:54am

Authors:

de Nooijer, N., Eindhoven university of technology
Gallucci, F., Eindhoven University of Technology
Van Sint Annaland, M., Eindhoven University of Technologhy

Introduction

An interesting development in the field of reforming reactors is the application of membrane reactors, where the in-situ selective extraction of hydrogen results in a shift in the equilibrium, which allows operation at lower temperatures while still achieving higher efficiencies. Moreover, no downstream separation is required since pure hydrogen is directly obtained from the membrane. Thin-film palladium-based membranes have the largest potential for integration in membrane reactors for hydrogen production due to their extremely high permeation fluxes and perm-selectivities. The permeation of hydrogen through the membrane depends only on the hydrogen partial pressure difference across the membrane. However, due to the high hydrogen flux, the mass transfer towards the membrane can become limiting, which effect is known as concentration polarization. In the case of fixed bed reactor configurations, the concentration polarization can significantly increase the required membrane area. The application of a gas-solid fluidized bed reactor configuration can reduce the concentration polarization (CP) effects due to its higher gas-solids mixing. Despite the decrease in the CP, it is still important to account for these mass transfer limitations, also for fluidized bed configurations; However, the prediction of CP in fluidized beds remains difficult. In this work the mass transfer rates in fluidized bed membrane reactors was studied, both the concentration polarization and the bubble-to-emulsion phase mass transfer rates.

Methods

Experiments were performed to quantify the extent of concentration polarization in a fluidized bed membrane reactor. The system used was a tube-in-shell system with a single thin-film (4 μm) PdAg membrane immersed into a fluidized bed of 250 μm catalyst particles, operated in the bubbling fluidization regime (3 to 7 times the minimum fluidization velocity). Experiments with mixtures, also under reactive conditions, were performed and used to develop a phenomenological description based on a stagnant film model to describe the CP in fluidized bed membrane reactors.

The bubble-to-emulsion phase and bulk-to-membrane mass transfer rates were studied in more detail with an in-house developed two-dimensional fluidized bed Infrared Radiation (IR) transmission system. This technique makes use of the IR absorbing properties of a tracer gas and is able to make instantaneous whole-field concentrations measurements of the dilute phase in the fluidized bed, from which the volumetric bubble-to-emulsion phase mass transfer coefficient can be determined.

Results and discussion

Unlike in fixed bed reactors, the gas passes through a fluidized bed consisting of Geldart B type particles via two phases, viz. the bubble and the emulsion phase, with different gas residence times, where there is mass transfer between the bubble and the emulsion phase. In the case of gas extraction through the membranes there is also mass transfer toward the membrane surface. Experimental results extracting hydrogen from a nitrogen hydrogen mixture show that fluidization can increase these mass transfer rates. A stagnant film layer description is shown to be able to predict these effects, once the film layer thickness has been estimated from experimental data.

The IR transmission technique has been applied to quantify bubble-to-emulsion mass transfer rates for single and multiple bubble injections into a fluidized bed at incipient fluidization conditions and bubble injections into a fluidized bed operated in the freely bubbling regime. The bubble-to-emulsion mass transfer is shown to be dominated by convective transport for larger particle sizes ranging from 300 to 700 μm. Extension of the image analyses by time averaging of instances allowed to study the mass transfer rates in beds consisting of smaller particles and to investigate the mass transfer rates from bubbles rising for longer times through the bed. The obtained results show that in beds consisting of smaller particles diffusive mass transfer becomes more pronounced and should be carefully considered.

Conclusions

Fluidized bed technology combined with membrane technology has proven to be an interesting concept for hydrogen production. The excellent heat and mass transfer of fluidized beds allow for an easy integration and improved performance combined with hydrogen extraction membranes. The mass transfer limitations towards the membrane can be described by a stagnant film model and are reduced due to the gas-solid mixing in the fluidized bed. A further understanding of the mass transfer in fluidized bed membrane reactors is, however, still required. With the use of a novel IR transmission technique the bubble-to-emulsion phase mass transfer could be studied in much more detail and the results have clearly indicated that the often-used existing correlations need to be revisited.

Acknowledgements

The presented work is funded within BIONICO. This project has received funding from the Fuel Cells and Hydrogen 2 Joint Undertaking under grant agreement No 671459. This Joint Undertaking receives support from the European Unionâ€™s Horizon 2020 Research and Innovation Programme, Hydrogen Europe and N.ERGHY.

Effect of Gas Switching on Powder Fluidization

Conference: Fluidization

Year: 2019

Proceeding: Fluidization XVI

Group: General Paper Pool

Session: 3C: Fluidized Bed Applications

Time: Tuesday, May 28, 2019 - 8:55am-9:07am

Authors:

Lu, H., Key Laboratory of Coal Gasification and Energy Chemical Engineering of Ministry of Education, Shanghai Engineering Research Center of Coal Gasification, East China University of Science and Technology
Guo, X., Key Laboratory of Coal Gasification and Energy Chemical Engineering of Ministry of Education, Shanghai Engineering Research Center of Coal Gasification, East China University of Science and Technology
Gong, X., Key Laboratory of Coal Gasification and Energy Chemical Engineering of Ministry of Education, Shanghai Engineering Research Center of Coal Gasification, East China University of Science and Technology
Cao, J., East China University of Science and Technology

In this paper, the effect of gas switching on powder fluidization were studied. Several powders, including Fluid Catalytic Cracking (FCC), glass bead and pulverized coal were used as experimental materials, and Heã€N_2 ã€Ar and CO_2 were used as fluidized gases while they were changed from one to another during the fluidization at a constant gas velocity. Experimental results showed that a decrease of the bed pressure drop might occur after switching the fluidizing gases, where the emulsion phase was compressed, bubbles disappear and channels were formed. The maximum relative decrease of bed pressure drop, the time duration of defluidization and the defluidization intensity were used to quantitatively describe this process, and the effect of gas, powder properties and gas switching sequence were discussed.

Keynote Talk: Fine Iron Powder as Renewable Dense Energy Carrier

Conference: Fluidization

Year: 2019

Proceeding: Fluidization XVI

Group: General Paper Pool

Session: 3C: Fluidized Bed Applications

Time: Tuesday, May 28, 2019 - 8:30am-8:55am

Authors:

Deen, N. G., Eindhoven University of Technology
Tang, Y., Eindhoven University of Technology
Homan, T., Eindhoven University of Technology
Hessels, C., Eindhoven University of Technology
Dam, N., Eindhoven University of Technology
de Goey, P., Eindhoven University of Technology

Metal fuels have been proposed as advanced storage and transport systems for clean energy, because of their inherently low safety risk and high energy densities. To ensure a clean metal fuel energy cycle, metal fuel utilization products, i.e., metal oxide powders, need to be captured and cyclically reduced back into metal fuel powder via advanced technology that is powered by renewable energy. In this work we focus on hydrogen reduction of iron oxides under fluidization conditions for recycling of iron powder as energy carrier. Iron is selected because of its favorable properties of heterogenic combustion, abundance and low price. Dynamic and thermodynamic behavior of this fluidization reduction process are studied via a combination of numerical modelling and experimental measurements. In this talk, we will discuss the main technical issues linked to the use of fine iron powders as a dense renewable energy carrier.

On-LINE Transition between Incompatible Catalysts

Conference: Fluidization

Year: 2019

Proceeding: Fluidization XVI

Group: General Paper Pool

Session: 3C: Fluidized Bed Applications

Time: Tuesday, May 28, 2019 - 9:55am-10:07am

Authors:

Enazi, N., SABIC

During the production of olefin polymers in a polymerization reactor it will often be necessary to transition from one type of catalyst system producing polymers having certain properties and characteristics to another catalyst system producing polymers having different specifications. The developed process relates to transition between two incompatible catalysts in continuous polymerization reaction, utilizing advanced Ziegler-Natta catalyst system and chromium-based catalyst. The transition process is performed with no discontinuity to the reactor operation, where, it starts with the first catalyst in continuous polymerization reaction of ethylene and alpha olefin co-monomer for producing a linear low density polyethylene (LLDPE) in a gas phase reactor to the second catalyst and different co-monomer for producing a high density polyethylene (HDPE). The transition process comprising initially a feed reducing of the first co-monomer into the reactor until the ratio of co-monomer to ethylene in the reactor is less than zero. Then, discontinuing the first catalyst of Ziegler-Natta while the introduction of a co-catalyst of the Ziegler-Natta catalyst is continued and subsequently discontinuing the introduction of the co-catalyst. The polymerization conditions in the reactor are kept with no change in this stage and permitting polymerization of LLDPE to continue for a time. Then, discontinuing the introduction of all feeds into the reactor, depressurizing, flow-purging the reactor and introducing more ethylene and other feeds to prepare the reactor for the new conditions for the second polymerization reaction, wherein the partial pressure of ethylene is increased to the pressure for the second polymerization reaction at such a speed that the reactor temperature is maintained at a temperature lower than the temperature of the first polymerization conditions. After that, reactor temperature to be increased to a temperature of the chromium-based catalyst polymerization conditions and start feeding the catalyst into the reactor. Then, the second co-monomer is introduced into the reactor to obtain a reactor composition for HDPE polymerization reaction.

Scale-up Considerations of the Ubfb Solar Receiver

Conference: Fluidization

Year: 2019

Proceeding: Fluidization XVI

Group: General Paper Pool

Session: 3C: Fluidized Bed Applications

Time: Tuesday, May 28, 2019 - 9:31am-9:43am

Authors:

Zhang, H., School of Life Science and Technology, Beijing University of Chemical Technology
Kong, W., School of Life Science and Technology, Beijing University of Chemical Technology
Flamant, G., CNRS
Ansart, R., Université de Toulouse, CNRS-Toulouse
Sabatier, F., Université de Toulouse, CNRS-Toulouse
Baeyens, J., European Powder and Process Technology

Concentrated Solar Power plants (CSP) are a promising technology for electricity generation. Indirect particle receivers can operate at high receiver temperatures and foster the application of advanced power generation cycles. The present work specifically considers the scale-up of the CNRS concept, referred to as Upflow Bubbling Fluidized Bed (UBFB) system. Previous papers have provided detailed information concerning the on-sun testing of a single tube (36 mm I.D., 0.5 m long) experiments at the solar furnace of PROMES-CNRS in Font Romeu (FR) [1-3] and of a 150 kW pilot module consisting of 16 parallel tubes[4] (29.7mm I.D., each 1 m long). SiC, a Geldart A type powder, was used and fluidized at superficial gas velocities of ~0.03 to 0.25 m s^{-1}. To scale-up these vertical UBFBs, some phenomena require additional attention. These phenomena are 5-fold, and related to (i) the pressure balance over the receiver and the achievable solids' flux and air velocity relationship; (ii) the efficiency of the vertical transport; (iii) the possible particle choking; (iv) the gas-solid hydrodynamics gradually transforming from a freely bubbling into a slugging mode in long UBFB tubes; and (v) attrition and erosion. Pressure Balance, Achievable Solid Flux and Air Velocity Relationship

The dispenser bed is operated at a fluidization velocity close to the minimum bubbling (U_{mb}) superficial velocity of the powder (~1.2 to 1.5 U_{mb}). The major part of the conveying gas is injected at the bottom of the tube (about 10 cm above its inlet). In the loop, different factors should be considered. The solid fraction, , in the upward part of the circuit (the riser) is lower than in the downcomer parts, being 0.35-0.40 and 0.45-0.50 respectively.

To operate the loop in a stable flow mode, driving downflow pressures (including the pressurization of hopper) should exceed pressure drops of the upflow branch including acceleration and friction losses[5]. The transport equation reveals that for solids circulating flux to be positive, several conditions should be met on the pressure balance, the gas velocity as a function on the terminal settling velocity of the particles, the bed voidage and the tube length.

Efficiency of the UBFB Conveyor System

The efficiency of the UBFB conveyor system can be calculated by comparing the compression and uplift work. To pressurize this air flow to the required pressure of the operating Î"P, above the atmospheric pressure, work is needed in the compressor The efficiency of the UBFB will be determined for two powders as a function of solids circulating flux. A higher powder density reduces the conveying efficiency through the higher the bed. For the same reason, a similar reduction of the efficiency will occur for deeper beds (longer tubes).

Slugging

Extensive research at ambient conditions demonstrated that wall slugging was found at a bed level of ~ 50 cm above the air inlet [6], whereas axi-symmetric slugging occurred at a height exceeding ~ 150 cm above the air inlet [7]. Slugs significantly reduce the bubble-induced particle mixing and associated excellent heat transfer. A long UBFB will suffer from slugging both towards the wall-to-bed heat transfer coefficient, as towards the considerable pressure fluctuations measured across the bed. At higher temperatures, and in-line with predictions by Kong et al. [7], the onset of slugging is deferred to higher bed levels. This finding is positive for the scaling up of the UBFB solar receivers that will involve multi-meter long tubes. Wall slugging is characterized by a slug frequency of 1 Hz. Axi-symmetric slugs move up the bed at a frequency of ~0.5 Hz. In this studies, a diagram of the bubble and/or slug frequency as a function of height at ambient temperature and 475°C temperature will be determined.

The Maximum Achievable Solid Flux

Common fluidized bed operations can be hampered in a specific superficial gas velocities (*U*) and solids circulation flux (*G*) range where choking occurs, being understood as the phenomenon where a small change in gas or solids flow rate prompts a significant change in the pressure drop and/or solids holdup: the stable upflow regime can no longer be maintained when -values exceed a certain limit for a low to moderate gas velocity. This choking can occur in dense upflow of particles when the superficial gas velocity and the driving pressure are no longer capable of entraining the particles. In the upflow bubbling fluidized bed concept, only -values up to 100 kg/m^2s were tested, where the stability of the operation was confirmed. This is expected since the pressure balance indicates that the loop will operate in a stable manner provided the external pressure in the dispenser compensates the upflow bubbling fluidized bed pressure drop. At high -values, acceleration and friction losses, both proportional to (n=1 to 2) [8] will increase and will finally hamper the system stability. To assess the impact of () combinations, a non-choking criterion is established when considering that the particle slip velocity must remain positive. To operate the upflow bubbling fluidized bed at -values in excess of 100 kg/m^2s, superficial air velocities should exceed 0.1 m/s, as commonly used in the single and multi-tube set-ups. In this work, a prediction of chocking limit at different superficial gas velocities and bed voidages will be determined.

Attrition of Particles

Since a fluidized bed is the key in operating the solar receiver at a high wall-to-bed heat transfer coefficient, attrition of particles was examined experimentally in order to select the type of Geldart A-powders less prone to attrition. This extensive research was fully reported by Zhang *et al* [9] and only essential features are summarized below. Although fast particle motion associates a high degree of mixing, it however causes inter-particle collision and bed-to-wall impacts, both leading to particle attrition. Attrition generates fines that can be lost in the dust collection system, whereas the particle size distribution of the bed will alter during the operation. Zhang *et al.* [9] clarified the influence of particle size and nature, bed height, fluidization velocity, action of jets and orifice diameter. An equation was developed and enables to predict attrition rates for different particles at different operating contribution. The total attrition rate combines the bubble-induced and jet-induced effects. It was moreover shown that particles with a high Abrasion Index (AI), as defined by CEMA [10], are less prone to attrition. The higher AI is however an indication of the expected wear of the equipment [11].

Conclusions

The UBFB is a key part of the novel solar power concept. The research has dealt with specific considerations in view of its scale-up, including: (i) the effect of the diameter of the receiver tubes and the use of internal fins; (ii) the applicable circulation flux; (iii) the possible extension of the pipe length from 0.5 or 1 m to beyond 3 m; (iv) the limited effect of powder attrition and equipment wear.

Acknowledgements

This work was supported by the Beijing Advanced Innovation Center for Soft Matter Science and Engineering of the Beijing University of Chemical Technology and funding was obtained from the European Union's Horizon 2020 research and innovation program under grant agreement 727762, project Next-CSP.

References

(1) Flamant, G.; Gauthier, D.; Benoit, H.; Sans, J.-L.; Garcia, R.; Boissière, B.; Ansart, R.; Hemati, M. Dense Suspension of Solid Particles as a New Heat Transfer Fluid for Concentrated Solar Thermal Plants: On-Sun Proof of Concept. *Chem. Eng. Sci.* **2013**, *102*, 567.

(2) Benoit, H.; Pérez López, I.; Gauthier, D.; Sans, JL.; Flamant, G. On-sun demonstration of a 750°C heat transfer fluid for concentrating solare systems: Dense particle suspension in tube. *Sol. Energy* **2015**, *118*, 622-633.

(3) Zhang, H. L.; Benoit, H.; Gauthier, D.; DegrÃ¨ve, J.; Baeyens, J.; LÃ³pez, I. P.; Hemati, M.; Flamant, G. Particle Circulation Loops in Solar Energy Capture and Storage: Gasâ€“solid Flow and Heat Transfer Considerations. *Appl. Energy* **2016**, *161*, 206.

(4) PÃ©rez-LÃ³pez, I.; Benoit, H.; Gauthier, D.; Sans, J. L.; Guillot, E.; Mazza, G.; Flamant, G. On-Sun Operation of a 150kW Th Pilot Solar Receiver Using Dense Particle Suspension as Heat Transfer Fluid. *Sol. Energy* **2016**, *137*, 463-476.

(5) BoissiÃ¨re B, Ansart R, Gauthier D, Flamant G, Hemati M. Experimental hydrodynamic study of gas-particle dense suspension upward flow for applications as new heat transfer and storage fluid. Canad J Chem

Eng. 2015;93(2):317â€“330.

(6) Kong, W. B.; Tan, T. W.; Baeyens, J.; Flamant, G.; Zhang, H. L. Bubbling and Slugging of Geldart Group A Powders in Small Diameter Columns. *Ind. Eng. Chem. Res.* **2017**, *56*(DOI:10.1021/acs.iecr.6b04798), 4136.

(7) Sabatier, F.; Ansart, R.; Zhang, H.; Baeyens, J.; Simonin, O. Experiments support simulations by the NEPTUNE_CFD code in an Upflow Bubbling Fluidized Bed reactor. Chemical Engineering Journal,**2018,** submitted.

(8) Kunii, D.; Levenspiel, O. *Fluidization Engineering*; Elsevier, 1991.

(9) Zhang, H.; DegrÃ¨ve, J.; Baeyens, J.; Wu, S.-Y. Powder Attrition in Gas Fluidized Beds. *Powder Technol.* **2016**, *287*, 1.

(10) CEMA Factors. In *Handbook of Powder Science and Technology*; Fayed, M., Otten, L., Eds.; Springer, 2013; pp 445â€“450.

(11) Vandewalle, C.; Baeyens, J. Attrition and Wear in Powder Circulation Loops. *Powder Handl. Process.* **2001**, *13* (1), 77.

SO2 Removal Efficiency in Fluidized Bed Combustor Burning Coals of Different Sulfur Content

Conference: Fluidization

Year: 2019

Proceeding: Fluidization XVI

Group: General Paper Pool

Session: 3C: Fluidized Bed Applications

Time: Tuesday, May 28, 2019 - 9:19am-9:31am

Authors:

Liang, B., Yankuang Group Co. Ltd.
Lan, T., Yankuang Group Co. Ltd.
Bai, H., Yankuang Group Co. Ltd.
Song, H., Yankuang Group Co. Ltd.

Coal represents over 94% of the fossil energy resources in China, and has played an important role in the countryâ€™s explosive economic development over the past decades. Unfortunately, due to various reasons, air pollution from utilization of coals has been a recognized problem for many years. In recent years, with the air emission standards becoming more stringent, coals with sulfur content higher than approx. 1% , which represents over 40% of the Chinese coal reserves, has been limited for use.

For clean and efficient utilization of high sulfur coals, sulfur removal experiments in the fluidized bed combustion were carried out for coals having sulfur content from 0.56% to 3.68%. Limestone with a preferred particle size distribution was used as a sulfur sorbent. Experiments were conducted for combustion temperatures from 800-950â„ƒ with a calcium-to-sulfur ratio (Ca/S) of 2.0-3.0. It is interesting to find out that the SO_2 removal efficiency increases with the sulfur content for given Ca/S and temperature. As the sulfur content is higher than 1.5%, the SO_2 removal efficiency can stably attain above 98% at Ca/S=2.5, resulting in a SO_2 emission of less than 50 mg/Nm3. For the coal with a sulfur content of 3.68%, the SO_2 emission was reduced to 18.2 mg/Nm3 (i.e. 99.6% removal efficiency) at Ca/S=2.5, which is lower than the Chinese new ultralow SO_2 emission requirement of 35 mg/Nm3. Therefore, it can be concluded that with advanced fluidized bed combustion technology, high sulfur coals can be utilized environmental friendly and cost effectively.

The Impact Fragmentation Tendency of Limestone Particles in Calcium Looping Cycles in the Presence of Steam and Sulphur Dioxide

Conference: Fluidization

Year: 2019

Proceeding: Fluidization XVI

Group: General Paper Pool

Session: 3C: Fluidized Bed Applications

Time: Tuesday, May 28, 2019 - 9:07am-9:19am

Authors:

Scala, F., Istituto di Ricerche sulla Combustione, Consiglio Nazionale delle Ricerche
Coppola, A., University of Naples Federico II
Esposito, A., University of Naples Federico II
Montagnaro, F., Università degli Studi di Napoli Federico II
Salatino, P., Università degli Studi di Napoli Federico II

This manuscript focuses on the calcium looping process, a flue gas post-combustion treatment aimed at carbon dioxide capture and its further release in a concentrated stream to be geologically stored or utilised in the chemical industry. Calcium looping is based on the reversible carbonation of Ca-based sorbents, such as those derived from natural (and cheap) limestone. It is a process carried out by means of two reactors (a carbonator and a calciner), typically in a dual interconnected fluidised bed configuration.

Processing sorbent particles in a fluidised environment unavoidably leads to attrition and fragmentation phenomena, with consequent changes in the particle size (and residence time) distribution, which in turn may influence the CO_2 capture capacity. Moreover, the possible presence, in the combustion flue gas entering the carbonator, of SO_2 and/or steam would have relevant effects on the sorbent reactivity. This and other research groups have widely investigated in the recent past the effect of the presence of SO_2/H_2O in the carbonator atmosphere, with reference to the effect of these species on the CO_2 capture capacity and lately on the attrition tendency of the sorbent particles, while the specific effect of sulphur dioxide and steam on the impact fragmentation tendency of the sorbent in calcium looping cycles has been so far mostly neglected.

Accordingly, the aim of this work was to give a contribution in this direction. Calcium looping experiments have been performed in a lab-scale twin fluidised bed apparatus (consisting of two identical interconnected bubbling fluidised bed reactors), able to reproduce a realistic particle thermal/chemical/mechanical history, using a reference high-Ca limestone. Calcium looping tests consisted of ten calcination/carbonation cycles. Carbonation was carried out at 650°C in an atmosphere containing 15% CO_2 to simulate a typical combustion flue gas, while calcination was operated at 940°C at 70% CO_2 to simulate oxy-combustion conditions. Six different operating conditions for carbonation were tested to study the effect of SO_2 and/or H_2O, where steam (when present) was fed at 10%, and SO_2 (when present) either at 75 ppm (a typical concentration in a pre-desulphurised combustion flue gas) or 1500 ppm (raw combustion flue gas). The CO_2 capture capacity was calculated for each carbonation stage and, at the end of the test, sorbent particles were further analysed for the determination of the degree of Ca sulphation.

After the tests, the exhausted sorbent particles were fed to an ex situ impact test apparatus, which is based on the well-established concept of entraining particles in a gas stream at controlled velocity, and impacting them against a target. The test rig consists of a vertical stainless steel eductor tube equipped with a particle feeding device. After feeding, the particles are accelerated by the air flow in the eductor tube. The particle velocity is controlled by regulating the air flow in the eductor tube, by means of a flowmeter. When the particles exit the eductor tube, they impact on a rigid target plate placed in a collection chamber 50 mm below the bottom end of the tube. The target is made of stainless steel and is inclined by 30Â° with respect to the horizontal. This inclination was chosen as a trade-off between the need of avoiding interference between impacting and reflected particles, and the need of minimising the departure of results from those obtained with a target perpendicular to the particle trajectory. The device is designed so as to minimise the loss of particles entrained by air in the chamber, enabling easy collection of the sorbent particles after impact for further analysis.

Samples (approximately 1.0 g in the particle size range 0.4â€“0.6 mm) of raw limestone (for reference), and of exhausted sorbent particles generated in the twin bed apparatus under the six different operating conditions, were used for fragmentation tests in the impact testing apparatus. The tests were carried out in air at ambient temperature with the following particle impact velocities: 10, 17, 24, 31, 38, 45 m sâ€“1. After each test, the samples were retrieved from the collection chamber and weighed. The collected particles were then sieve-analysed to obtain their particle size distribution (with related probability density functions), and to calculate the fraction of fragments (<0.4 mm) cumulatively generated. The analysis of data, as a function of the impact velocity, allowed to discuss the impact fragmentation tendency of each sample in the light of the broad particle breakage mechanisms â€œchippingâ€, â€œsplittingâ€ and â€œdisintegrationâ€, as defined in literature by this and other research groups. The behaviour was then critically related to the presence, in the calcium looping cycles generating the samples, of steam and/or SO_2, which produced sorbent particles containing CaO, $CaCO_3$ and

$CaSO_4$ (materials which have different mechanical properties) to various extents as inferred from the analysis of the CO_2/SO_2 degree of capture.

The raw sample resulted to be the most resistant among all those tested, as it is rich in hard $CaCO_3$. Samples obtained in presence of SO_2 resulted to be harder than in absence of it, as higher SO_2 concentrations in the carbonator determined thicker $CaSO_4$-based shells around the particles. This is in particular relevant at low impact velocities, where chipping is the main fragmentation mechanism (while splitting/disintegration are active at higher velocities) and the impact tendency of the particles is governed by the mechanical properties of their outer peripheral layer. The presence of steam in the carbonator, orientating the reactivity of CaO towards CO_2 rather than SO_2 (when present), on the one hand determines particles less resistant than those obtained in absence of steam and in presence of the sole SO_2 together with CO_2. On the other hand, as steam indeed favours the carbonation reaction of CaO, we have observed particles with a larger $CaCO_3$ fraction and therefore more resistant than those obtained in conditions where both steam and SO_2 were absent. From the quantitative point of view, the fraction of fragments resulted, at 10 m sâ€“1, in the range from 1.5% (raw limestone) to 4.7% (sample exhausted in absence of both steam and SO_2). At 45 m sâ€“1, these values increased in the range from 5.4% to 23.2%, depending on the sample.

Thermochemical Energy Storage Employing Fluidized Bed Technology: Experimental Investigations with CaO/Ca(OH)2 on a 21kWh Reactor

Conference: Fluidization

Year: 2019

Proceeding: Fluidization XVI

Group: General Paper Pool

Session: 3C: Fluidized Bed Applications

Time: Tuesday, May 28, 2019 - 9:43am-9:55am

Authors:

Wuerth, M., Technical University of Munich
Spliethoff, H., Technical University Munich
Becker, M., Technical University of Munich
Gleis, S., Technical University of Munich
Vandersickel, A., Technical University Munich
Talebi, E., Technical University of Munich

Thermochemical energy storage (TCES) represents one of the most promising energy storage technologies, currently investigated. It uses the heat of reaction of reversible reaction systems and stands out due to the high energy density of its storage materials combined with the possibility of long-term storage with little to no heat losses.

Amongst possible reversible reaction systems, gas-solid reactions, especially the reaction system $CaO/Ca(OH)_2$, have come to the fore.Until now, the focus for the technical implementation of TCES has been pointed towards fixed bed reactors. However, due to the limitations of fixed bed reactors, the application of fluidized bed reactors for TCES has gained increased attention. Fluidized bed reactors do not only enable the favorable option of operating the system continuously, which allows for a decoupling of storage power and capacity. They also offer significant advantages for the scale-up of the system due to the improved heat and mass transfer. The heat transfer between immersed heat exchangers and the fluidized bed is of particular interest for the herein investigated reaction system $CaO/Ca(OH)_2$, as preliminary experiments have identified it to be the main aspect limiting the hydration/dehydration reaction. In addition to the importance of the heat transfer characteristics, a deeper insight into the fluidization behavior of the storage material itself is of major importance. Even though gas-solid fluidized beds are nowadays implemented in a wide range of industrial operations, the fluidization of cohesive materials, such as the aforementioned metal oxides/hydroxides, still represents a sparsely investigated field. The consequent lack of knowledge of physical, chemical and technical parameters of the processes on hand is currently a hindering aspect for a proper design and scale-up of fluidized bed reactors for MW-applications of TCES.

Preliminary experimental work at TUM on a lab scale fluidized bed reactor has been carried out to gain first insights into the topic of chemical cycle stability of storage materials. In addition, the fluidization behavior of the bulk material has been investigated in a fluidized bed cold model, containing a heat flux probe and operating at ambient conditions.

In order to provide in depth information on the operation of a fluidized bed reactor for TCES at industrial operation conditions, a state-of-the-art pilot fluidized bed reactor has been set up at TUMâ€™s research facilities. While being able to be operated continuously, the initial experiments presented herein focus on batch operation at conditions of up to 700Â°C and -1 to 6 bar_g. In this mode, the system itself possesses a thermal storage capacity of up to 21 kWh. The storage material used for the first experiments holds a mean particle diameter of about 350 Âµm. Solid samples, for further characterization of the process, are taken throughout operation with an online sampling system. This characterization is focused on changes in particle size distribution due to effects such as attrition, breakage and agglomeration, on conversion rates, and on heat transfer coefficients. The initial batch experiments focus on low cycle numbers, where one cycle is defined as the execution of one hydration and one dehydration step of the material. The results, obtained from these experiments, help to prepare for multiple cycle studies and give first insight for a successful scale-up to MW-scale.

CFD Simulation of the Coal Pyrolysis in a Fluidized-Bed Reactor

Conference: Fluidization

Year: 2019

Proceeding: Fluidization XVI

Group: General Paper Pool

Session: 3D: Modeling and Simulation

Time: Tuesday, May 28, 2019 - 10:07am-10:19am

Authors:

Guan, Y., North China Electric Power University
Qi, N., North China Electric Power University
Zhang, K., North China Electric Power University

The coal topping process in circulating fluidized beds (CFBs) is beneficial for efficient conversion and utilization of high-volatile solid fuels, such as biomass and low-rank coal. The hydrodynamics and pyrolysis behavior of coal are explored by experimental measurement and computational fluid dynamic (CFD) simulation in this study. A fluidized bed system is built to explore the effect of drag model on the CFD simulation of gas-solids flow. The distribution of solids velocity and pressure are obtained by the wire-mesh electrodes and pressure transducers, respectively. A hydrodynamic model of the fluidized bed is established based on the Eulerian-Eulerian two-fluid model with the kinetic theory of granular flow. Effect of the drag model, including the empirical correlation based on experiment, the correlation based on direct numerical simulation and the correlation based on energy minimization multi-scale method, on the computational results is investigated and validated by the experimental data. Both qualitative and quantitative results indicated that the drag model had a significant effect on the flow behavior. Furthermore, a comprehensive CFD model, involving continuity equations, momentum equations, energy equations, species transport equations and related constitutive equations, is established to predict the hydrodynamics and pyrolysis behavior of low-rank coal in the fluidized bed.

Considerations for Practical Industrial CFD Simulations of Fluidized Systems

Conference: Fluidization

Year: 2019

Proceeding: Fluidization XVI

Group: General Paper Pool

Session: 3D: Modeling and Simulation

Time: Tuesday, May 28, 2019 - 10:19am-10:31am

Authors:

Blaser, P., CPFD, LLC
Cocco, R., Particulate Solid Research, Inc.
Akhavan, A., CPFD Software

Computational Fluid Dynamics (CFD) is based upon both fundamentals and empirical models. The fundamental principles, such as conservation of mass, momentum and energy, are known. Empirical models, on the other hand, have more limited scopes of applicability, and thus are less fully understood.

CFD for fluidized systems, such as fluidized beds or circulating systems, has additional inherent complexity, compared with pure fluid flows. Results can be highly dependent upon empirical models related to fluid-particle, or particle-particle interaction terms, which take the form of drag, collisional, stress, granular temperature, or similar models. The development and validation of these models, which is often undertaken at small test scales, is a topic of significant, current research.

Many industrial systems utilizing fluidization, on the other hand, are constructed and operated at scales which may be orders of magnitude larger than the lab-scale or pilot-scale units upon which they are based. Industrial processes such as Fluidized Catalytic Cracking (FCC), polyolefins production, gasification, pyrolysis, or cement manufacture, to name a few, often involve characteristic length scales which can be up to ten meters or greater.

This presentation examines the relationship between empirical model development and practical CFD simulations of industrial systems. The role of test data, and the scale at which the data is obtained, is explored. Impacts of test data on model calibration for industrial-scale simulations are discussed. Multiple industrial case studies with practical outcomes are presented.

Coupled Lbm/DEM Simulations with a Multi-Particle Partially Saturated Method- Theoretical and Computational Details

Conference: Fluidization

Year: 2019

Proceeding: Fluidization XVI

Group: General Paper Pool

Session: 3D: Modeling and Simulation

Time: Tuesday, May 28, 2019 - 9:31am-9:43am

Authors:

Tsigginos, C., Daresbury Laboratory STFC
Meng, J., Daresbury Laboratory STFC
Gu, X. J., Daresbury Laboratory STFC
Emerson, D. R., Daresbury Laboratory STFC

A coupled Discrete Element (DEM) - Lattice Boltzmann scheme was developed for modelling particle transports in fluid flows. The fluid particle interaction is modeled with the multi-particle partially saturated method (MPPSM). In the MPPSM, the fluid-solid collusion operator of the PSM scheme is split in two parts, where the solid operator is implemented as a boundary condition that properly enforces the rigid body motion in the particle interior. The original PSM scheme that is based on the cell concept, does not properly resolve the fluid flow in the vicinity of contacting particles. By introducing a solid collision operator and weight parameter B for each particle, the flow in the vicinity of contacting particles can be sufficiently resolved. An assessment of the schemeâ€™s convergence rate and the effect of its parameters on performance is conducted. Finally, a necessary condition that allows the timestep factor (Î»=Î”t^{LBM}/Î”t^{DE}) to attain values larger than one was found.

Drag Model Modifications for Gas-Solids Two-Phase Flows with Clusters in Circulating Fluidized Bed Risers

Conference: Fluidization

Year: 2019

Proceeding: Fluidization XVI

Group: General Paper Pool

Session: 3D: Modeling and Simulation

Time: Tuesday, May 28, 2019 - 9:07am-9:19am

Authors:

Sun, Z., Western University
Zhang, C., Western University
Zhu, J., The University of Western Ontario

Drag model modifications for gas-solids two-phase flows with clusters in circulating fluidized bed risers Introduction

CFD modeling has been considered as an effective tool for researchers to design and study the multiphase flows inside circulating fluidized bed (CFB) risers. The Eulerian-Eulerian two-fluid method (TFM) is widely applied into simulations of gas-solids CFB risers. However, there are always discrepancies between the numerical results and the experimental data. The main reason might be the inaccurate drag models used between the gas and solids since single particles tend to agglomerate to clusters in the gas-solids CFB riser. The cluster usually consists of a group of single particles, so that it has higher solids concentration than the surrounding gas-solids suspension. The intensive gas-solids interactions result in the continuous formation and breakup of particle clusters in the CFB riser. When it comes to CFD modelling on the gas-solids flow in the CFB riser, the dynamic nature of the clusters makes it difficult to derive an accurate correlation for the drag force between the gas and solids phases.

Researchers have put a lot of effort to the modification of the drag models to include the cluster effect and all the current drag models that include the cluster effect can be generally classified into three groups based on different theories. In the first group, the drag models were modified based on the pressure drop data from the packed beds to include the cluster effect, such as Wen and Yu model, Gidaspow model, and Huilin-Gidaspow model. In the second group, drag model modifications were obtained based on the concept of the Richardson-Zaki equation, but with a correlated Vr based on the multiparticle system, such the popular used Syamlal-OIn both groups, the properties of single particle were used in the modified drag models. In the third group, the meso-scale heterogeneity theory of gas-solids system was used to modify the drag model, such as EMMS model and CSD model. Although â€œnumericalâ€ clusters were calculated based on the minimum energy dissipation theory form EMMS/CSD model, it has not been validated that those â€œnumericalâ€ clusters are the same as those real clusters existing in a fluidized bed.

Because of lacking experimental data on clusters, most of the modified drag models try to include the cluster effect by adding a factor to the drag models developed for the uniformly distributed particles, However, the properties of single particles were still used in those modified drag models. Some of those modified drag models work well under certain flow conditions in fluidized beds, but they are not suitable for general cases. A general drag model that includes the cluster effect in gas-solids fluidization systems is developed in this study based on the statistical data of the clusters with the help of image analytical experiments.

The drag model including the clustering effect

There are two types of clusters in CFB riser: one is â€œcoreâ€ cluster in which particles are tightly packed by the fluid and can be hardly breakup when rising along the riser; another is â €œcluster of core clustersâ€ which are many loosely aligned â€œcoreâ€ clusters and they are frequently formed and breakup inside the CFB riser. With the help of the experimental data, the statistical results of the diameter and density of the â€œcoreâ€ clusters can be obtained from the image analysis. The solids in a CFB riser can be divided into a cluster phase which consists of â€œcoreâ€ clusters and a single particle phase which only contains individual particles. Therefore, the drag force between gas and solids phases in CFD simulations can be determined from the summation of the drag force between the cluster phase with the cluster properties and the gas phase, and the drag force between the single particle phase and gas phase.

The Sparsely Distributed Solid Particles drag model (Schiller & Naumann) is used for the calculations of the drag force for both the cluster phase and single particle phase, but with different input parameters for those two phases. . When calculating the drag force for the cluster phase, the clusters are assumed as stable spheres (â€œcoreâ€ clusters) with a constant diameter, d_{cl} and density, , based on statistical data from the experiments. The volume fraction of the clusters phase is also defined from the statistical data by the image analysis. The slip velocity of the clusters is assumed based on different operating conditions.The effects of the cluster relaxation time and the surrounding solids holdup are also considered and compared by two other cases in the results section to better present the effectiveness of the proposed cluster drag model.

Three CFD cases as shown in Tab. 1 are setup to compare the cluster-driven drag model and the commonly used Syamlal and O'Brien drag model in TFM method. A typical operating condition of Ug = 5m/s, Gs = 100 kg/m^{2}s for FCC particles in the CFB riser is selected for numerical simulation. The corresponding statistic data for the cluster properties under the same operating condition are: cluster diameter, d_{cl} = 0.0058m, cluster density is 0.052, and cluster phase volume fraction to the solids phase is 0.5.

Table 1: Cases setup details

Case #	Drag calculation	Relaxation time	Other considerations
1	Syamlal and O'Brien drag model	Single particle	
2	Schiller Naumann drag correlation on clusters and single particles	Single particle	
3	Schiller Naumann drag correlation on clusters and single particles	Cluster/single particles	

Results and discussion

The overall solids holdup profiles form Case #1-3 do not have much difference between the commonly used Syamlal and O'Brien drag model and the proposed drag model and they both have a slight deviation from the experimental data. The overall solids velocity profiles show that Case #2 overestimated the particle velocity compared with Case #3 because the effects from cluster relaxation time and surrounding solids holdup are not included. However, there is almost no difference among Case #1 and Case #3, which indicates that the impacts from cluster relaxation time and the surrounding solids holdup are not as significant as the cluster size and density. More details will be provided in the full paper.

Conclusions

A new drag model which includes the cluster effect is proposed in this work. With the help of the image analysis, the statistical data from the experiments for the properties of the clusters such as the cluster diameter, density, and its total volume fraction of the solids phase are employed in the proposed drag model. The total drag force then can be determined by the summation of the drag force between cluster phase and gas phase, and the drag force between the single particle phase and gas. Although the overall flow structures don not have a significant difference between the proposed drag model and the commonly used drag model, the flow details did reflect the impacts of the clusters on the drag force. More frequent clustering phenomenon can be found in the numerical results by the proposed drag modification, which indicates the advantage and accuracy by implementing the cluster properties into drag model although more detailed future work needs to be done.

Mechanistic Analysis of Air-Induced Segregation during Powder Flow into a Confined Space

Conference: Fluidization

Year: 2019

Proceeding: Fluidization XVI

Group: General Paper Pool

Session: 3D: Modeling and Simulation

Time: Tuesday, May 28, 2019 - 8:30am-8:55am

Authors:

Wu, C., University of Surrey
El Hebieshy, J. F., University of Surrey

Die filling is a typical process step in the manufacturing of a wide range of particulate products (e.g. pharmaceutical tablets, catalyst pellets and powder metallurgical components) in various industries, which normally takes place in the presence of air. Segregation of powder mixtures during die filling due to the buildup of air pressure in the confined space can have a detrimental effect on the quality of the end product. In this study, segregation mechanisms of powder mixtures during die filling were experimentally and numerically investigated. Binary mixtures of powders with either similar particle density but different mean particle sizes or similar particle size but different particle densities were considered to investigate segregation due to size or density differences. Using a segmented shoe and die, the size induced segregation and density induced segregation during die filling was examined quantitatively. The results indicated that, during die filling, depending on the shoe speed, different segregation mechanisms came into effect; the sieving segregation mechanism often occurring in hoppers was observed during all condition where air was able to freely escape and the fluidization segregation mechanism was observed when the air inside the die, while being displaced by the powder was only able to escape by permeating through the powder bulk in the shoe It is also found that the composition ratio of the mixtures had only a minor effect on the segregation patterns for the materials considered in this study. Numerical analysis was also performed using a coupled DEM-CFD, from which the air-induced segregation mechanisms were explored.

Numerical Simulation of Gas-Solid Flow in Downer with a New Drag Model Based on the Fluidized Spatial Superposition

Conference: Fluidization

Year: 2019

Proceeding: Fluidization XVI

Group: General Paper Pool

Session: 3D: Modeling and Simulation

Time: Tuesday, May 28, 2019 - 9:43am-9:55am

Authors:

Pan, X., Taiyuan University of Technology

Nowadays,
substantial progress has been made in simulating downers by using two-fluid model (TFM). Li et al. pointed out that the accuracy of the TFM model mainly depended on the drag model. After that, the drag model continued to be developed. Currently, three drag models that based on the homogeneous structure or heterogeneous structure are often used by many researchers. However, these models either simulation results
deviated from the experimental data greatly or considered most comprehensive and sufficient so that need solve many structure parameters and equations. Therefore, it is needed to construct a novel drag model that simultaneously considering the two existing forms of particles in the downer and reducing the complexity of the model for easy solution.

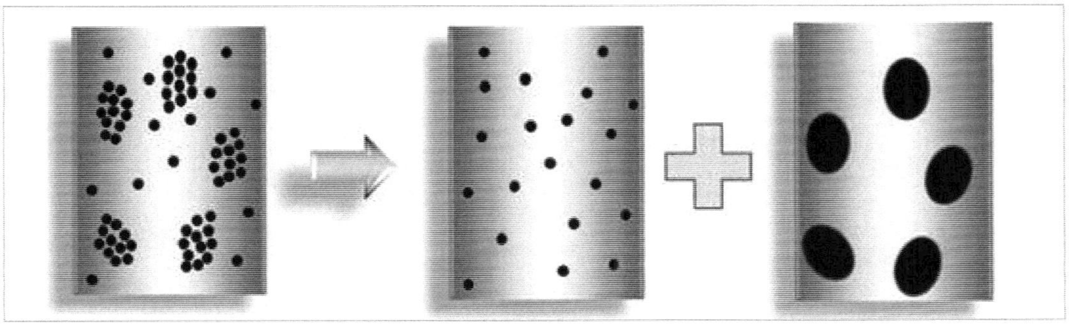

Fig.1. Decomposition of the control volume

In this work, the heterogeneous multiphase flow structure is divided into dilute phase and agglomerate phase, as shown in figure.1. Moreover, the dilute and agglomerate particles are uniformly dispersed in the control volume and are fluidized by all gas. Therefore, the real flow structure is formed by spatial superposing the flow behavior of the dilute and agglomerate phase. According to this new approach, five structure parameters were obtained and then used to calculate the drag coefficient for a control volume, which is further incorporated into the TFM. Finally, comparisons of the simulation results between this new drag modification and the Wen & Yu/Ergun Equations are carried out to demonstrate that present work have the capacity to capture major hydrodynamic features of downer reactor.

Numerical Simulation of Hydrodynamic Behaviours in a Novel Gas-Solids Moving Bed Reactor Under a Downer

Conference: Fluidization

Year: 2019

Proceeding: Fluidization XVI

Group: General Paper Pool

Session: 3D: Modeling and Simulation

Time: Tuesday, May 28, 2019 - 9:19am-9:31am

Authors:

Guan, G., Hirosaki University
Zhao, Z., Hirosaki University
Situmorang, Y. A., Hirosaki University
Fushimi, C., Tokyo University of Agriculture and Technology
Tsutsumi, A., The University of Tokyo
Hao, X., Taiyuan University of Technology
Yoshida, A., Hirosaki University
Abudula, A., Hirosaki University

The downer can be used in the circulating fluidized bed as a pyrolyzer of coal and/or biomass because of its characteristics of gas-solids contact in a short time, uniform distribution, and so on. In a practical process, some escaping particles of coal or biomass which are not pyrolyzed completely and produced tar could get in the char gasification unit and hinder the gasification of char. To solve this problem, a gas-solids moving bed rector is considered to be added at the bottom of downer which can let coal/biomass pyrolyzed completely and simultaneously tar cracked effectively on the char. In this study, FLUENT 19.0 commercial software, a 3D grid, and the Eulerianâ€"Eulerian two-fluid model were applied to simulate the hydrodynamic behaviours in this gas-solids moving bed reactor. The results of numerical simulation showed that this moving bed could increase solids holdup extremely, reaching a value of above 0.35. There was a peak of solids holdup near the wall while many particles concentrated at the center of the moving bed. This phenomenon should be caused by the radial velocity of particles. In addition, the axial velocity distribution was the similar as that in the downer, where the first acceleration section, the second acceleration section and the constant velocity section were observed.

[Acknowledgements] This is supported by the JSPS KAKENHI Grant-in-Aid for Scientific Research B (Kiban B, 17H03451), Japan and the National Natural Science Foundation of China (U1710101)

Review of Fluid-Particle Drag Experimental Validation

Conference: Fluidization

Year: 2019

Proceeding: Fluidization XVI

Group: General Paper Pool

Session: 3D: Modeling and Simulation

Time: Tuesday, May 28, 2019 - 9:55am-10:07am

Authors:

Freireich, B., Particulate Solid Research, Inc.
LaMarche, C. Q., Particulate Solid Research, Inc.
Cocco, R., Particulate Solid Research, Inc.

Predicting accurate fluidized particle behavior with computational fluid dynamics (CFD) is beneficial to industry via improving operation scaling up speed and reliability, fluidized bed operational efficiency, troubleshooting existing operations, as well as exploring new equipment design, etc. Modeling the drag force, a major component of the fluid-particle momentum exchange, is essential to establishing a relevant toolbox for predicting fluidized particle behavior. Currently, a gamut of drag models exist, ranging from models that are based on the assumption of homogenous distributions of particles with each fluid grid to models that utilize sub-grid formulations to account for solids volume fractions gradients (i.e., heterogeneous particle spatial distributions). A critical, comprehensive review of these models is presented with a focus on the parameter space relevant to development and validation. Key findings with respect to the limited range of experiments used for validating models, and the mismatch in drag model validation across different fluidization regimes are reported. Furthermore, the potential influence of neglecting, or incorrectly modeling, physical interactions between particles (e.g. cohesion, friction) are explored along with the effect of erroneously attributing all inaccuracies of the simulation to the drag model. Gaps associated with the experimental data available for model calibration and validation are discussed.

Simulation Research on CO2 Capture Process By Chemical Method Based on Self-Heat Recuperation Technology

Conference: Fluidization

Year: 2019

Proceeding: Fluidization XVI

Group: General Paper Pool

Session: 3D: Modeling and Simulation

Time: Tuesday, May 28, 2019 - 8:55am-9:07am

Authors:

Chen, D., Taiyuan University of Technology

To
reduce the high energy consumption of the CO_2 capture process, the self-heat regeneration technology is applied to optimize the traditional chemical CO_2 capture process. Simulation analysis of the capture process by Aspen plus£¬established
traditional chemical of CO_2 capture model based on flue gas emitted from the triple-bed combined circulating fluidized bed(TBCFB). After that, the self-heating regeneration technology is applied to optimize the energy saving of the capture system. The simulation
results show that the energy consumption of the optimized capture process is 1.66GJ/ (t CO_2), and the energy consumption can decrease 37%.The economic evaluation of the process was carried out through the calculation of investment cost and operating cost. The optimized capture integration was reduced by 25.84% compared with the traditional chemical process. The improved capture process comprehensively utilizes the residual heat of the system, and realizes that the recycling heat of the absorption heat of the absorption tower and the steam condensation heat of the stripping tower, which significantly reduces the energy consumption of the system, reduces the need for additional heat sources and increases energy efficiency simultaneously.

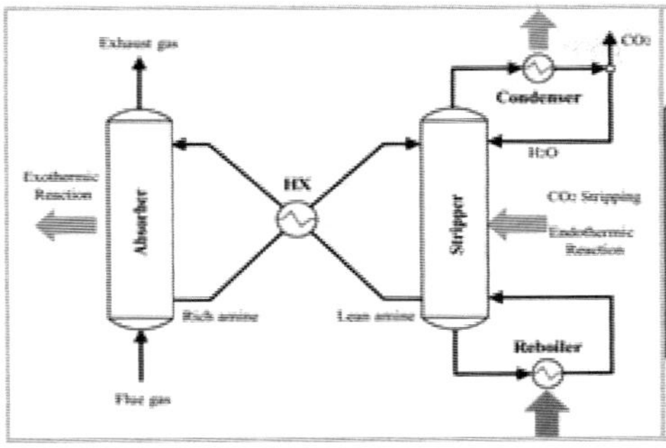

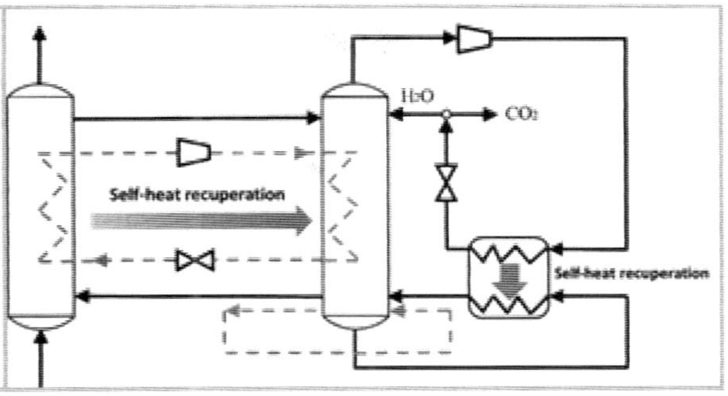

Fig.1 Traditional chemical CO_2 capture process Fig.2 Chemical CO_2 capture process by self-heat recuperative

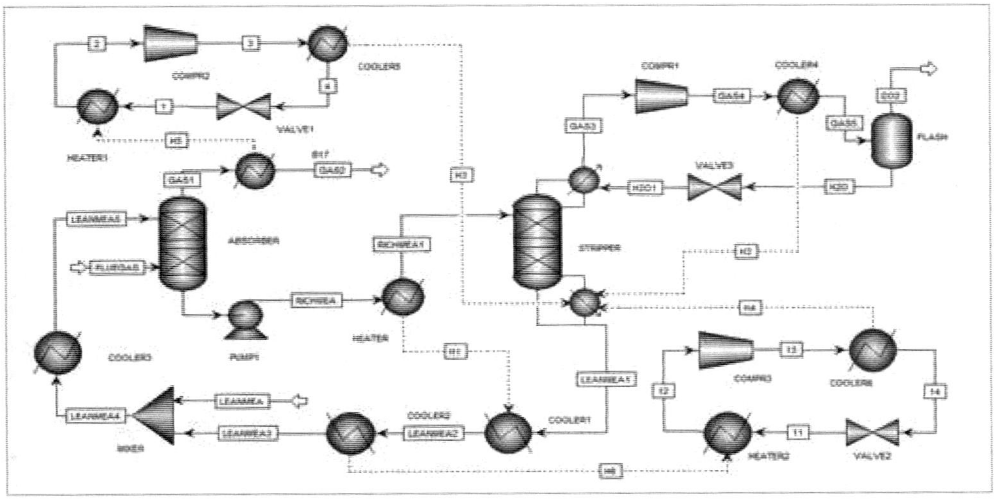

Fig.3 System model
CO2 capture process by self-heat recuperative

Plenary Talk: Fluidization in 100 Years - Future Perspectives

Conference: Fluidization

Year: 2019

Proceeding: Fluidization XVI

Group: General Paper Pool

Session: Plenary Talk & Panel Discussion

Time: Tuesday, May 28, 2019 - 1:30pm-2:00pm

Authors:

Zhu, J., The University of Western Ontario

Fluidization shall have its 100 anniversary next year and it is very useful to review the past and then start to look for the future. What would fluidization look like in another 100 years, or in some other name?

Since the inception, the â€œoriginalâ€ gas-solid â€œconventionalâ€ fluidized bed has been expanded, to liquid-solid and three-phase fluidization by changing the fluidizing medium, to circulating fluidization by changing (increasing) the fluidizing velocity beyond terminal velocity, to downflow fluidization by changing the fluid flow direction, and to inverse fluidization by changing the relative density of the fluid. Taking advantages of its large fluid-particle interfacial area, vigorous mixing, and excellent heat and mass transfer, fluidized bed processes have had major successes in the past 100 years, such as fluid catalytic cracking of crude oil, the combustion of coal, and many catalytic chemical reactions.

Recent developments have been pointing to intensifying and integrating fluidization processes, instituting new processes in other â€œnon-conventionalâ€ process industries, comprehensive process modelling, and more in-depth fundamental studies to reveal the underlining principles fully utilizing the rapid developments of powerful measurement tools and computational capabilities. Utilizing fine group C particles and establishing micro flow fluidized bed reactors shall also come to play. New fluidization theories, especially those related to the above-mentioned developments, can be anticipated.

Adsorption Desulfurization Performance and Adsorption-Diffusion Study of B2O3 Modified Ag-CeOxTiO2-SiO2

Conference: Fluidization

Year: 2019

Proceeding: Fluidization XVI

Group: General Paper Pool

Session: Poster Session

Time: Tuesday, May 28, 2019 - 4:00pm-6:00pm

Authors:

Zheng, M., Fuzhou University

Adsorption desulfurization performance and adsorption-diffusion study of B_2O_3 modified Ag-CeO_x/TiO_2-SiO_2

Meiqin Zheng[1]*

[1] Fujian Province University Key Laboratory of Green Energy and Environment Catalysis, College of Chemistry and Materials, Ningde Normal University, Ningde, Fujian 352100, China

* Correspondence author E-mail: 15280002581@163.com.

In this work, the adsorption desulfurization performance and adsorption diffusion study of B_2O_3 modified Ag-CeO_x/TiO_2-SiO_2 adsorbent were investigated. The adsorption desulfurization performance was studied by batch and fixed bed tests. The homogeneous surface diffusion model (HSDM) [1] were employed to investigate the adsorption and diffusion behavior of 4,6-dimethyldibenzothiophene (4,6-DMDBT) in diesel. It was found that the addition of B_2O_3 promotes the dispersion of CeO_x species and then further facilitates the dispersion and oxidation of Ag species resulting in higher adsorption desulfurization activity[2]. Ag species are in a state of Ag, Ag_2O and Ag_2O_2, among which, Ag_2O and Ag_2O_2 were found to be the active centers[2,3].

The kinetics of adsorption desulfurization of model diesel fuel was investigated to provide guiding significance for the prediction of breakthrough curves of fixed-bed adsorption columns.

The batch kinetic experiment modeled by HSDM model indicates that surface diffusion controls the main rate. The surface diffusion coefficient D_s determined by batch adsorption experiments is independent of operation conditions, which can be used to directly predict the breakthrough behavior in fixed bed adsorption. The modified HSDM model is proposed to describe the breakthrough behavior. Results indicate that the breakthrough time is affected by bed height, flowrate and influent concentration.

Simultaneously, a 2D porous media CFD model is developed for a fixed bed adsorber to remove the 4,6-dimethyldibenzothiophene (4,6-DMDBT) from diesel. The CFD simulation results demonstrated that porous media CFD model can describe successfully the adsorption mass transfer process for removal of 4,6-DMDBT from diesel. The mass transfer rate increased with increasing feed concentration at the beginning due to the higher driving force at higher feed concentration, conversely, the lower mass transfer rate was obtained at higher feed concentration after a period of adsorption time because of the decrease in the saturation time at the fixed bed column entrance gradually at higher feed concentration.

Key words: adsorption desulfurization; B_2O_3; diffusion; HSDM; CFD.

REFERENCES

1. Zheng M, Xu C, Hu H, Ye Z, Chen X. A modified homogeneous surface diffusion model for the fixed-bed adsorption of 4,6-DMDBT on Agâ€“CeO_x/TiO_2â€“SiO_2. *RSC Adv.* 2016;6(114):112899-112907.
2. Zheng M, Hu H, Ye Z, Huang Q, Chen X. Adsorption desulfurization performance and adsorption-diffusion study of B2O3 modified Ag-CeOx/TiO2-SiO2. *J. Hazard. Mater.* 2019/01/15/ 2019;362:424-435.
3. Xu C, Zheng M, Chen K, Hu H, Chen X. CeO_x Doping on a TiO_2-SiO_2 Supporter Enhances Ag Based Adsorptive Desulfurization for Diesel. *J. Fuel. Chem. Techno.* 2016;44(8):943-953.

Characterizing Fluidization of Cohesive Particles in Fluidized Bed By Axial Distribution of Particle Motions

Conference: Fluidization

Year: 2019

Proceeding: Fluidization XVI

Group: General Paper Pool

Session: Poster Session

Time: Tuesday, May 28, 2019 - 4:00pm-6:00pm

Authors:

Lin, W., Zhejiang University
Yang, Y., College of Chemical and Biological Engineering
Huang, Z., Zhejiang University
Sun, J., Zhejiang University
Wang, J., Zhejiang University
Yang, Y., Zhejiang University

To characterize the fluidization of cohesive particles in the fluidized bed, particle motions distribution along the axial direction was investigated by acoustic emission detection. Results showed that the energy of acoustic emission signals increased rapidly at minimum fluidization gas velocity. This phenomenon provided a new direction of cohesive particle fluidization characterizing in fluidized bed. Meanwhile, the wavelet analysis results showed that the particle-wall interaction changed at minimum fluidization gas velocity, which regarded as the reason of the change of acoustic energy. When particles fluidized, particle-wall interaction transformed from friction dominated to collision dominated. The transformation of particle motion type indirect characterized the change of fluidization behavior of cohesive particles, which may be further applicate in cohesive particle fluidized bed characterizing.

Cluster Velocity and Slip Velocity in the CFB Riser: Experiment and Simulation

Conference: Fluidization

Year: 2019

Proceeding: Fluidization XVI

Group: General Paper Pool

Session: Poster Session

Time: Tuesday, May 28, 2019 - 4:00pm-6:00pm

Authors:

Chen, H., Chongqing University
Gu, S., Chongqing University

Cluster is a common phenomenon in the circulating fluidized bed, which influences obviously the transfer efficiency of mass and heat. Experimental studying on the cluster velocity was carried out in a thin 2D fluidized bed and Lagrange approach was employed to computing the velocity of the clusters which captured by a high speed camera. The statistical analysis was carried out based on the moving direct of the clusters. The experimental results show that the cluster velocity can be accurately computed by the Lagrange method. The velocity of the upward clusters is slightly greater than that of the downward cluster. The cluster velocity increases with increasing the superficial gas velocity and most of cluster velocity ranges from -3 to 3 m/s. The number of the downward clusters is greater than that of the upward clusters in the peripheral region, while the upward clusters were slighter more than the downward cluster in the central region of the riser. The number of the downward cluster is about 60-70% of the total clusters. The x-direct component of cluster velocity is low, which is the larger in the central region and the smaller in the peripheral region of the riser. Comparison of cluster slip velocity (between cluster velocity and gas velocity) and particle slip velocity (between particle velocity and gas velocity) has been implemented with two-fluid model (TFM). The simulation results show that the cluster slip velocity is nearly equal to the particle slip velocity in the bottom region, whereas it is obviously larger than the particle slip velocity in the top region. The downward cluster slip velocity is greater than that of the upward cluster because of its inverse direct to the gas flow, which results in the shorter during time of the downward cluster.

Coupled Modeling of Hydrodynamics and Kinetics to Determine Single-Pass Conversion in a Heavy Oil Hydrotreater

Conference: Fluidization

Year: 2019

Proceeding: Fluidization XVI

Group: General Paper Pool

Session: Poster Session

Time: Tuesday, May 28, 2019 - 4:00pm-6:00pm

Authors:

Olivo, L., IPN
Donaldson, A., Dalhousie University
Matinez, J., IPN

This work focuses on the phenomenological characteristics of transient 3D multiphase flow in an ebullated bed reactor operating under high severity conditions. The model presented incorporates simultaneous heat and momentum transference within porous media through an Euler-Lagrange approach, with the kinetics of the hydrotreating reactions considered by means of a multi-phase porous media model. The impact of mass diffusion within the gamma-alumina catalyst pores and the translation of the solid phase from the surface to the center of the catalyst is accounted for, along the evolution of trajectories in the continuous phase. The simulations are used to explore pressure drop and velocity distribution in the fluidization system, exploring the impact of particle shape and density on fluidization behavior and pseudo-steady-state flow patterns within the system.

CPFD Simulation for Variables Affecting the Flow Regime of Pneumatic Conveying

Conference: Fluidization

Year: 2019

Proceeding: Fluidization XVI

Group: General Paper Pool

Session: Poster Session

Time: Tuesday, May 28, 2019 - 4:00pm-6:00pm

Authors:

Sung, W. C., Sungkyunkwan University
Chung, S. W., Institute for Advanced Engineering
Jung, J. S., Seintec Corporation
Lee, D. H., Seintec Corporation

The study was carried out with Computational Particle Fluid Dynamics(CPFD) using Multiphase Particle in Cell (MP-PIC) method for predicting pneumatic conveying characteristics. Simulation was performed on horizontal, vertical, vertical upward, and vertical downward circular pipes. Particle diameter was varied 50 to 500 Î¼m and particle density was varied 1000 to 3000 kg/m^3. Gas velocity was varied from 1 to 15 m/s and solid loading was varied from 1 to 100. Simulation results were compared with the modeling data presented in the previous literature and the optimal drag model was confirmed. Relation of Gas velocity, solid loading and shape of conveying line(horizontal, vertical, vertical upward, and vertical downward) to pressure drop was confirmed and the flow regime according to each condition was confirmed.

Direct Production of Rare Earth Oxides from a Fresh REE Ore through Thermal Cracking in a Fluidized Bed Reactor

Conference: Fluidization

Year: 2019

Proceeding: Fluidization XVI

Group: General Paper Pool

Session: Poster Session

Time: Tuesday, May 28, 2019 - 4:00pm-6:00pm

Authors:

Latifi, M., Ecole Polytechnique Montreal
Chaouki, J., Ecole Polytechnique Montreal
Carrillo Garcia, A., Polytechnique Montreal

The rare earth elements (REE) are demanding metals with a vast number of applications in advanced technologies. The rare earth elements such as neodymium, dysprosium, europium, terbium and yttrium are the critical metals in the rapidly growing green energy and high technology sectors whereas China is the main supplier of the rare earth elements in the world (more than 85 %). In the past seven years, rare earthsâ€™ market has been volatile due to the strategic dependency of developed countries on China. That reliance caused a market crisis of the rare earths elements that was induced by the new policies on Chineseâ€™s export quotas implemented in 2011. The event pushed the rest of the world (ROW) to invest in their own supplies, as it can be seen by the recently emerging commercial plants or researches.
There is no known technology and adequate infrastructure for recovery of the REE from the ores in ROW due to definite dependency on China. Due to special characteristics of the REE deposits, the REE mineral processing is not as easy as mineral processing of other metals such as gold and copper. Also, although there are quite a lot of the REE-bearing minerals, only a few of them are viable to be processed for the REE recovery.

Carbonatite type bastnäsite (REE.F.CO3) and phosphate type monazite (REE.Th.PO_4) are two REE bearing minerals from which extraction of the rare earth elements is viable. Conventional production of the rare earth elements, in the form of oxides, from an ore of bastnÃ¤site and monazite usually includes physical beneficiation, minerals cracking, hydrometallurgy, and calcination steps.

Physical beneficiationâ€™s objective is to produce a concentrate of the above-mentioned REE bearing minerals, which is usually a complex array of unit operation of magnetic, gravity and electrostatic separation as well as froth flotation.

The acid/alkali baking of the concentrate is another challenge in conventional REE production process because the consumption of acid or alkali imposes economic, e.g. capital cost and operational cost, and safety challenges to the process. The presence of fluorine in bastnäsite is another challenge since toxic fluoride gases can be produced during acid baking of the rare earth concentrate. Therefore, an improper handling of these substances can result in extensive environmental damages. A serious drawback in sulfuric acid baking is the production of SO_2 and SO_3 that would cause serious environmental issues like acid rain advent. In addition to the formation of SOx, applying sulfuric acid baking to such an ore mixture (bastnästie and monazite) is related to the production of hazardous gases such as HF and SiF_4 off the bastnäsite and sulfates of thorium and uranium off the monazite.

To overcome serious issues of physical beneficiation and minerals cracking steps of the conventional REE production processes, authors developed a direct production of the rare earth element oxides from bastnäsite and monazite REE bearing minerals through a process that is much simpler, more environmentally friendly and economical. Therein, the fresh ore **directly undergoes thermal cracking** with air to produce rare earth oxides in a fluidized bed reactor. Therefore, physical beneficiation, i.e. concentrate production, and minerals cracking, i.e. liberation of elements by acid/alkali baking of the conventional processes are eliminated.

Operating in a bubbling regime, the fluidized bed reactor provided a uniform heat and mass transfer in the bed of ore particles of Geldart's group A so that a maximum temperature of 800 ^{o}C was required under which particles agglomeration was not observed, and full conversion was achieved.

Distribution of Particles from Inlet Pipe in Fluidized Beds with Different Inlet Structures

Conference: Fluidization

Year: 2019

Proceeding: Fluidization XVI

Group: General Paper Pool

Session: Poster Session

Time: Tuesday, May 28, 2019 - 4:00pm-6:00pm

Authors:

Li, J., China University of Petroleum (Beijing)
Liu, L., China University of Petroleum (Beijing)
Yao, X., China University of Petroleum
Lu, C., China University of Petroleum-Beijing

For actual industrial process, catalyst particles usually flow into the fluidized bed reactor from one side by inlet tube, which is a crucial part to transport the particles from one reactor to another one. However, the particles are not uniform distribution in inlet region of fluidized bed due to too large size of reactor and limit of internals, particularly for the catalyst cooler in FCC. The cooler is a fluidized bed with several vertical tubes to exchange heat. The uneven particles distribution in the inlet region not only impacts the heat exchange between cold tubes and hot particles but also causes the erosion and even leak of tubes. Therefore, the distribution of particles in inlet region of fluidized bed is investigated for three inlet structures in this paper. For the original structure without internal, most particles rapidly flow to the opposite region with inlet region due to particles with a high velocity. In order to improve the distribution of particles, two new structures are developed, which are made of a decelerate plate and a rectifier plate. Their decelerate plates has different slant angle, which are 15 ° and 0 ° respectively. Moreover, the rectifier plate of structure 2 is grooved. For two new structures, the particles distribution becomes more uniform and the erosion is weakened after the two inlet structures were installed, while the structure 2 has superior effect.

Dynamics of Cluster in a Pilot High-Density Circulating Fluidized Bed Riser

Conference: Fluidization

Year: 2019

Proceeding: Fluidization XVI

Group: General Paper Pool

Session: Poster Session

Time: Tuesday, May 28, 2019 - 4:00pm-6:00pm

Authors:

Su, X., State Key Laboratory of Heavy Oil Processing, China University of Petroleum
Wang, C., China University of Petroleum
Lan, X., China University of Petroleum
Gao, J., China University of Petroleum

Circulating fluidized bed (CFB) has been utilized in many industrial processes, such as chemical engineering, petroleum, coal combustion and so on. The fluid catalytic cracking (FCC) processing is a typical commercial application, and it is usually operated with very high solids circulation rate of 400-1200 kg/m^{2}s, which is different from other commercial process. To distinguish to those process, the circulating fluidized bed with high solids circulation rate of more than 200 kg/m^{2}s along with more than 10% solids concentration is defined as high-density circulating fluidized bed. The gas-solids flow in high-density circulating fluidized bed is characterized by highly heterogeneous and nonlinear flow structure about local solids holdup and particle velocity. However, a set of investigations indicate that a common hydrodynamic feature in circulating fluidized bed is the present of cluster. The continue formation and destruction of cluster structure influence the solids holdup and particle velocity and then effect on the gas-solids flow structure, heat transfer and reactor performance. In two decades the characteristics of cluster in circulating fluidized bed has been investigated according to pressure or fiber optic data using different analysis methods. But the characteristics of cluster focus on the circulating fluidized bed with less than 200 kg/m^{2}s solids circulation rate. So it is necessary to research on the dynamics of cluster in high-density circulating fluidized bed riser.
In this work, experiments are mainly carried out in a riser with 150 mm diameters and 18 m height using the FCC particles with mean diameter of 85 Î¼m and particle density of 1500 kg/m^3. The experimental setup includes a downcomer with 450 mm diameters and a solids

storage tank with 660 mm. The downcomer and storage tank can provide enough drive force to supply enough solids particle to the riser. The experimental operating condition range from 50 kg/m^2s to 600 kg/m^2s with the superficial gas velocity from 5 to 9 m/s. The reflective fiber optic probe of model PV6D is used to measure local solids concentration and particle velocity in this work developed by the Institute of Processing Engineering, Chinese Academy of Science, Beijing, China. The clusters are identified using a suitable analysis methods according to the instantaneous solids concentration data. The time fraction, frequency, mean concentration and during time of cluster at different local axial and radial position in the circulating fluidized bed are obtained and displayed in the work. This result can further account for the heterogeneous flow structure in the macro-scale and contribute to understand the mechanism of the high-density circulating fluidized bed risers.

Effect of a Mechanical Vibration on Particle Migration Behavior in a Gas-Solid Fluidized Bed for Fine Cohesive Powders

Conference: Fluidization

Year: 2019

Proceeding: Fluidization XVI

Group: General Paper Pool

Session: Poster Session

Time: Tuesday, May 28, 2019 - 4:00pm-6:00pm

Authors:

Mawatari, Y., Kyushu Institute of Technology
Nakamura, R., Kyushu Institute of Technology
Sonoda, K., Kyushu Institute of Technology

This study aimed to investigate the particle migration behavior under mechanical bed vibration in a gas-solid fluidized bed for fine cohesive powders. In this study, the vertical vibration was added to the particle bed, and the characteristics of the pressure drop curve, particle flow patterns, and the migration velocity on the bed surface were examined under various vibration parameters (vibration frequency and vibration amplitude). The powders used in this study were spherical shaped powders of silica, copper dioxide, zirconia, and alumina. The primary size and density were ranged from 6 to 20μm and from 2500 to 6040 kg/m3 , respectively. When the vibration amplitude was lower than a certain critical value (vibration frequency was constant), the dominant particle flow pattern changed from the formation of the stable gas channels to the bubble formation with continuously channel breakage as the gas velocity increased. In such vibration conditions, the pressure drop curve against the gas velocity was a common pressure drop curve that consists of the fixed bed and fluidized bed (the value of the bed pressure is almost constant) was obtained. However, when the vibration amplitude was larger than a certain critical value, vigorous particle migration appeared in the whole bed. From the observation of the vigorous particle migration behavior, the particles and/or agglomerates had a regular migration direction, hence it is considered that a high particle mixing state was achieved under such a vigorous particle migration behavior. This vigorous particle migration behavior was observed between the fixed static state and the bubble formation with channel breakage state. Furthermore, we found that the bed pressure drop became lower while the vigorous particle motion occurred, and the degree of the reduction of the bed pressure drop became larger as the particle migration became vigorous. The experimental result of the particle migration velocity had a strong interrelation with the reduction degree of the bed pressure drop, i.e., as the vibration amplitude increased the particle migration velocity increased while the bed pressure drop was decreased.

Effect of Solids Viscosity on Unary and Binary Gas-Solid Fluidization: ECT Measurements and CFD Simulations

Conference: Fluidization

Year: 2019

Proceeding: Fluidization XVI

Group: General Paper Pool

Session: Poster Session

Time: Tuesday, May 28, 2019 - 4:00pm-6:00pm

Authors:

Buwa, V. V., Indian Institute of Technology Delhi
Parvathaneni, S., Indian Institute of Technology Delhi
Singh, B. K., Indian Institute of Technology Delhi
Roy, S., Indian Institute of Technology Delhi

Industrially relevant applications like coal gasification (involving fluidization of coal and ash), catalytic polymerization, waste incineration, powder granulation, etc. involve fluidization of mixture of solids with different properties (size, shape, density, etc.). Gas-solids flow in fluidized bed reactors is inherently dynamic and performance of such reactors largely depends on the temporal and spatial distribution of the different solids phases in the bed. Further, depending on the operating conditions and solids properties, segregation of solids phases can also occur, which in turn influences the local heat and mass transport and consequently the overall reactor performance. Prediction of time-averaged flow behavior in multiphase reactors (e.g. gas-solid fluidized beds) using Eulerian multi-fluid models is possible to a reasonable accuracy, but most of the models seem less promising in their ability to predict the time-evolution (dynamics) of phase distribution in the reactors. The ability to predict the dynamics is imperative since most transport and kinetic effects in multiphase reactors occur on time scales which are significantly smaller than the time-scale of averaging. While long recognized, relatively fewer research efforts have focused on establishing the prediction of the dynamics using Eulerian multi-fluid models owing to the paucity of reliable time- and spatially- resolved experimental data.

The gas-solid interphase momentum exchange models (drag force) are considered to be very crucial in Eulerian multi-fluid simulations, and most often modified by researchers to reduce the discrepancies between the results of simulation and their corresponding measurements. In addition to the drag force, the force due to solid shear stresses also play an important role in the bubbling behavior [1,2]. It was seen that the previous study on the effect of solid viscosities are carried out on pseudo-3D beds for unary gas-solid fluidization, and the effect of solid viscosities on the segregation and mixing phenomena of binary gas-solid fluidization was not reported in the previous study.

In the present work, a cylindrical fluidized bed of internal diameter of 9 cm and height of 150 cm was used for measurements and the same geometry was used for the Eulerian multi-fluid simulations. Air and glass beads of different sizes (96 μm, 430 μm, and 922 μm; density = 2500 kg/m^3) were used for the gas and solid phases. Also, different compositions (30%, 50%, and 70% by weight) of 96 μm and 922 μm particles were used to make different mixtures of solids phases for binary fluidization. A commercial measurement instrument Electrical Capacitance Tomography (ECT) system was used to measure temporally and spatially resolved solid volume fraction. In case of Eulerian multi-fluid simulations, the conservation equations (mass and momentum) of gas and solid phases, and the Kinetic Theory of Granular Flow (used to model solid stresses and energy losses due to particle-particle collisions) are solved in commercial software Ansys Fluent 18.0. The details of the measurements (e.g. sensor calibration, image reconstruction algorithms, etc.) and simulations (e.g. grid size, boundary conditions, solution techniques, discretization schemes, etc.) will be given in the full-length manuscript.

The dynamics of segregation and fluidization of unary particles and binary mixtures (with varying compositions of unary particles) with different size ratio (d_{pr} = ~2- 9) in a cylindrical fluidized bed were investigated using the temporally- and spatially-resolved measurements of solid volume fraction performed using the ECT. In the segregating bed conditions, the segregation-time measured using the ECT was compared with that measured using the high-speed imaging and it was shown that ECT data can be analyzed to serve as a quantitative measure of the segregation behavior. The ECT measurements were used further to quantify dynamics of unary and binary fluidization behavior using the time-evolution of spatial solid volume fraction fluctuations, corresponding frequency distribution, and bubble size distribution. Further, a relation between the measured variance of solid volume fraction fluctuations with the spatial solid volume fraction distribution and corresponding flow structures under different fluidization conditions was established. These measurements helped to provide a database to test the Eulerian multi-fluid CFD simulations of segregation and fluidization of binary mixtures.

The Eulerian multi-fluid simulations were performed for unary and binary gas-solids fluidization, and the time-averaged, as well as instantaneous solid volume fraction distribution, were compared with their corresponding measurements. The drag force and force due to solid shear stresses in a cell were compared. It was found that besides the drag force, the force due to solid shear stresses strongly influence the bubbling behavior by influencing the compactness in the solid phases. Higher the compactness in the solids phase, the less gas is transported through the boundary of the bubbles and therefore bigger gas bubbles were observed. It was also seen that among all the solids viscosities (kinematic, collisional, bulk, and frictional), the effect of frictional viscosity on the bubbling behavior was found more dominating than the other viscosities. Generally, three types of frictional pressure models have been used widely in the literature i.e. KTGF-based, Johnson and Jackson [3], and Syamlal et al. [4]. However, the frictional viscosity values obtained using the KTGF-based, Johnson and Jackson [3], and Syamlal et al. [4] were significantly different and therefore choosing the right model for frictional pressure became a challenge in the Eulerian multi-fluid simulations especially in the case of binary gas-solid fluidization.

It was found that the use of KTGF-based frictional pressure leads to produce relatively small-sized but large number of the bubbles per unit time. With an increase in the solid viscosity caused by the use of Johnson and Jackson [3] or Syamlal et al. [4] frictional pressure models at different critical solid volume fractions, the bubbles size increased in comparison to the size of the bubbles observed using the KTGF-based frictional pressure model. Also, the number of bubbles per unit time decreased in comparison to the KTGF-based model. Due to the change in the size, and number of bubbles per unit time in the binary gas-solid fluidization, there was a significant change observed in the segregation rate. It is worthwhile to mention that the agreement between the measured and simulated bubbling characteristics (size and frequency) and segregation rate can be improved by changing the frictional viscosity of the solids in the bed.

In the full length manuscript, the detailed comparison of time-evolution of solid volume fraction for unary and binary gas-solid fluidization measured using the ECT will be reported. Also, through the Eulerian multi-fluid simulations, the effect of solid viscosities on the time-evolution of solid volume fraction will be discussed. The bubble chord length distribution and power spectra of bubble chord length fluctuations will be reported with their corresponding experimental validation. In addition, the change in segregation rate due to the change in the bubbling characteristics for different mixture compositions have been analyzed and it will be reported in the full length manuscript. The present work will be helpful in understanding the role of solid viscosities on the segregation and bubbling behavior of binary gas-solid fluidization in a cylindrical fluidized bed.

References:

[1] D.J. Patil, M. Van Sint Annaland, J.A.M. Kuipers, Critical comparison of hydrodynamic models for gas-solid fluidized beds - Part I : Bubbling gas-solid fluidized beds operated with a jet, Chem. Eng. Sci. 60 (2005) 57â€“72. doi:10.1016/j.ces.2004.07.059.

[2] S. Cloete, S.T. Johansen, A. Zaabout, M. van Sint Annaland, F. Gallucci, S. Amini, The effect of frictional pressure, geometry and wall friction on the modelling of a pseudo-2D bubbling fluidised bed reactor, Powder Technol. 283 (2015) 85â€“102. doi:10.1016/j.powtec.2015.04.060.

[3] P.C. Johnson, R. Jackson, Friction collisional constitutive relations for granular materials,

with application to plane shearing, J. Fluid Mech. 176 (1987) 67â€“93. doi:10.1017/S0022112087000570.

[4] T.J. Syamlal, M., Rogers, W., Oâ€™Brien, Mfix documentation: volume I, theory guide. Technical Report DOE/METC-9411004, NTIS/DE9400087, 1993.

Enhancement of Gas-Liquid Two-Phase Flow and Mass Transfer in the Baffled Microchannel

Conference: Fluidization

Year: 2019

Proceeding: Fluidization XVI

Group: General Paper Pool

Session: Poster Session

Time: Tuesday, May 28, 2019 - 4:00pm-6:00pm

Authors:

Yin, Y., Tianjin University
Ma, D., Tianjin University
Guo, R., Tianjin University
Zhu, C., Tianjin University
Fu, T., Tianjin University
Ma, Y., Tianjin University

The enhancement of gas-liquid two-phase flow and mass transfer in the baffled microchannel were investigated experimentally for CO_2 absorption into monoethanolamine/1-butyl-3-methylimidazolium tetrafluoroborate (MEA/ [Bmim][BF_4]) aqueous solutions. Three flow patterns were observed in the baffled microchannel: Taylor-bubbly flow, Taylor flow and broken-Taylor flow. The influences of baffle size on gas-liquid flow, mass transfer rate and pressure drop were explored via varying the MEA/[Bmim][BF_4] concentration ratio, gas-liquid flow rate and blockage ratio. The mass transfer enhancement by baffles relied dramatically on the gas-liquid two-phase flow characteristics. For the Taylor-bubbly flow occurring at low gas flow rates, the mass transfer performance was less affected by varying the liquid flow rate, the concentration ratio, or the blockage ratio. Once the flow entered the Taylor flow regime at moderate gas flow rates, mass transfer performance would be obviously affected. As the gas flow rate was further increased, the flow fell into a broken-Taylor regime, the higher fluid disturbance and bubble breakup would take place in this case, and the mass transfer would be greatly enhanced. Under higher two-phase flow rates, lower MEA/[Bmim][BF_4] concentration ratios and larger blockage ratios, the mass transfer intensification is more pronounced and the enhancement factor could reach to 1.5 with an acceptable increase in pressure drop 0.3 kPa (20% in relative term) compared to the unobstructed microchannel.. This work could serve as guidance for chemical process intensification and the design of equipment miniaturization.

Fluidized Bed Drying of Fruits and Vegetables

Conference: Fluidization

Year: 2019

Proceeding: Fluidization XVI

Group: General Paper Pool

Session: Poster Session

Time: Tuesday, May 28, 2019 - 4:00pm-6:00pm

Authors:

Hari Aditya, K., Sepuluh Nopember Institute of Technology

The search for novel technologies in the area of food processing and preservation has led to the application of fluidization technique for enhancing the process capabilities, energy efficiencies and overall quality retention in the final products. However, this paper reviews the various fruits and vegetables dried in the fluidized bed dryer and their retention of sensory attributes in the dried products. This review states that fluidized bed drying indeed has proved better compare to conventional drying in respect to drying time of fruits and vegetables especially for better product structure and its shorter reconstitution time. Fluid-bed drying of different fruits and vegetables particles has been investigated. The results have shown that drying of fruit and vegetable in a fluidized bed produces dry fruit and vegetable particles of improved quality in a much shorter time, with improved rehydration times and qualities, than in continuous belt-driers that are generally used

Gas-Liquid Two-Phase Flow and Mass Transfer of CO2 Absorption into Amino Acids Sodium Aqueous Solution in Microchannel

Conference: Fluidization

Year: 2019

Proceeding: Fluidization XVI

Group: General Paper Pool

Session: Poster Session

Time: Tuesday, May 28, 2019 - 4:00pm-6:00pm

Authors:

Chu, C., Tianjin University
Sun, X., Tianjin University
Zhu, C., Tianjin University
Fu, T., Tianjin University
Ma, Y., Tianjin University

Carbon dioxide (CO_2) is a typical greenhouse gas and the primary contributor to the global warming and climate change. It is imperative to remove huge amounts of CO_2 from the exhaust gases in many industrial processes. The amino acid salt solutions is a new type of CO_2 absorbent with advantages such as low volatility, low toxicity, antioxidant degradation and good absorption performance, which accordingly have received increasing attention. The gas-liquid two-phase flow and absorption performance of CO_2 into sodium glycinate, sodium L-alaninate, sodium L-valinate and sodium L-threoninate were investigated by a high-speed camera in a microreactor. Under Taylor-bubbly and Taylor flows, the CO_2 absorbed molar amount of four kinds of sodium amino acids at same condition decrease in the order: n(sodium glycinate) > n(sodium L-alaninate) > n(sodium L-valinate) > n(sodium L-threoninate). Furthermore, the mass transfer characteristics of CO_2 absorption into sodium glycinate aqueous solution were also investigated experimentally. The effects of gas phase flow rate, liquid phase flow rate and liquid phase concentration on specific surface area (a), liquid-side volumetric mass transfer coefficient (k_La) and liquid-side mass transfer coefficient (k_L) were studied. The results show that the larger gas phase flow rate, the smaller liquid phase flow rate and liquid phase concentration could lead to larger specific surface area. The liquid side volumetric mass transfer coefficient (k_La) increases with increasing the gas phase flow rate, liquid phase flow rate or concentration of sodium glycinate solution. The liquid side mass transfer coefficient (k_L) decreases slightly with rising gas phase flow rate, while increases with increasing the liquid phase flow rate and liquid phase concentration of sodium amino acids. A dimensionless empirical correlation was proposed to predict liquid side volumetric mass transfer coefficient considering the enhancement of chemical reaction on mass transfer.

Impact of Horizontal Internals on the Hydrodynamics in a Gas-Solid Fast Fluidized Bed

Conference: Fluidization

Year: 2019

Proceeding: Fluidization XVI

Group: General Paper Pool

Session: Poster Session

Time: Tuesday, May 28, 2019 - 4:00pm-6:00pm

Authors:

Kim, J. Y., Korea Institute of Energy Research
Jin, G. T., Korea Institute of Energy Research
Park, Y. C., Korea Institute of Energy Research
Moon, J. H., Korea Institute of Energy Research
Nam, H. S., Korea Institute of Energy Research
Yi, C. K., Korea Institute of Energy Research
Jo, S. H., Korea Institute of Energy Research

In this study, the effect of horizontal internals on the hydrodynamics in a gas-solid fast fluidized bed has been investigated at the bench-scale dry-sorbent CO_2 capture process which consists of a fast fluidized-bed type sorption reactor and a bubbling fluidized-bed type regeneration reactor. In order to investigate the hydrodynamics as well as visual confirm of solid flow in a gas-solid fast fluidized bed with horizontal internals such as heat transfer tubes, a lower part of the sorption reactor has been made of acrylic plexiglass plates. The hydrodynamics were analyzed by changing the solid inlet port into the lower part of the sorption reactor to two or three heights. Based on the experiments, the fast fluidized-bed type sorption reactor in the 10 MWe-scale dry-sorbent CO_2 capture has been designed.

Influence of Geometric Factors of Confined Conical Spouted Beds on Fine Particle Entrainment

Conference: Fluidization

Year: 2019

Proceeding: Fluidization XVI

Group: General Paper Pool

Session: Poster Session

Time: Tuesday, May 28, 2019 - 4:00pm-6:00pm

Authors:

Tellabide, M., University of the Basque Country
Estiati, I., University of the Basque Country
Pablos, A. Sr., University of the Basque Country
Altzibar, H. Sr., University of the Basque Country
Aguado, R. Sr., University of the Basque Country
Olazar, M. Sr., University of the Basque Country

INFLUENCE OF GEOMETRIC FACTORS OF CONFINED CONICAL SPOUTED BEDS ON FINE PARTICLE ENTRAINMENT

Idoia Estiati, Mikel Tellabide*, Aitor Pablos, Haritz Altzibar, Roberto Aguado and Martin Olazar
University of the Basque Country; Dept. Chemical Engineering
PO Box 644, 48080 Bilbao, Spain
*T: +34-94-601-5994; F: 94-601-3500; E:mikel.tellabide@ehu.es

The spouted bed is a fluid-particle contact technique that has been successfully applied to systems where fluidization has yielded unsatisfactory results. Conical spouted beds are highly versatile in the gas flow rate, allowing operating with particles of irregular texture, fine particles and those with a wide size distribution and sticky solids, whose treatment is difficult using other gas-solid contact regimes. These features of the conical spouted beds make them attractive for different applications.

The ratio between the inlet diameter and particle diameter limits the scaling up of spouted beds (the inlet diameter should be smaller than 20-30 times the particle diameter). The insertion of a draft tube is the way to overcome this limitation.

Many applications of conical spouted beds require the use of inert materials and/or catalysts. Therefore, they involve operations using mixtures made up of coarse (waste) and fine (inert and/or catalyst) particles, which require rather high gas velocities due to the presence of coarse particles, thereby leading to fine particle entrainment. As a result, this loss of material (catalyst) involves high cost, damage to the environment and a poor gas-solid contact. In order to avoid these problems, a modification of the conventional spouted bed regime has been proposed by our research group (1). This modification consists in a new internal device called fountain confiner, which stabilises the operation and changes the trajectory of the gas leading to an additional contact between the gas and the solid in the fountain.

The main aim of this study is to assess the potential of the fountain confiner for avoiding fine particle entrainment in conical spouted beds. Accordingly, runs have been carried out using different contactors, fountain confiners and types of draft tubes. The performance of the different alternatives has been compared in order to ascertain the optimum configuration for reducing fine particle entrainment. The materials used for operation are glass beads of 4 mm diameter and bread crumbs of 0.41 mm diameter.

The results show that the fountain confiner reduces fine particle entrainment and ensures a better contact between the gas and the solid. Furthermore, particle entrainment is minimized using the longest confiner and a nonporous draft tube provided with the lowest height of the entrainment zone.

REFERENCES

1. Altzibar, I. Estiati, G. Lopez, J.F. Saldarriaga, R. Aguado, J. Bilbao, M. Olazar, Fountain confined conical spouted beds, Powder Technol. 312, (2017) 334â€“346.

Investigation of Loop Seal Operation By Pulse Air Injection for Improving the Performance and Stability

Conference: Fluidization

Year: 2019

Proceeding: Fluidization XVI

Group: General Paper Pool

Session: Poster Session

Time: Tuesday, May 28, 2019 - 4:00pm-6:00pm

Authors:

Sudo, Y., Gunma University
Masfuri, I., Agency for the assessment and application of technology
Bagus, T., Agency for the assessment and application of technology.
Wakana, K., Gunma Universuty
Rosyadi, E., Agency for the assessment and application of technology.
Noda, R., Gunma University

In this study, the solid recirculation characteristics of loop seal have been tested in a Fluidized Bed cold model system, and the hydrodynamic behavior in downcomer and riser section in various operating conditions. The aim of this work is to fix the appropriate structure and operation of actual scale loop seal installed in a cold model of dual bubbling fluidized beds(DBFBs) system.

The hydrodynamics of solids material recycle through a loop seal (size 70 x 28 cm) in the downcomer and riser section of the DBFBs cold model have been measured in various operation conditions. Silica sand was applied as bed material and tested the fluidization characteristics in a reactor chamber (gasifier: 80 cm, combustor: 93 cm) with 70 cm in height. A reactor chamber was connected to stand pipe then to a loop seal with several riser pipes without mechanical valves. Solids circulation rate (G_s) experiment have been performed in various outlet riser pipe size (6 inches and 4 inches) and various aeration velocity in the loop seal in this study. Some experiments were carried out by flowing solids particle from reactor chamber through standpipe by continuous aeration to obtain the solids particle behavior in the circulating fluidized bed system. Solid flow rate through the loop seal increases linearly with increasing aeration rate of loop seal gas. At the same aeration rate, the solid circulation rate can be obtained by applying 10 or 6 inches riser pipes. The obtained solid circulation rate has been calculated from total amount of solid flow out from loop seal with the operation time. According to the result, the most influence factor to reach high solid circulation rate is aeration velocity of the loop seal, especially in the downcomer section. The higher circulation rate can be obtained by increasing in gas velocity in downcomer in this experiment, influence of fluidization regimes in riser was also investigated by changing standpipe size from 10 inches to 4 inches. It can be concluded from the experimental and theoretical considerations that superficial velocity should be more than the transport velocity to make particle flow steady. The higher aeration velocity in loop seal can be obtained by changing standpipe size in outlet of riser section, from 10 inches initially to be 6 inches and 4 inches. And also in this study, it was proposed a pulse operated loop seal (POLS) using pulse injective flow to loop seal for to establish simple operation method. Performance of continuous flow and pulse flow of loop seals was examined to check on the validity. It was observed that averaged particle circulation rate could be achieved by application of POLS in similar aeration gas injection. The result of particle circulation in POLS will be reported in the presentation.

Investigation of Loop Seal Performance in Cold Model Circulating Fluidized Bed

Conference: Fluidization

Year: 2019

Proceeding: Fluidization XVI

Group: General Paper Pool

Session: Poster Session

Time: Tuesday, May 28, 2019 - 4:00pm-6:00pm

Authors:

Sudo, Y., Gunma University
Noda, R., Gunma University

In this study, the particle recirculation characteristics of loop seal have been tested in a Fluidized Bed cold model system, and the hydrodynamic behaviour in downcomer and riser section in various operating conditions. The aim of this work is to develop a hydrodynamic model for a cold model of fluidized bed gasification incorporating with loop seal and stand pipe to predict the basic hydrodynamic properties such as pressure drop, minimum fluidization velocity and bed material circulation rate. The hydrodynamics of solids material recycle through a loop seal (size 70 x 28 cm) in the downcomer and riser section of CFB cold model have been determined. Fluidization testing was conducted in a reactor chamber (gasifier ï¦ 80 cm, combustor 93 cm) and filled by solids particle up to around 70 cm in height. The reactor chamber was connected to stand pipe and a loop seal as non-mechanical valve. Particles circulation rate (G_s) experiment have been performed in various outlet riser pipe size (6 inches and 4 inches) and various aeration velocity in the loop seal ($V_{J,RD}$and $V_{J,RU}$) in this study. Some experiments were carried out by flowing solids particle from reactor chamber through standpipe and a loop seal, to obtain the solids particle behavior in the circulating fluidized bed system. Solid flow rate through the loop seal increases linearly with increasing aeration rate of fluidizing gas. At the same aeration rate, the solid circulation rate can be obtained by using standpipe of downcomer 10 inches and 6 inches for riser pipe. A number factors influenced to the solids material behaviour, such as solid inventory, fluidizing gas velocity, and outlet standpipe size. The obtained solid circulation rate has been correlated with bed inventory height in the reactor and correlation between total amount of solid flow out from loop seal with the operation time. The solids circulation rate through the loop seal increases linearly to a maximum value with increasing aeration rates of downcomer up to 4Umf and riser up to 120Umf, due to the increase of frictional drag on the particles. Therefore, the most influenced factor to reach high solid circulation rate is aeration velocity of the loop seal, especially in the riser section. The higher aeration velocity in loop seal can be obtained by changing standpipe size in outlet of riser section, from 10 inches initially to be 6 inches and 4 inches. And also in this study, it was proposed a pulse operated loop seal (POLS) using pulse injective flow to loop seal for to establish simple operation method. Performance of continuous flow and pulse flow of loop seals was examined to check on the validity. As a result, it was found that the operability was improved by POLS. It was also suggested that the particle circulation rate can be controlled by low pulse average rate.

Lattice Boltzmann Simulation of Two-Dimensional Buoyant Impinging Streams with the Soret and Dufour Effects

Conference: Fluidization

Year: 2019

Proceeding: Fluidization XVI

Group: General Paper Pool

Session: Poster Session

Time: Tuesday, May 28, 2019 - 4:00pm-6:00pm

Authors:

Zhang, W., Ningbo University
Li, Y., Ningbo University
Wang, L., China University of Geosciences
Kuang, S., Monash University
Wang, Y., Ningbo University

Impinging streams is the flow form in the gastifier of entrained-flow coal gasification, which has the vital practical significance to greatly increase the utilization efficiency of coal and dramatically reduce the pollution originating from coal. The impinging streams is typically dense gas-particle turbulent flows with heat and mass transfer and chemical reaction under high temperature and pressure. In this work, the regularized lattice Boltzmann model is used to investigate the buoyancy-driven impinging streams with the Soret and Dufour effects. Under the Soret and Dufour effects, the influences of configuration parameters (H/W, H and W are the jetsâ€™ separation and jet inlet width), Reynolds number (Re), buoyancy ratio, Prandt number (Pr), Schmidt number (Sc) on the structures of flow, temperature and concentration fields are studied. The effects of Soret number (Sr) and Dufour number (Df) on the variation of average Nusselt number and Sherwood number on the walls are also investigated.

Lbm-DEM Simulation of Particle-Laden Flows in the Presence of Liquid Bridge Forces

Conference: Fluidization

Year: 2019

Proceeding: Fluidization XVI

Group: General Paper Pool

Session: Poster Session

Time: Tuesday, May 28, 2019 - 4:00pm-6:00pm

Authors:

Chen, H., Tsinghua University

LBM-DEM simulation of particle-laden
flows in the presence of liquid bridge forces

Hongsheng Chen [1],
Sheng Chen [1], Shuiqing Li [1,*], and Zhaoli Guo [2]

1

Key Laboratory for
Thermal Science and Power Engineering of Ministry of Education, Department of Energy and Power Engineering, Tsinghua University, Beijing 100084, China

2

State Key
Laboratory of Coal Combustion, School of Energy and Power Engineering, Huazhong University of Science and Technology, Wuhan
430074, China

Abstract: Air flows, carrying micron-sized particles, occur extensively in numerous fields of interest such as aerosols, dust removing and pharmaceuticals. As well known, micron-sized particles become extremely cohesive due to the presence of liquid bridge forces, being responsible for various complex phenomena involved in gas-solid flows, e.g., clustering, aggregation and clogging. In this study, the dynamics of microparticles passing over a circular cylinder in a channel flow is numerically investigated at mesoscale by combining a multi-relaxation-time lattice Boltzmann method with a multiscale discrete element method. Sliding, rolling and twisting resistances between colliding particles are taken into consideration, and the presence of liquid bridge introduces additional capillary cohesion, liquid viscous damping and liquid viscous torque to the particle-particle and particle-wall interactions. The fluid-solid interaction is achieved by the momentum exchange method, considering solid surfaces as bounce-back boundaries. The Weber number is applied to discuss the capture of wet microparticles by the circular cylinder, as well as the behavior of wet microparticles in the wake behind the circular cylinder. Therefore, this research is of certain importance for a better understanding of the roles of liquid bridge forces playing in the dynamics of microparticles carried by viscous fluids and in the capture of microparticles by an obstacle.

Keywords: particle-laden flows, wet particles, fluid-solid interaction, LBM-DEM

Minimum Spouting Velocity for Conical Spouted BEDS of Agroindustrial Waste Biomasses

Conference: Fluidization

Year: 2019

Proceeding: Fluidization XVI

Group: General Paper Pool

Session: Poster Session

Time: Tuesday, May 28, 2019 - 4:00pm-6:00pm

Authors:

Saldarriaga, J. F. Sr., Universidad de los Andes
Estiati, I., University of the Basque Country
Aguado, R. Sr., University of the Basque Country
Olazar, M. Sr., University of the Basque Country

Spouted beds are an interesting alternative for the treatment of agroindustrial biomass for energy purposes, given that the insertion of a draft-tube allows large-scale handling of solids with a wide range of sizes and diverse properties. The hydrodynamic correlations proposed in the literature have not been validated with particles of low density and sphericity. Therefore, a detailed experimental study has been conducted using rice husk and olive pit of agroindustrial biomasses and different combinations of the contactor geometric factors. The correlations in the literature for plain conical spouted beds and for those provided with nonporous and open-sided draft-tubes have been taken as a starting point and, based on the experimental results, they have been modified in order to account for the properties of the solids. Two strategies have been considered in the methodology for modifying the coefficients in the literature correlations. Both are based on the introduction of coefficients one by one based on a hypothesis test. Accordingly, new correlations based on existing ones have been proposed, with the number of coefficients modified being minimum.

Numerical Simulation of a Three-Phase Fluidized Bed Reactor: Hydrodynamic Analysis of the Expansion Zone

Conference: Fluidization

Year: 2019

Proceeding: Fluidization XVI

Group: General Paper Pool

Session: Poster Session

Time: Tuesday, May 28, 2019 - 4:00pm-6:00pm

Authors:

Casillas, N., Instituto Politécnico Nacional
Olivo, L., IPN
Elizalde, I., Instituto Politécnico Nacional

The present work focuses on the hydrodynamic analysis of a three-phase fluidized bed reactor and the influence of several key aspects, such as volume fractions, dynamic pressure, magnitude of gas velocities, liquid and solid phases (hydrogen, extra heavy oil and gamma alumina), and phase interactions, through drag models with the *k-epsilon* turbulence approach.

The simulation uses an Euler-Euler method, coupled with different correlations with materials characterized by physical and thermal properties, the transport processes that are developed within the transient regime.

The proposed model with a three-phase fluidized bed reactor was solved numerically (by ANSYS-Fluent software) with different mesh refinements on bed expansion zone, to observe the influence of the hydrodynamic parameters and convergence criteria.

With our model, and supported by numerical simulations, the involved hydrodynamics parameters could be predicted from defined boundaries conditions.

Numerical Study on Segregation Characteristics of Biochar in a Bubbling Fluidized Bed with Varying Aspect Ratio

Conference: Fluidization

Year: 2019

Proceeding: Fluidization XVI

Group: General Paper Pool

Session: Poster Session

Time: Tuesday, May 28, 2019 - 4:00pm-6:00pm

Authors:

Park, H. C., Yonsei University
Choi, H. S., Yonsei University
Choi, M. K., Yonsei University

In this study, computational particle fluid dynamic (CPFD) was applied to investigate the segregation characteristics of biochar in a bubbling fluidized bed. The aspect ratio of the bubbling fluidized bed was changed and the effects of aspect ratio on the segregation characteristics were investigated. In the present study, the segregation characteristics of a mixture of biochar and sand particles were analyzed in terms of pressure fluctuation, bubble size distribution, and mixing index. It was observed that as increasing the aspect ratio, the bubble size decreased which led to better segregation between biochar and sand particles. The mixing index of biochar and sand particles decreased as the aspect ratio increased. In addition, the effects of pyrolysis temperature on the physicochemical properties and structure of biochar were investigated.

Particle Dynamics of Irregular Sand Particles Flow within a Rotating Drum

Conference: Fluidization

Year: 2019

Proceeding: Fluidization XVI

Group: General Paper Pool

Session: Poster Session

Time: Tuesday, May 28, 2019 - 4:00pm-6:00pm

Authors:

Yang, H., University of Shanghai for Science and Technology
Wang, C., University of Shanghai for Science and Technology
Zhang, Y. J., University of Shanghai for Science and Technology
Wang, S. S., University of Shanghai for Science and Technology
Wang, Q., University of Shanghai for Science and Technology
Li, R., University of Shanghai for Science and Technology

Rotating drums are important due to their industrial application but also as a simple model system for studying transient avalanche process. The speckle visibility spectroscopy (SVS), spatial filtering velocimetry (SFV), and alongside simple imaging techniques were used to measure the time-resolved dynamics of sand particles in a rotating drum operating in discrete and continual avalanching regime. The SVS measurement of discrete avalanching shows that the fluctuation velocity, so called granular temperature, increased sharply before plateauing then decreases sharply to zero values during a typical avalanche event which is different to spherical particle behavior where these transitions are more gradual. Furthermore, the granular dynamics of avalanching flow is influenced by visually observed flowing mound down the avalanching surface so there are more characteristics peaks in plateau region of the signal for measurement point in lower part of the bed. The SFV measurement of continual avalanching shows that there are two velocity peaks in the velocity curve of irregular glass sand at a relatively lower rotating speed. Furthermore, with the increase of the drum speed, the middle valley gradually moves to the upper right and merges with the second peak and there is only one peak in the velocity profile along stream wise direction of the bed surface consequently. And the variation of the velocity distribution of irregular particles with particle sizes, and drum diameters are also studied as well.

Research on the Relationship between Flow Patterns of FCC Catalyst and Open Rate of Butterfly Valve in the Inclined Pipe

Conference: Fluidization

Year: 2019

Proceeding: Fluidization XVI

Group: General Paper Pool

Session: Poster Session

Time: Tuesday, May 28, 2019 - 4:00pm-6:00pm

Authors:

Wei, Y., China University of Petroleum
Yan, C., China University of Petroleum
Song, J., China University of Petroleum

To investigate the flow rate characteristics of FCC conveying along with the inclined pipe, the experiment was carried out in a 150 mm diameter 45°inclined pipe. Not only were the flow patterns observed but also the particle mass flux was measured during the conveying process of solids by changing opening rate of butterfly valve in order to study the relationship between particle mass flux and opening rate of butterfly valve. Experimental results showed that the opening rate of butterfly valve can be divided into one controlled range and two uncontrolled ranges. When the opening rate of butterfly valve is in the range from 3 to 7, the butterfly valve is in the controlled range whose flow patterns are undulated flow and gas-solid stratified flow, which has the function of particle mass flux regulation. If opening rate of butterfly valve is smaller than 3 or bigger than 7, the valve has no action on particle mass flux regulation, whose flow patterns are creeping flow and fluidized flow. In different opening rates of butterfly valve, the fluctuation profile of particle mass flux is of great difference during the whole particle feeding process. In a specific opening rate of butterfly valve, there are remarkable differences concerning the fluctuation profile of particle mass flux during the different time periods. In general, the profile of particle mass flux of undulated flow has the most severe fluctuations.

Role of Temperature on TAR Characteristics in a Fountain Enhanced Conical Spouted BED Reactor

Conference: Fluidization

Year: 2019

Proceeding: Fluidization XVI

Group: General Paper Pool

Session: Poster Session

Time: Tuesday, May 28, 2019 - 4:00pm-6:00pm

Authors:

Tellabide, M., University of the Basque Country
Cortazar, M., University of the Basque Country
Alvarez, J., University of the Basque Country
Amutio, M., University of the Basque Country
Santamaria, L., University of the Basque Country
Fernadez, E., University of the Basque Country
Lopez, G., University of the Basque Country
Olazar, M., University of the Basque Country

Role of temperature on tar
characteristics IN A Fountain enhanced conical SPOUTED BED reactoR

M. Cortazar, J. Alvarez, M. Amutio,
L. Santamaria, E. Fernandez, G. Lopez, M. Olazar

Univ. of the Basque Country, Dpt. Chem.
Engineering, PO Box 644– 48080 Bilbao, Spain

maria.cortazar@ehu.eus

Biomass
gasification processes have considerably gained attention for the sustainable
production of syngas which could be used as fuel or
as an intermediate product in the production of other fuels and chemicals.

However, the major challenge of this process is still the high level of tar presented in the gas, which hinders the improvement of its industrial viability [1, 2]. This work addresses the influence the temperature has on tar characteristics, including a detailed tar characterization by means of Fourier transform infrared spectroscopy (FTIR), gas chromatography coupled with mass spectrometry (GC/MS) and simulated distillation.

The steam gasification of sawdust (1-2 mm) was carried out in a gasification pilot plant provided with a fountain confined conical spouted bed reactor. The influence of gasification temperature on tar concentration and composition was studied in the 800-900 °C temperature range. The bed was initially made up of 100 g of olivine (90-150 um), and a water flowrate of 1.5 mL/min was used to attain fountain enhanced regime. All the runs were carried out in continuous mode by feeding 0.75 g/min of biomass, so steam to biomass (S/B) ratio was maintained at 2. The analysis of the volatile stream leaving the reactor was conducted on-line, using a CG Agilent 7890 provided with a flame ionization detector (FID) and a Varian 4900 micro-gas chromatograph for permanent gases. The identification of the tars collected in the condensation system was determined in a GC/MS (Shimadazu UP-2010S). Fourier transform infrared spectroscopy (FTIR) measurements were performed in a Thermo Nicolet 6700 instrument to determine the organic functional groups in the tar samples. Moreover, tars collected at different temperatures were fractionated by simulated distillation using a GC (agilent 6890) provided with FID detector and according to the ASTM-D2887-84 standard.

Not only does the temperature affect the amount of tar produced, but also the composition of tar. Tar concentration fell from 49.2 g Nm^{-3} (on a dry basis) at 800 °C to 6.7 g Nm^{-3} operating at 900 °C. Tar analysis (FTIR and GC) revealed that its composition was significantly influenced by temperature, opening a pathway for the evolution of its composition to more stable species (enhanced when temperature was increased) in the range of secondary and tertiary tars (of higher molecular weight) due to rearrangement reactions. The major fraction in the 800-900 °C range was that of light PAHs, although tar composition shifted from phenolic compounds and alkyl-substituted PAHs to more stable and heavier PAHs (non-substituted species) by increasing temperature, with naphthalene being the major compound at 850 °C and 900 °C. Simulated distillation confirmed that the average boiling point of the tars increased (especially of those obtained from 800 to 850 °C) when the gasification was performed at high temperatures.

References

[1] Alauddin,
Z., Lahijani, P., Mohammadi, M., and Mohamed, A. R. (2010). Renewable & Sustainable Energy Reviews,
14(9), 2852-2862.

[2] Parthasarathy, P., and Narayanan, K. S. (2014). Renewable Energy, 66, 570-579

Scale-up of the Circulating Fluidized Bed Methanation Reactor: A Numerical Study

Conference: Fluidization

Year: 2019

Proceeding: Fluidization XVI

Group: General Paper Pool

Session: Poster Session

Time: Tuesday, May 28, 2019 - 4:00pm-6:00pm

Authors:

Yuli, Z., Hohai University
Xiao, R., Southeast University
Ye, M., Dalian Institute of Chemical Physics, Chinese Academy of Sciences

Methanation of the syngas from biomass gasification is one of the key processes in the biomass-to-SNG technology. Circulating fluidized bed reactor turns to be suitable for the operation of methanation reaction with high exothermicity. The methanation reaction involves a sufficient reduction in the gas volumetric flow. Thus the scale-up of a circulating fluidized bed methanation reactor may have some differences from the traditional scaling-up criterion. In this research, the fluidization characteristics in circulating fluidized bed methanation reactors with different sizes are investigated with the numerical study. With the 'multi-scale visualization' simulation results, reaction, hydrodynamics and mass transfer can be separately analyzed from both cluster scale and reactor scale, which can provide a theoretical basis for the optimization and scale up of methanation circulating fluidized bed reactor.

Solar Thermal Treatment of Non-Metallic Minerals: Powder Flowability As Critical Application Parameter

Conference: Fluidization

Year: 2019

Proceeding: Fluidization XVI

Group: General Paper Pool

Session: Poster Session

Time: Tuesday, May 28, 2019 - 4:00pm-6:00pm

Authors:

Zhang, H., School of Life Science and Technology, Beijing University of Chemical Technology
Baeyens, J., European Powder and Process Technology
Ansart, R., Université de Toulouse, CNRS-Toulouse
Sabatier, F., Université de Toulouse, CNRS-Toulouse
Flamant, G., CNRS

SOLPART intends to decompose non-metallic minerals by concentrated solar irradiance of a fluidized bed reactor. Two different types of particles will be treated, i.e. free-flowing and cohesive particles. This is already evident from the particle size, when the Sauter mean particle size is considered, being â‰¤ 10 Âµm for cement raw meal (CRM), 50-130 Âµm for limestone and dolomite, and < 350 Âµm for milled phosphate and other ores. The combination of mean particle size and density of most of the minerals and their products (e.g. CaO from $CaCO_3$) classifies them as Geldart A or B-type powders, easily handled, stored, discharged and conveyed. CRM is even visually seen as highly cohesive. This cohesiveness was hence examined in detail at temperatures between 20 and 850 Â°C according to various criteria, being the Geldart classification: C/A; the Hausner ratio = Ï$_{tapped}$/Ï$_{loose;}$ the Yield limit: consolidation vs. shear; the Angle of Repose (AoR); and the balance of forces. A Hausner ratio in excess of 1.25 is characteristic for a cohesive powder. Results clearly illustrate the difference between CRM and 58 Î¼m limestone, with values for CRM ranging from1.25 to 1.38, against 1.05 for limestone. Bulk densities are moreover slightly higher at higher temperature, meaning that the powder is more consolidated, with typical values of 1191 kg/mÂ³ (850Â°C) and 1078 kg/mÂ³ (20Â°C) for CRM, and 1560 kg/mÂ³ (850Â°C) against 1492 kg/mÂ³ (20Â°C) for 58 Âµm limestone.The

Angle of Repose (Geldart et al.) provides additional information and was determined in an electrical furnace by pouring the powders on a flat plate and measuring the angles of the slopes. Criterion of assessment are: AoR < 30 (flowable); AoR 30 - 45 (some cohesiveness); AoR 45 - 55 (true cohesiveness) and AoR > 55 (high cohesiveness, limited flowability). Results are again confirming the cohesiveness of CRM*,with AoR increasing from62-65Â° at 20Â°C, to 70-72Â° at ~850Â°C. Limestone values vary from 27Â° at 20Â°C to 31Â° at ~800Â°C.Initially developed by Smolders and Baeyens[1], a total force balance for the powder behaviour can be established. Whereas the CRM powder will be characterized by very strong Van der Waals cohesive forces, coarse particles such as limestone-lime will show no cohesiveness. The experimental work on the Yield limit (Jenike criteria[2]) assesses the effect of the consolidation on the incipient flow. The flow factor ratio ff_c = Ïƒ$_1$/Ïƒ$_c$ characterizes the flowability of bulk solids, where greater values of the ratio represent a greater flowability of the powder. Ïƒ$_1$ is the normal consolidation stress, whereas Ïƒ$_c$ is the stress at failure. The numerical classification is the following [Schulze[3]]: ff_c < 1 (non-flowing); 1 < ff_c < 2 (very cohesive); 2 < ff_c < 4 (cohesive); 4 < ff_c < 10 (easy-flowing); and 10 > ff_c (free-flowing). Results obtained in a shear cell at different temperatures will be illustrated. With a ff_c â‰¤ 2, CRM is considered very cohesive. Limestone and dolomite have a ff_c > 10, and are freely flowable. It can hence be concluded that Limestone/dolomite and other A/B-type minerals pose no problem, since perfectly flowable, whereas Cement raw meal is cohesive to very cohesive according to all criteria. Shear is needed (air flow) to reduce the cohesive Van der Waals forces, and this explains the use of a lean phase pre-calciner and aerated hopper design in CRM storage silos.

Acknowledgements: â€œThis project has received funding from the European Unionâ€™s Horizon 2020 research and innovation programme under grant agreement No 654663, SOLPART project."

References

(1) K. Smolders K, J. Baeyens. A charaterisation of cohesive (C) and free-flowing (A) powders. Powder Handling and Processing 2005ï¼Œ17(4), 196-199.

(2) W. Jenike, Storage and flow of solids, Bulletin nÂ° 123 Utah Engineering Experiment Station, 11th edition, 1986.

(3) D. Schulze, in Powder technology and pharmaceutical processes, Handbook of powder technology, volume 9, Elsevier, 1994

Solid Circulation and Reduction Characteristic of Mass Produced Particles in a 0.5 Mwth Chemical Looping Combustion System

Conference: Fluidization

Year: 2019

Proceeding: Fluidization XVI

Group: General Paper Pool

Session: Poster Session

Time: Tuesday, May 28, 2019 - 4:00pm-6:00pm

Authors:

Jin, G. T. J., Korea Institute of Energy Research
Ryu, H. J., Korea Institute of Energy Research
Jo, S. H., Korea Institute of Energy Research
Lee, S. Y., Korea Institute of Energy Research
Lee, D., Korea Institute of Energy Research
Nam, H., Korea Institute of Energy Research
Baek, J. I., KEPCO Research Institute

In Korea, KIER(Korean Institute of Energy Research) has been doing research on chemical looping combustion technology since 2000 and KIER and KEPCO RI(Korea Electric Power Research Institute) started a collaborative research from 2006 to develop chemical looping combustion process and oxygen carriers, respectively. We started new research project to develop a 0.5 MWth pressurized chemical looping combustor from 2015. In this project, our target is development of pressurized chemical looping combustor up to 5 bar-abs and achievement of high CO_2 selectivity (> 98%) using natural gas and syngas as fuels. Recently we have finished the installation of the 0.5 MWth chemical looping combustor. In this study, continuous solid circulation test at high temperature and high pressure conditions and batch type reduction-oxidation tests were performed to check feasibility of a 0.5 MWth chemical looping combustion system. Pressure drop profiles were maintained stable during continuous solid circulation up to 16 hours. Therefore, we could conclude that the solid circulation between an air reactor and a fuel reactor could be smooth and stable. The measured fuel conversion and CO_2 selectivity were high enough even at high capacity and even after cyclic tests. Therefore, we could expect high reactivity of oxygen carrier at real operation condition.

Stability Analysis of Gas Solids Flow through Parallel Paths

Conference: Fluidization

Year: 2019

Proceeding: Fluidization XVI

Group: General Paper Pool

Session: Poster Session

Time: Tuesday, May 28, 2019 - 4:00pm-6:00pm

Authors:

Zhang, C., Tsinghua University
Qian, W., Tsinghua University
Wei, F., Tsinghua University

The core point for gas solids flow through parallel paths is to ensure the stability of desirable gas solids distribution, especially when each path is non-identical. In this study, a framework for studying the effect of non-identity on the stability of gas solids distribution is established. After introduce of a parameter, i.e. $\hat{I}'$ = (C_{d2} â€“ C_{d1})/C_{d2}, reflecting of deviation from identity, the criterion, based on the concavity of total pressure drop surface, for detecting instability of gas-solid distribution through parallel system with non-identical paths is developed. Moreover, experimental studies are also carried out to verify the theoretical analysis.

Study on Gas Phase Flow Characteristics and Particle Separation Efficiency of Three-Stage Internal Cyclone Separator in Fluidized Bed Reactor (FBR) of in Cold Hydrogenation Reaction

Conference: Fluidization

Year: 2019

Proceeding: Fluidization XVI

Group: General Paper Pool

Session: Poster Session

Time: Tuesday, May 28, 2019 - 4:00pm-6:00pm

Authors:

Hua, C. Jr., CAS Key Laboratory of Green Process and Engineering, Institute of Process Engineering, Chinese Academy of Sciences, Beijing 100190, China
Lu, P. Sr., CAS Key Laboratory of Green Process and Engineering, Institute of Process Engineering, Chinese Academy of Sciences, Beijing 100190, China

In fluidized bed reactor(FBR) of cold hydrogenation reaction of silicon tetrachloride(SiCL4), three-stage internal cyclone separator is an important equipment for separating silicon powder and chlorosilane mixed gas. To optimize the separation efficiency and energy loss of the internal cyclone separator in FBR, the main structural parameters affecting the performances are deeply determined. Based on ansys fluent(CFD) software code, the trace of solid particle size is numerically simulated by the response surface model and Lagrange model. The effect of the feed location, the feed velocity, the pipe diameter of gas outlet and leg height on separation efficiency was well studied. The results show that the diameter of the inlet pipe has little effect on the pressure drop(â–³p)and the separation efficiency. The diameter and velocity of the exhaust pipe have great effects on the pressure drop and the separation efficiency while the interaction between the diameter of the inlet pipe and the feed velocity is obvious for internal cyclone separator. As for silicon particle size of 0.5~10 Î¼m of the reactant, the optimal parameters are De/D=0.35ã€Dd/D=0.37ã€V=12 m/s: the outer diameter(D) the dust outlet diameter (D_d), exhaust port diameter (D_e), and inlet velocity (V).The optimal operation parameters are applied into one certain three-stage internal cyclone separators which diameters are separately 1.2m,0.8m and 0.5m. There is little error of < 5.0% between the industrial data and the simulated data while the separation efficiency of 0.1Î¼m particle size is increased to >99.5% and the pressure drop is decrease to <3.2KPa by half drop. The established response surface model can accurately describe the relationship between design variables and objective functions Therefore the industrial equipmentâ€™s efficiency of cold hydrogenation reaction was greatly improved. These studies provide an effective guidance for the separation of 0.5-10 Î¼m particle size in three-stage internal cyclone separation.

Thermal Behaviour of Granular Materials in Directly Irradiated Fluidized Beds

Conference: Fluidization

Year: 2019

Proceeding: Fluidization XVI

Group: General Paper Pool

Session: Poster Session

Time: Tuesday, May 28, 2019 - 4:00pm-6:00pm

Authors:

Tregambi, C., Università degli Studi di Napoli Federico II
Bevilacqua, C., Istituto di Ricerche sulla Combustione, Consiglio Nazionale delle Ricerche
Chirone, R., Consiglio Nazionale delle Ricerche
Salatino, P., Università degli Studi di Napoli Federico II
Solimene, R., Consiglio Nazionale delle Ricerche

The production and storage of renewable energy is a top challenge for mankind. Extensive R&D is in progress to exploit the huge amount of solar energy falling on Earth (~100,000 TW). Thanks to the easy integration with Thermal Energy Storage (TES), Concentrating Solar Power (CSP) systems stem out as one the most promising technology to take advantage of the solar energy. TES allows to separate the process of usage and absorption of solar energy: in this way, the energy coming from the sun can be exploited at will, even during the night or in cloudy sky conditions. The current benchmark in the CSP technology is represented by the solar tower receiver with molten salts working as heat transfer fluid and sensible heat storage medium. The main drawback of this system is the relatively low working temperature of the molten salts (about 565 °C), which affects the efficiency of the subsequent Rankine cycle for energy production. Particle receivers in CSP systems are gaining ever increasing interest, as the solid particles can simultaneously act as receiver, heat transfer fluid and heat storage medium. Dense gasâ€'solid suspensions can work at higher temperature with respect to molten salts, even between 1000â€"1500 Â°C, without any technological problem. Fluidized beds (FBs) as dense gasâ€'solid suspensions are considered as affordable and reliable systems for CSP. In solar FB, the interaction between the incident radiative flux and the FB can occur in an indirect or direct way. In the indirect heating, the concentrated radiation is focused onto a cavity or an exposed surface, and heat is eventually transferred to the solid suspension by convection/conduction. In the direct heating, the concentrated radiation enters the reactor by the use of optically accessible windows located in the ceiling of the receiver, or through the use of transparent walls. The direct heating configuration permits higher operating temperatures as the high concentrated solar radiation directly heats the dense gasâ€‘solid suspension, but care must be paid to avoid the bed surface overheating.

In this work, different materials were investigated to scrutinize their potential use as solid suspension in directly irradiated FB reactors. The experimental apparatus consists of a FB reactor with both the internal bed diameter and height equal to nearly 0.1 m, so that the aspect ratio (internal diameter-height ratio) approaches the unit value. The windbox section, below the reactor zone, is 0.15 m high. The upper part of the FB reactor is connected to a conically-shaped section, 0.4 m high, which represents the freeboard and hosts, at is extremity, a 4 mm transparent window resistant to high temperature, essential to let the solar simulated radiation enter. The conical shape of the freeboard section is required to both not hinder the pathway of the solar simulated radiation and to further protect the transparent window against particles impact, thanks to the gas slowdown due to the cross-section increase. Gas exit is provided at half the conical section through four 1 inch pipes. The FB reactor surface is exposed to a 12 kW_{el} simulated solar furnace made by an array of three short-arc Xe-lamps coupled with elliptical reflectors. The peak flux produced on the FB surface is of nearly 3000 kW m$^{â€“2}$, while the total power irradiated over the whole FB surface is of about 3 kW_{th} when all the lamps are turned on. On the other side, when only two lamps are used, the value of peak flux and total power becomes 2000 kW m$^{â€“2}$ and 2 kW_{th}, respectively. The reactor is also surrounded by two semi cylindrical radiant heaters in ceramic fiber, which are used to both heat and insulate the reaction chamber. An electronic mass flow controller is used to supply the air required for the reactor operation. Concerning the diagnostic tools, five K-type thermocouples are located inside the FB to measure the temperature at different locations.

The experimental tests were aimed at studying the dynamics of the directly irradiated FB reactor with specific reference to temperature distribution at the surface and in the bulk of the bed as a function of the inlet gas velocity. The tests were performed by powering only two of the three Xe-lamps because the temperature level reached with this configuration were already sufficiently high for the exploitation of thermodynamic cycles for energy production. In each test, the bed was initially charged with an amount of material so that the static bed height was equal to 7.8 cm. To speed up the initial warm up of the reactor, the system was heated by using the radiant heaters up to a temperature of 400 Â°C while fluidizing with an inlet gas velocity of 10 cm s^{-1}. Then, the radiant heaters were turned off and simultaneously the solar simulator was powered on. It was then waited until all the temperature values read by the K-type thermocouples become stationary. Once the steady state condition was obtained, the inlet gas velocity was reduced by 1 cm s$^{â€“1}$to eventually reach a new stable operation condition. Typically, a time of 20 min was always sufficient to achieve a new steady state operation condition. Following this method, the inlet gas velocity was progressively reduced to the lowest possible stable value according to the fluidization behaviour of the investigated material. During the tests, the temperature data were continuously acquired at 1 Hz using a purposely developed script under Labview environment. Data of steady-state bed temperature as a function of the inlet gas velocity were collected for the different materials and eventually used to compare the material performance and to estimate an efficiency of absorption of the solar energy. More into detail, the data were used to solve an energy balance model on the FB reactor whose solution returns the absorption thermal efficiency, as a function of the inlet gas velocity, for all the investigated material.

Experimental results highlighted that steady-state temperatures tend to decrease when the inlet gas velocity was increased. This behaviour was attributed to both an increase in the thermal losses due to convective flow of the fluidizing gas and to the reduction of absorption capacity of the bed surface, as probably clouds of particles ejected by bubble bursting could hinder the penetration of the concentrated radiation to the FB. However, by reducing the inlet gas velocity, temperature gradients inside the fluidized bed are increased due to a reduction of the axial thermal diffusivity in the FB reactor. By solving the energy balance equation on the FB reactor under steady state conditions, the overall absorption coefficient for the investigated materials was estimated. It was then possible to rank the materials according to their efficiency in absorbing the incident solar energy. It was also observed that, as for the steady state temperature, the absorption coefficient increased when the fluidization velocity was reduced. However, to properly choose the best material to be used as granular material in directly irradiated fluidized beds, it is important to take into consideration also the lowest fluidization velocity at which the material can work to withstand the high concentrated solar flux. Indeed, gas pumping and compression represent one of the major costs in FB reactors, thus a lower inlet gas velocity brings to lower operational costs.

Wall-to-Bed Heat Transfer Characteristics in a Fluidized Bed of CNT Particles

Conference: Fluidization

Year: 2019

Proceeding: Fluidization XVI

Group: General Paper Pool

Session: Poster Session

Time: Tuesday, May 28, 2019 - 4:00pm-6:00pm

Authors:

Lee, M., Korea National University of Transportation
Kim, S., Korea National University of Transportation

Carbon nanotubes (CNTs) with high thermal and electrical conductivities are very promising materials in a wide range of potential applications. In recent years, fluidized bed reactors have been increasingly applied for mass production in the CNT synthesis using catalytic chemical vapor deposition technology. Heat supply to the catalyst or CNT particles in the bed for the CNT synthesis is achieved by using heated reactor wall. The wall-to-bed heat transfer coefficients are needed to design the reactor, where temperature control plays an important role.
The effect of gas velocity on the average and local heat transfer coefficients between a heat transfer probe (20 mm-OD) at wall and a fluidized bed has been firstly determined in a fluidized-bed (0.15 m i.d. X 2.6 m high) of the CNT particles (d_p=0.485 mm; bulk density=92 kg/m^3). The average heat transfer coefficient exhibits a maximum value (692 W/m^{2}K) at 2.8 U_{mf} with variation of gas velocity. The variation of instantaneous heat transfer coefficients reflects well the entangled CNT aggregates behavior near wall region in the dense phase. The obtained average heat transfer coefficients have been correlated with the experimental parameters from present results and previous literature.

A Generalized Model on Prediction of Solid Flow Rate Though L-Valve

Conference: Fluidization

Year: 2019

Proceeding: Fluidization XVI

Group: General Paper Pool

Session: 5A: Fundamentals of Fluidization

Time: Wednesday, May 29, 2019 - 10:19am-10:31am

Authors:

He, P., The Ohio State University
Fan, L. S., The Ohio State University

Abstract

L-valve serves as an essential non-mechanical part to control solid flow rate in circulating fluidized bed (CFB). A number of correlations were suggested in last few decades by various authors to predict solid flow rate providing aeration rate or pressure drop across L-valve, and other parameters such as properties of particle or L-valve geometry, etc. The accuracy of these empirical correlations are questionable especially when operating condition is beyond the tested range. In this work, a simple model was proposed for gas and solid flow in L-valve section. By closing the model with literature data, a generalized dimensionless correlation was found to intrinsically account for the momentum transfer between the gas and solid flow. With the correlation as additional independent equation, the model is then able to predict solid flow rate providing other measurable parameters. The correlation was obtained as a single curve based on the data generated at various experimental configurations, such as particle properties and L-valve pipe size, etc. Thus, this model is considered as potentially generalized estimation for solid flow rate through L-valve, i.e. solid circulation rate in CFB, although it requests further data validation from other conditions, e.g. at elevated temperature and pressure.

A Reflection of Gas-Solid Fluidization Regimes

Conference: Fluidization

Year: 2019

Proceeding: Fluidization XVI

Group: General Paper Pool

Session: 5A: Fundamentals of Fluidization

Time: Wednesday, May 29, 2019 - 8:30am-8:55am

Authors:

Bi, X., University of British Columbia
Grace, J. R., The University of British Columbia

Flow patterns and flow regimes in gas-solid fluidization systems have attracted a great attention in the fluidization research community since 1940s. A number of flow regime maps have been proposed to distinguish different unique flow patterns, such as bubbling, slugging, turbulent and fast fluidization, and dilute phase transport, but controversies still exist on turbulent fluidization, which was first proposed in late 1960s, fast fluidization, which was first proposed in late 1970s, and high density suspension upflow, which was proposed in late 1990s. We will present the recent advances in understanding the transition among turbulent fluidization, fast fluidization and dense suspension upflow regimes. The future research needs in developing a generalized flow regime diagram are proposed based on improved understanding of high velocity fluidization and dense suspension flow in the presence of equipment configuration and operational constraints.

Effect of Elevated Pressure on Gas-Solid Flow Characteristics in a Circulating Fluidized Bed

Conference: Fluidization

Year: 2019

Proceeding: Fluidization XVI

Group: General Paper Pool

Session: 5A: Fundamentals of Fluidization

Time: Wednesday, May 29, 2019 - 9:07am-9:19am

Authors:

Liu, D., Southeast University
Ma, J., Southeast University
Song, J., Southeast University
Chen, X., Southeast University

Gas-solid fluidized beds are extensively applied because of their excellent heat and transfer properties. The pressurized Circulating Fluidized Beds (CFB) are found in some applications, e.g, the pressurized oxy-fuel combustion for CO_2 capture. The particle concentration profile and solid circulating rate are crucial parameters, which affect the design of pressurized CFB reactor. In the current study. A cold pressurized CFB experimental device is established. The inner diameter of the riser is 50 mm and the height is 3300 mm. Geldart B glass beads with an average particle size of 250 Î¼m and a density of 2550 kg/m^3 is used as material. The operation pressures ranges from 0.1 MPa to 0.6 MPa. The distribution of the particle concentration along the axial and radial directions in the riser is studied. The result shows that, as the pressure is increased, the particle concentration presents a slight increase in the top and middle parts of the riser, while the particle concentration in the bottom part of the riser does not change obviously. The particle concentration is generally higher near the wall under different pressures, but the particle concentration changes slightly along the radial direction under different pressures.

A further analysis of pressure fluctuation shows that the domain frequency of the particle concentration sequence signal is around 5 Hz based on the power spectrum density curves. The domain frequency increases with increasing pressure and decreases with increasing circulating flow rate. The average gas-solid slip velocity decreases at the same particle concentration conditions and particle circulation flow rate.

Eulerian Simulations of Mixing and Segregation of Binary Gas-Solid Flow of Particles with Different Densities

Conference: Fluidization

Year: 2019

Proceeding: Fluidization XVI

Group: General Paper Pool

Session: 5A: Fundamentals of Fluidization

Time: Wednesday, May 29, 2019 - 9:19am-9:31am

Authors:

Parvathaneni, S., Indian Institute of Technology Delhi
Buwa, V., Indian Institute of Technology Delhi

Introduction

Several industrial processes employing gas-solid fluidized beds often involve a mixture of particles differing in size, density, and shape (e.g. coal and inerts, coal and ash, biomass and inert used in gasification process). The difference in the physical properties (size, density, and shape) of these particles give rise to non-uniform mixing and segregation which leads to the formation of hot-spots and agglomerates (e.g. clinker formation), thus deteriorating the performance of the gasifier. Therefore, the performance of these industrial processes is influenced by the local mixing of the gas and solid phases, which primarily is governed by the local fluidization/bubbling behaviour. Therefore, it is important to develop CFD models capable of predicting the bubbling characteristics, mixing and segregation phenomena of binary gas-solid flow accurately. Such models can then be extended to predict the local temperature and concentration distributions in gasifiers and eventually to optimize performance of such fluidized bed gasifiers/reactors.

The focus of the present work is on simulating the binary gas-solid flow differing in density and predicting the mixing and segregation phenomena. However, in order to predict the mixing and segregation phenomena, it is important to predict the bubbling characteristics (bubble size distribution, bubble rise velocity, and bubble frequency) as these are responsible for the mixing and segregation phenomena at a given superficial gas velocity (U_g) and bed composition. In our research group, we have been working on simulations of unary gas-solid flow in fluidized beds (Singh et al. (2018)). The Two-Fluid Eulerian simulations performed in our research group using different gas-solid drag models, for e.g. widely used model of Gidaspow et al. (1992), the multi-zonal drag on different values of local gas volume fraction (Gao et al. (2009), Singh et al. (2018)) and EMMS-based drag models (Hong et al. (2013), Li and Yang (2017)) that account for the heterogeneous flow structures. We found that the predicted time-averaged radial solid hold-ups for all drag models considered in the present work were in a reasonably good agreement with the experiments, whereas the predictions of bubbling characteristics (bubble size distribution, bubble frequency) were significantly different from the measurements. Therefore, such models are unlikely to predict the mixing and segregation phenomena as the latter are governed by the bubbling characteristics. As the existing gas-solid drag models mentioned above do not predict the bubbling characteristics well, it becomes important to investigate the effects of other closures and parameters such as particle frictional stress model, minimum and maximum frictional packing limits which are shown to have a significant impact on the bubble formation, growth and size (Passalacqua and Marmo (2009)).

The present work is carried in a pseudo 3-D (rectangular) fluidized bed because of its advantage in visualization and measurements of several parameters simultaneously. The objectives of the present work are (i) to measure the bubble size distribution, bubble rise velocity, time-evolution of average dispersed bed heights of individual solid phases and % segregation for binary systems differing only in density (same size), (ii) to analyze the contribution of particle frictional stress models (frictional pressure and frictional viscosity), maximum and minimum frictional packing limits on bubble size distribution and bubble rise velocities of two unary systems (both having same size but differing in density) and, (iii) to predict and to validate the time-evolution of individual solid-phase average dispersed bed heights, % segregation, bubble size distribution and bubble rise velocities for binary gas-solid flow differing in density (size ratio = 1, density ratio = 3.47) for a wide range of U_g and bed compositions.

Experimental and Computational methodology

Unary and binary gas-solid fluidization experiments were performed in a pseudo 3D (rectangular) column with height (H) of 1 m, width of 0.2 (m) and depth of 0.02 (m). Air was used as the fluidizing medium and glass beads (GB) (d_p = 2.26 mm; Ï$_p$ = 2500 kg/m^3), mustard seeds (MS) (d_p = 2.18 mm; Ï$_p$ = 720 kg/m^3) were used as the solid phases. Dynamics of fluidization of unary_GB, unary_MS and binary system (GB: MS) in the weight ratio of 70%:30%, 50%:50%, and 30%:70% solids was investigated for superficial gas velocities in the range 1.1-2 U_{mf} of the dense phase. The experimental images were recorded with a highâ €“speed monochromatic camera at 640 Ã— 800 pixel^2 resolution and at a rate of 400 fps. Further details of the experimental methodology will be presented in the full manuscript.

The Two-and Multi-Fluid Eulerian models were used to simulate the unary_GB, unary_MS and binary system_GB: MS under transient flow conditions. The Kinetic Theory of Granular Flow (KTGF) model was used to model the solid stresses. The model equations were solved using the commercial software â€œAnsys Fluent 19.1.0â€. No-slip and Johnson-Jackson wall boundary conditions were used for gas and solid phases, respectively. The user defined functions were used to implement different closures models (gas-solid drag, frictional pressure, and frictional viscosity). Further details on the effects of the grid and time step-independence studies and simulation parameters will be presented in the full manuscript.

Results and discussion

For the binary gas-solid experiments performed for a system composed of GB: MS = 70:30 wt. % at a U_G of 1.73 m/s (1.1 U_{mf} of GB), it was observed that the time-evolution of average dispersed bed height of GB was found to reach 15 cm from the initial static bed height of 25 cm and the % segregation was found to reach 60 % after 25 sec. At U_G = 2.78 m/s (1.78 U_{mf} of GB), a complete mixing of the solid phases was observed as the operating U_G was much higher than the U_{mf} of GB and therefore segregation was not observed. From the information of the movement of bubble with respect to time and axial location, the bubble size distribution and bubble rise velocity are calculated for unary_GB, unary_MS and binary systems.

The Two-Fluid Eulerian simulations were performed using the gas-solid drag models of Gidaspow et al. (1992), Gao et al. (2009), Singh et al. (2018), and Li and Yang (2017) for unary_GB at U_g = 2.08 m/s (1.3 U_{mf}), 2.78 m/s (1.78 U_{mf}). Although the aforementioned gas-solid drag models predicted the dispersed bed heights in a satisfactory agreement with the measurements, the bubble size distribution predicted using Singh et al. (2018) was qualitatively in a better agreement with the experiments in comparison with the corresponding predictions of the other drag models. The drag model of Singh et al. (2018) predicted significantly higher solid shear stress over solid normal stress and this led to a higher compaction of solids around the bubble, thereby leading to less dispersion of gas into the solid phase. To further improve the predictions of bubble size, frictional pressure models of Syamlal et al. (1993) and Johnson and Jackson (1987) were used at U_g = 2.78 m/s for unary_GB system. At a minimum frictional packing limit of 0.61 and maximum packing limit of 0.63, the formation of large bubbles was observed in both the cases in comparison to the KTGF based frictional pressure model. Further investigations on the effect of frictional stress models (frictional pressure and frictional viscosity) at different frictional packing limits on bubbling characteristics and multi-fluid Eulerian simulations will be performed for a binary system (GB: MS) differing in density and the detailed analysis will be presented in full manuscript.

Conclusions

In this study the effect of different closures models for gas-solid drag and frictional stress models, minimum and maximum frictional packing limits on predictions of bubble size distribution, bubble rise velocity and time-evolution of % segregation at different U_G and bed compositions were investigated and compared quantitatively with corresponding experimental measurements. The frictional pressure model and frictional packing limit plays a significant role in the prediction of bubble size quantitatively in comparison to the gas-solid drag models. This study will provide a comprehensive analysis of the effect of gas-solid drag and particle frictional stress models on predictions of the bubbling characteristics, mixing and segregation phenomena.

References

Gao, J., Lan, X., Fan, Y., Chang, J., Wang, G., Lu, C. and C. Xu, *Chemical Engineering Journal,* 64 (2009), 4302-4316.

Gidaspow, D., Bezburuah, R. and J. Ding, *Fluidization VII, Proceedings of the 7th Engineering Foundation Conference on Fluidization.* (1992), 75-82.

Hong, K., Shi, Z., Wang, W. and J. Li, *Chem. Eng. Sci.* 99 (2013), 191â€“202.

Johnson P.C. and R. Jackson, *J. Fluid Mech.* 176 (1987), 67â€“93.

Li, J. and B. Yang, *Chemical Engineering Journal,* 329 (2017), 100-115.

Passalacqua, A. and L. Marmo, *Chem. Eng. Sci.* 160 (2009), 2795-2806.

Singh, B. K., Roy, S. and V.V. Buwa, *Internal Report, Indian Institute of Technology Delhi, New Delhi* (2018).

Syamlal, M., Rogers, W. and T.J. O'Brien, *MFIX Documentation: Volume I, Theory guide. Technical Report DOE, Springfield, VA,* (1993).

Experimental Study of a 3D-Printed 3-Phase Miniaturized Fluidized Bed for Bioprocessing Screening

Conference: Fluidization

Year: 2019

Proceeding: Fluidization XVI

Group: General Paper Pool

Session: 5A: Fundamentals of Fluidization

Time: Wednesday, May 29, 2019 - 9:31am-9:43am

Authors:

Zhang, Y., Newcastle University

Experimental Study of a 3D-Printed
3-Phase Miniaturized Fluidized Bed for Bioprocessing Screening

Yi
Zhang[1], Yuen Ling Ng[2], Kheng-Lim Goh[2], Steven
Wang[1], Vladimir Zivkovic[1]*

[1] School of Engineering, Newcastle
University, Newcastle Upon Tyne, NE1 7RU, UK

[2] Newcastle Research & Innovation
Institute Singapore (NewRIIS), 80 Jurong East Street 21, #05-04, Singapore
609607

Abstract:

Fluidized bed as one type of reactors has been
commonly used in chemical and process industry due to its excellent
multi-phases contact, minimum diffusional resistance, and good heat and mass
transfer [1]. Miniaturized fluidized
beds (MFBs) are attracting a lot of attention as a cost-effective and process
intensification (PI) tool for fast screening of solid processes and bioprocesses

in recent years, out-performing many conventional large-scale fluidic systems [2, 3].

In
this research, the miniaturized fluidized beds were innovatively fabricated using 3D printing technique. This novel micro fluidized bed reactor highlights the application of rapid design and optimization utilizing 3D printing technique. The performance of the flow of multiple microbubbles under different parameters (i.e. air sparger inner diameter, air-inlet numbers, and liquid & gas velocity) was studied. The effects of air sparger distance on multiple bubble flow were also investigated. Furthermore, the 3D-printing approach was used to manufacture cell-mimicking particles, based on density characteristics, to study the dynamics of cell in micro-fluidized bed reactor, providing a possibility for fast screening of cells during cell cultivation and bioprocessing processes. Specifically, the 3D-printed polymer particles (1.12g/cm3 in density) with diameters of 1 mm, 2 mm and 3 mm were respectively introduced into the fluidized column to form a liquid-solid-gas fluidized system, to mimic the fungal cell cultivation within the fluidized bed. The different flow regime including dispersed bubble flow, coalesced bubble flow and slug flow were mapped as function of operating conditions.

Reference

ADDIN EN.REFLIST 1. Doroodchi, E., et al., Fluidisation and packed bed behaviour in capillary tubes. Powder Technology, 2012. 223: p. 131-136.

2. Li, X., M. Liu, and Y. Li, Bed expansion and multi-bubble behavior of gas-liquid-solid micro-fluidized beds in sub-millimeter capillary. Chemical Engineering Journal, 2017. 328: p. 1122-1138.

3. Wang, H., et al., A review of process intensification applied to solids handling. Chemical Engineering and Processing: Process Intensification, 2017. 118: p. 78-107.

---

* Corresponding
author:

E-mail
address: vladimir.zivkovic@ncl.ac.uk

Measurement of Residence Time Behavior in a Continuously Operated Spouted Bed

Conference: Fluidization

Year: 2019

Proceeding: Fluidization XVI

Group: General Paper Pool

Session: 5A: Fundamentals of Fluidization

Time: Wednesday, May 29, 2019 - 8:55am-9:07am

Authors:

Pietsch, S., Hamburg University of Technology
Kieckhefen, P., Hamburg University of Technology
Schönherr, M., BASF SE
Kleine Jäger, F., BASF SE
Heinrich, S., Hamburg University of Technology

Continuous gas-solid processes are applied in many industries with particulate products as e.g. for agglomeration, granulation, coating or drying. An important criterion for these apparatuses, which often combine several unit operations in one plant, is a narrow residence time distribution and low back-mixing between the stages in order to obtain a constant product quality. A novel tracking method with magnetizable particles was established in order to measure the residence time distribution in continuously operated fluidized or spouted beds. In this work, a continuously operated horizontal prismatic spouted bed with four unseparated chambers (ProCell 25, Glatt) was investigated. The tracer was produced by coating the bed material, namely Cellets®500 particles, with a magnetizable paint. The applicability of the particles as tracer material was evaluated regarding the requirements and confirmed. The residence time distribution was determined by applying an impulse of tracer material during continuous process operation at the inlet and taking samples of the outflow in defined time intervals. The tracer material was separated by a magnetic rod and the measurement of the weight concentration in every sample allowed the fast calculation of the exit age distribution.

Experiments were performed with different bed masses and gas volume flow rates. Besides the original chamber configuration, plates with different opening geometries, as e.g. a mousehole, were inserted to physically separate the four chambers. It was found that the sealing of the plates is an important step in order to force the particles to flow through the opening geometry. In addition, experiments with closed in- and outlet boundary conditions were performed to characterize the back-mixing and the flow behavior of the particles dependent on the separation plates. Therefore, the tracer particles were inserted in one chamber and the distribution after a defined process duration was used for quantification of back-mixing. It was found that the back-mixing can be reduced by decreasing the size of the mousehole in the separation plates. Nevertheless, the size reduction is limited as a minimum hole diameter has to be provided for reaching the desired mass flow rate through the apparatus. The experimental data are used for validation of simulations (e.g. CFD-DEM, rCFD) of continuous spouted bed operation.

Measuring the Velocity of Gas and Particles in and Around a Single Bubble in a 3D Fluidised Bed

Conference: Fluidization

Year: 2019

Proceeding: Fluidization XVI

Group: General Paper Pool

Session: 5A: Fundamentals of Fluidization

Time: Wednesday, May 29, 2019 - 9:43am-9:55am

Authors:

Holland, D. J., University of Canterbury
Rice, N., University of Cambridge
Boyce, C. M., Columbia University
Sederman, A. J., University of Cambridge
Gladden, L. F., University of Cambridge

Understanding the flow of gas and particles is critical to predicting the behaviour of gas-solid fluidised beds, and is not yet fully resolved. The bubbles in a fluidised bed drive the mixing behaviour of the particulate phase, while the flow of gas through the bubbles and the particulate phase determines the gas-solid contact. In this paper, we investigate some of the classical models of fluidised beds using Magnetic Resonance Imaging (MRI) measurements of the velocity of the gas and solid phases in a three dimensional gas-solid fluidised bed. In order to obtain sufficient signal from the gas phase, sulphur hexafluoride gas was used at a pressure of 7.5 barg; poppy seeds were used for the particulate phase. Even with this gas at pressure, MRI measurements of the velocity are typically slow, especially when measuring the gas phase owing to the low signal intensity of the gas. Therefore to enable images of the gas and solid velocity field, a system was developed to inject single, isolated bubbles reproducibly into an incipiently fluidised bed. Images of the flow of gas and solid are then obtained by averaging measurements from a large number of these isolated bubbles. The MRI measurements were triggered such that data were acquired at specific times after the injection to ensure that all of the bubbles that were measured were in the same position for each time. Here we first quantify the reproducibility of the gas injection system and then demonstrate that quantitative measurements of the gas and particle velocities can be obtained. The resulting images clearly show a recirculation pattern in the gas phase that is consistent with the potential flow solution of Davidson and Harrison, for these group D particles. From the gas velocity field it is also possible to estimate the through-flow of gas through the bubble and show that this is consistent with the estimate of three times the minimum fluidisation velocity, U_{mf}. The velocity of the gas in the particulate phase is found to be less than expected in the classical theory owing to the net downward velocity of the particles relative to the rise velocity of the bubble. If the particle velocity is used to correct the expected gas velocity, it is in good agreement with the estimated interstitial velocity U_{mf}/e_{mf}, where e_{mf} is the voidage at minimum fluidisation.

Simultaneously Production of High Quality Tar and CH4 from Catalytic Bituminous Coal Hydrogasification in a Pressurized Fluidized Bed

Conference: Fluidization

Year: 2019

Proceeding: Fluidization XVI

Group: General Paper Pool

Session: 5A: Fundamentals of Fluidization

Time: Wednesday, May 29, 2019 - 10:07am-10:19am

Authors:

Yan, S., State Key Laboratory of Coal Conversion, Institute of Coal Chemistry, Chinese Academy of Sciences, China
Qu, X., State Key Laboratory of Coal Conversion, Institute of Coal Chemistry, Chinese Academy of Sciences
Bi, J., ENN Group State Key Laboratory of Coal-based Low Carbon Energy

Coal catalytic hydrogasification (CCHG) is an attractive technique of coal clean utilization, which can simultaneously produce SNG (Substituted natural gas) and aromatic hydrocarbon liquids under mild reaction conditions. This study works on achieving high CH_4 yield, light tar yield and coal conversion through the method of coal catalytic hydrogasification (CCHG) in a pressurized fluidized bed. The effects of catalyst loading (cobalt-calcium dual catalyst), temperature (600â€"850 Â°C) and H_2 pressure (0.1â€"3 MPa) on the formation of gaseous and liquid products were explored. Experimental results showed that carbon conversion and CH_4 yield increased markedly at a given ratio of Co-Ca. The addition of Ca retards Co sintering and more importantly triggers the catalytic conversion of graphite carbon in coal structure. Increasing temperature, especially higher than 750 Â°C, facilitated the activity of Co-Ca catalyst, making more carbon catalytic hydrogasified with lower activation energy. Nevertheless, elevated temperature was not beneficial for the yield of tar products because of the intensified thermal cracking effect. Elevating H_2 pressure facilitated the depolymerization effect of Co-Ca catalyst on coal during hydropyrolysis and promoted Co-Ca-C-H_2 interaction during hydrogasification, and thus encouraged CH_4 and tar formation. Through one stage CCHG at 850 °C and 3 MPa H_2, the yields of light aromatic hydrocarbons (LHs) and gaseous hydrocarbons (GHs) were 1.47 wt.%

(BTX, PCX and naphthalene) and 81.2% wt.% (CH_4 and C_2-C_3), respectively, when using 5%Co-1%Ca as catalyst. Through two-stage CCHG (firstly catalytic hydropyrolysis at 600 °C and 3 MPa H_2, and then catalytic hydrogasification at 850 °C and 3 MPa H_2), good results with LHs yield of 3.36 wt.% and GHs yield of 74.2 wt.% were achieved, which could serve as an alternative way for simultaneously production of LAHs and CH_4 from CCHG.

Solid Circulation Control with L-Valves: A Parametric Study

Conference: Fluidization

Year: 2019

Proceeding: Fluidization XVI

Group: General Paper Pool

Session: 5A: Fundamentals of Fluidization

Time: Wednesday, May 29, 2019 - 9:55am-10:07am

Authors:

Tebianian, S., IFP Energies nouvelles
Bertholin, S., IFP Energies nouvelles
Gauthier, T., IFP Energies nouvelles
Yazdanpanah, M. M., TOTAL

In processes such as Chemical Looping Combustion with high operating temperatures, the solid circulation control with mechanical valves would result in complex technologies and significant cost. Non-mechanical devices such as L-valves represent simple, cheap and reliable means to control the solid circulation rate around the flow loops. However, their design, especially for large units, must take into account the operating conditions such as particle properties, pressure profile and solid circulation rate in order to be able to control the flow in the range of operation.

In this work, a parametric study has been performed in order to study the effect of operating conditions on the design and operability of L-valves. The importance of pressure balance, solid flowrate and particle physical properties has been investigated and suggestions for providing a flexible design have been made.

A Novel Method to Investigate Lateral Mixing of Solids in Bubbling Fluidized Beds

Conference: Fluidization

Year: 2019

Proceeding: Fluidization XVI

Group: General Paper Pool

Session: 5B: Fluidized Bed Applications

Time: Wednesday, May 29, 2019 - 9:06am-9:18am

Authors:

Martinez Castilla, G., Chalmers University of Technology
Lundberg, L., Chalmers University of Technology
Pallarès, D., Chalmers University of Technology
Larsson, A., Bioshare AB
Johnsson, F., Chalmers University of Technology

The fluidized bed technology is widely used in a large number of industrial processes due to its high heat and mass-transfer capability. When applied to combustion and gasification of solid fuels, the mixing of solids is of high importance since it controls the mass and heat transfer. Numerous works have been published in the last decades investigating the solids mixing phenomena, in particular with respect to its influence on heat conductivity. However, interpreting and extrapolating results from published data on solids dispersion in fluidized beds is often challenging since data were obtained under different conditions. Moreover, there is a lack of data measured under conditions representative for operation under industrial conditions. Fluid-dynamically down-scaled units can be a cost-efficient tool to perform studies under ambient conditions, while maintaining similar fluid-dynamic properties as under industrial conditions at elevated temperatures.

This work presents a novel experimental method to investigate the lateral mixing in bubbling fluidized beds. The method, which is applied to a fluid-dynamically down-scaled bed, is based on placing a heat source in the bed and monitoring the temperature field over the bed surface with a high-precision thermographic camera placed above the bed. Based on the temperature field recorded by the camera and the heat input into the system, effective heat conductivity values can be computed, which are directly related to the dispersion coefficients of the bed material.

With this, the method is used to evaluate lateral mixing of solids under different operational conditions. An adjustable internal division wall is used to estimate how the lateral mixing depends on axial position along the bed height. The division wall provides an adjustable gap between the air-distributor and the division wall allowing particles, and thereby, heat to be transferred in the horizontal direction. Therefore, the axial mixing can be blocked over different parts of the bed height.

The fluid-dynamically down-scaled model used for the experiments resemble the bed of the Chalmers 12 MW boiler. The bed has a cross sectional area of 0.26m x 0.26m and consists of 60-Î¼m bronze particles fluidized with ambient air, applying a bed height of 0.11 m. The bed is operated without the division wall and with the division wall applying gap heights of 0.02 m, 0.05 m and 0.11 m. The fluidization velocities range from 3 to 5 times the minimum fluidization velocity using two different air distributors with different pressure drop.

The results show that the lateral solids dispersion coefficients in the bed operated without division wall fall within the order of magnitude of values reported in literature, which is considered to validate the method presented in this paper. The dispersion coefficient increases with fluidization velocity in agreement with previous findings. Applying the division wall, it is seen that the dispersion coefficient is reduced with reduced gap between air distributor and division wall, i.e. there is an axial dependence in the lateral solids mixing.

Combustion Behavior of High Ash Bituminous Coal in a Pilot Scale Circulating Fluidized Bed Rig

Conference: Fluidization

Year: 2019

Proceeding: Fluidization XVI

Group: General Paper Pool

Session: 5B: Fluidized Bed Applications

Time: Wednesday, May 29, 2019 - 8:54am-9:06am

Authors:

Sarbassov, Y., PI "National Laboratory Astana", Nazarbayev University
Tokmurzin, D., National Laboratory Astana
Aiymbetov, B., National Laboratory Astana
Inglezakis, V. J., Nazarbayev University
Anthony, E., Cranfield University
Suleymenov, K., Energy Technology R&D Center, ERG Group

In contrast to pulverized coal combustion systems, fluidized bed technologies offer significant advantages such as a low temperature environment and lower NOx, SOx emissions. Another advantage is the potential utilization of a wide range solid fuels such as biomass and low rank coals with high ash content. In this study, a bituminous coal from the northern part of Kazakhstan with ash content of around 40% was tested in a pilot-scale circulating fluidized bed rig under air environment. The pilot-scale circulating fluidized bed has a thermal output of 200 kW, a riser height of 7000 mm and inner diameter of 150 mm. The combustion and fluidization behavior of coal were determined by measuring the pressure drop profiles along the height of the rig and flue gas measurements with FTIR system. The coal and bed material are introduced from separate silos equipped with an electrical motor connected to the feed screw to achieve a controlled feeding rate. The rig is also equipped with primary and secondary air supply as well as solid separation cyclones. Part of the exit stream of flue gas is directed to a FTIR for gas analysis system (Gasmet dx4000). The level of bottom ash in the dense zone of CFB rig was controlled with a bottom ash screw removal system. Typically, roughly 40% of feed coal was converted to the ash, hence tests were also performed using the ash as the bed material. In addition to combustion tests, the chemical composition of the fly ash and bottom ash were examined by X-ray fluorescence and the results are presented in the paper.

Design and Proof of Concept of a Continuous Pressurized Multistage Fluidized Bed Unit for Deep Sour Gas Removal Using Adsorption

Conference: Fluidization

Year: 2019

Proceeding: Fluidization XVI

Group: General Paper Pool

Session: 5B: Fluidized Bed Applications

Time: Wednesday, May 29, 2019 - 8:30am-8:42am

Authors:

Driessen, R. T., University of Twente
Knaken, B., University of Twente
Buzink, T., University of Twente
Brilman, D. W. F., University of Twente

Design and proof of concept of a continuous pressurized multistage fluidized bed unit for deep sour gas removal using adsorption

Rick T. Driessen, Benno Knaken, Tim Buzink,
Derk W.F. Brilman

Sustainable Process Technology, Faculty of Science and Technology, University of Twente, PO Box 217, 7500 AE Enschede, The Netherlands Introduction

Adsorption is a widely used separation technology in the (petro)chemical industry. The conventional process for an adsorption process are multiple fixed beds, operated in a certain sequence to enable continuous processing while using adsorption and desorption. Although pressure swing adsorption (PSA) using fixed beds is an industrially established process, it has some disadvantages:

1. Ineffective
use of sorbent: In fixed bed adsorption only the sorbent located within the

mass transfer zone is active. The sorbent upstream and downstream of the mass transfer zone are either saturated with adsorbates or don't see any adsorbate yet. As a consequence, the effective use of the sorbent, during adsorption and at breakthrough, is relatively low resulting in large adsorption beds.

2. Slow heat transfer: When using temperature swing adsorption (TSA), heat has to be supplied to the fixed bed. However, fixed beds show slow heat transfer. This implies that (a) heating the fixed bed to the desired temperature takes a considerable amount of time or (b) additional investment is needed to provide sufficient heat transfer area. Both lead to higher costs.

This work proposes a process which overcomes these two disadvantages. First, to overcome the ineffective use of the sorbent, a continuous adsorption process could be designed in which the adsorber length matches the length of the mass transfer zone, and where the solid circulates between the adsorption zone and a desorption zone. This implies that almost every sorbent particle is either adsorbing or desorbing, minimizing the amount of sorbent in an inactive zone. Second, with respect to heat transfer, gas-solid fluidization can be applied, since fluidized beds are known for their excellent heat transfer because of the vigorous movement of the particles. This work combines the two mentioned characteristics by using multistage fluidized beds (MSFBs) to adsorb sour gas compounds (carbon dioxide and hydrogen sulfide) on supported amine sorbents (SAS) from natural gas. For this application, deep removal is needed because the pipeline specifications of natural gas are tight: <3 ppm H_2S and 1-2% CO_2 (50-150 ppm CO_2 for liquefied natural gas).

Recently, continuous adsorption processes employing MSFBs received renewed attention in view of post-combustion carbon dioxide (CO_2) capture. For a full review of these papers we refer to our earlier work, nevertheless we want to highlight some [1].
Veneman et al. use a gas-solid trickle flow reactor to adsorb CO_2 and use a multistage fluidized bed for desorption [2].
The group of Meikap investigated multistage fluidized beds experimentally with various sorbents [3,4].
The group of Hofbauer built a bench scale, ambient pressure, multistage fluidized bed and presented their first experimental results [5].
All these researchers use continuous MSFB processing with TSA to adsorb CO_2 from gases. However, to our knowledge, the use of PSA alone, let be pressure-temperature adsorption (PTSA) in a continuous sorbent process has not been investigated.

This paper presents the design of our experimental continuous, pressurized multistage fluidized bed setup for PTSA operation. Furthermore, we present some first results to prove that this process can be used for the removal of sour gases from natural gas. Experimental setup

The lay-out of the setup is too extensive to be described here in detail, but a summary will be given. The core components of the experimental setup are, in order of circulation: a MSFB adsorber (at elevated pressure), a high pressure/low pressure (HP/LP) lock to depressurize the sorbent, a MSFB desorber (at atmospheric pressure), a riser to transport the solids (at atmospheric pressure), and a low pressure/high pressure (LP/HP) lock to pressurize the sorbent. The MSFB adsorber (three stages, 50 mm inner diameter and a shallow bed height of 13 cm) is typically operated at 20-60 °C and at elevated pressures up to 10 bara. The HP/LP lock is a vessel operated in a sequential manner: (1) pressurization with nitrogen (N_2) to adsorber pressure; (2) filling with sorbent from adsorber; (3) depressurization to desorber pressure; (4) emptying the sorbent to the desorber. The MSFB desorber (seven stages, 200 mm inner diameter) is operated at approximately 100 °C and at atmospheric pressure. The riser transports the sorbent pneumatically with N_2 to a cyclone where the riser gas is separated from the sorbent. The sorbent is fed to the LP/HP lock, which is operated in a sequential manner similar to the HP/LP lock. Each main unit operation (adsorber, HP/LP lock, desorber and LP/HP lock) is equipped with a buffer vessel to facilitate a full continuous flow of sorbent.

A commercial ion exchange resin functionalized with benzylamine (Lewatit VP OC 1065, Lanxess) is used as amine sorbent. This amine sorbent is capable of adsorbing both H_2S and CO_2. Both methane, the main compound in natural gas, and N_2 do not adsorb on this amine sorbent. Therefore, N_2 is used to mimic methane because of safety reasons. A LI-COR LI840A CO_2 analyzer (0-20 000 mol ppm) is used to measure the inlet, outlet and concentrations of CO_2 per stage. [6,7] First results

The first round of experiments was conducted with mixtures of CO_2 and N_2. CO_2 concentrations of a few parts per million were achieved, while having an inlet concentration of approx. 10 000 mol ppm. Deep removal is therefore shown to be feasible possible, proving the concept of this process. In the first stage already the majority of CO_2 is adsorbed: the concentration is lowered from 9500 mol ppm to 1500 mol ppm. The second and third stages enable the deep removal of CO_2. The residence time of the gas and solid phase in the adsorber are about 6 s and 45 s respectively, underlining the fast rate of the process.

Gas-based Murphree tray efficiencies, representing the degree to which adsorption equilibrium is reached, were calculated and show high values. The measured tray efficiencies are high: generally the tray efficiency is larger than 0.8 and frequently they are higher than 0.9. This again underpins that the adsorption process is fast: the gas and the sorbent are approaching equilibrium. Conclusion

A new process for continuous adsorption is presented employing MSFBs to (a) maximize the effective use of sorbent and (b) speed up heat transfer. To our knowledge, we are the first researchers who apply a pressure swing in a continuous sorbent flow adsorption process. The first results show that the adsorption process is fast: gas and solid residence times are low, in the order of seconds, while deep removal down to a few mol ppm (over 99.8% removal) is possible. This is a prove that this process could be used for the deep removal of H_2S and CO_2 from natural gas. In addition, tray efficiencies indicate that the gas and the sorbent are near their equilibrium.

In view of the targeted application, sour gas removal from natural gas, activities are ongoing to investigate the influence of various process parameters (such as gas velocity, solid flow and inlet concentrations). In the next step, hydrogen sulfide will be introduced to mimic natural gas sweeting for sour gas mixtures. Acknowledgements

This research was carried out in the context of the Compact Advanced Sour gas Processing (CASPer) project, coordinated by the Institute for Sustainable Process Technology and co-financed by the Ministry of Economic Affairs of the Netherlands (RVO.nl project number: TEEI115008). References

[1] R.T. Driessen, M.J. Bos, D.W.F. Brilman, A Multistage Fluidized Bed for the Deep Removal of Sour Gases: Proof of Concept and Tray Efficiencies, Ind. Eng. Chem. Res. 57 (2018) 3866–3875. doi:10.1021/acs.iecr.7b04891.

[2] R. Veneman, T. Hilbers, D.W.F. Brilman, S.R.A. Kersten, CO_2 capture in a continuous gas-solid trickle flow reactor, Chem. Eng. J. 289 (2016) 191–202. doi:10.1016/j.cej.2015.12.066.

[3] D. Das, D.P. Samal, B.C. Meikap, Removal of CO_2 in a multistage fluidized bed reactor by diethanol

amine impregnated activated carbon, J. Environ. Sci. Heal. Part A. 51 (2016) 769–775. doi:10.1080/10934529.2016.1170462.

[4] S. Roy, C.R. Mohanty, B.C. Meikap, Multistage Fluidized Bed Reactor Performance Characterization for Adsorption of Carbon Dioxide, Ind. Eng. Chem. Res. 48 (2009) 10718–10727. doi:10.1021/ie901133r.

[5] F. Dietrich, G. Schöny, J. Fuchs, H. Hofbauer, Experimental study of the adsorber performance in a multi-stage fluidized bed system for continuous CO_2 capture by means of temperature swing adsorption, Fuel Process. Technol. 173 (2018) 103–111. doi:10.1016/J.FUPROC.2018.01.013.

[6] S. Sutanto, J.W. Dijkstra, J.A.Z. Pieterse, J. Boon, P. Hauwert, D.W.F. Brilman, CO_2 removal from biogas with supported amine sorbents: First technical evaluation based on experimental data, Sep. Purif. Technol. 184 (2017) 12–25. doi:10.1016/j.seppur.2017.04.030.

[7] E. Sonnleitner, G. Schöny, H. Hofbauer, Assessment of zeolite 13X and Lewatit VP OC 1065 for application in a continuous temperature swing adsorption process for biogas upgrading, Biomass Convers. Biorefinery. (2017) 1–17. doi:10.1007/s13399-017-0293-3.

Elucidating Bubble Deformation in Fluidized Beds with Vertical Tube Banks Using X-Ray Tomography and Optical Probes

Conference: Fluidization

Year: 2019

Proceeding: Fluidization XVI

Group: General Paper Pool

Session: 5B: Fluidized Bed Applications

Time: Wednesday, May 29, 2019 - 9:54am-10:06am

Authors:

Riechmann, P., Paul Scherrer Institut
Schildhauer, T. J., Paul Scherrer Institut
Schillinger, F., Paul Scherrer Institut
Wagner, E. C., Delft University of Technology
Mudde, R. F., Delft University of Technology
van Ommen, J. R., Delft University of Technology

Proper understanding of the hydrodynamics in fluidized beds is the basis for successful design and safe scale-up of such reactors. This also holds for bubbling fluidized beds for strongly exothermic reactions such as methanation reactors in Power-to-Gas applications for energy storage. In such reactors, immersed heat exchanger tubes allow removing the heat of reaction which avoids the formation of hot spots and favors higher conversions in this equilibrium limited reaction system [Schildhauer2016].Voids or bubbles play a key role: they have a positive impact on the heat transfer and the movement of catalyst particles which was shown to avoid catalyst deactivation by managing the surface coverage with carbonaceous species [Seemann2010]. On the other hand, higher gas flow rates and the inherently large bubbles can lead to mass transfer limitations [Beetstra2009] and the increase of particle elutriation [Chew2015].

To better understand the influence of heat exchanger tubes on the size, growth rates and rise velocity of bubbles, Paul Scherrer Institut and TU Delft applied a number of methods in cold flow models and pilot scale, pressurized reactors with Geldart B type materials. This way, it could be shown that vertical heat exchanger tubes avoid the local defluidisation observed between dense banks of horizontal tubes [Schillinger 2015]. Further, vertical tubes structure the fluidized bed such that the local hydrodynamics is independent of the bed diameter [vanOmmen2007, RuÌˆdisuÌˆli2012b]. Therefore, the sectorial scale-up approach enables to maintain the hydrodynamic situation when the reactor is scaled up by increasing the number of heat exchanger tubes (with constant tube diameter) with the square of the bed diameter ratio [RÃ¼disÃ¼li 2012a, Maurer 2014].

Pressure fluctuation measurements showed that bubbles grow significantly less in reactors with vertical tubes than in beds without immersed internals and also more slowly than expected from common correlations in literature [Karimipour2011]. This highlights the fact that the immersed heat exchanger tubes cause dominant â€œwallâ€ effects for which reason none of the typical equations an correlations for fluidized beds can be used a priori for beds with immersed heat exchanger tubes.

Bubble hold-up and bubble distribution in the bed as well as bubble sizes and rise velocities were investigated in detail in beds of different diameters and vertical tubes arrangements [Schillinger 2017]. Both optical sensors and X-ray tomography were applied. The former method can be applied in hot and pressurized systems, but delivers information of the local situation at the sensor tip only while the latter gives full insight of the situation over the complete cross section of the bed [Maurer 2016]. It was shown that square and concentric arrangements of vertical tube banks deliver similar results and that different bed diameters lead to nearly identical bubble properties if the number of tubes was adapted according to the sectorial scale-up approach [Schillinger 2017].

The strictly local information obtained by optical sensors has to be interpreted correctly with respect to statistical effects, because optical probes have to be used at several radial positions and pierce large bubbles with a higher probability than small bubbles, but do not necessarily pierce them at the largest chord length. It has been shown by Monte-Carlo simulations that these two errors balance out for spherical bubbles yielding in close to correct average bubble diameters [Rüdisüli 2012c]. Already for the bubble rise velocity, the higher piercing probability for larger bubbles causes a bias in the measurements which necessitates the use of statistical methods to derive the original average bubble rise velocity [Maurer 2105].

For chemical reactors, average bubble sizes and rise velocities are however not sufficient to predict the reactor performance because e.g. the mass transfer will be limited most by large and fast bubbles. Therefore, knowledge of the distributions of bubble sizes and rise velocity are needed. Moreover, none of the above mentioned statistical methods work a priori, if the bubbles are not spherical as it is the case in beds with immersed internals.

The results from a fictitious optical sensor signal synthesized from the X-ray data were compared to the results of the 3D reconstruction (two spatial dimensions in the cross section, one time dimension) based on X-ray tomography data [Schillinger 2018]. Interestingly, when the bubble rise velocities are mapped as function of the measured geometrical dimension (pierced chord lengths for optical sensors, volume-equivalent diameters for the X-ray data evaluation), both methods seem to give deviating messages. Optical sensors measure high bubble rise velocities for bubbles with high pierced chord lengths which fits to typical bubble rise velocity correlations where the rise velocity increases with the square root of the bubble diameter . Evaluation of the reconstructed X-ray data show, however, smaller velocities for large-volume bubbles and both fast and slow bubbles with smaller volume, but hardly fast bubbles with large volume [Maurer 2016, Schillinger 2017].

The seeming inconsistency of both methods could now be explained by plotting the rise velocities of bubbles as function of their aspect ratio (largest cross section compared to maximum length in axial direction). While fast bubbles are slender with high aspect ratios (leading to high pieced chord length even with small volume), slow bubbles have a tendency to low aspect ratios which causes smaller pierced chord lengths even at high bubble volumes.

This shows that in systems far from standard assumptions such e.g. spherical bubbles, strong deviations from common correlations have to be expected which necessitates the combined use of several measurement methods to obtain the full picture. This then opens a pathway to derive design approaches also for non-standard reactor systems.

[Beetstra2009] Beetstra, R., Nijenhuis, J., Ellis, N., & van Ommen, J.R. (2009). The influence of the particle size distribution on fluidized bed hydrodynamics using highâ€throughput experimentation. *AIChE journal*,*55*(8), 2013-2023.

[Chew2015] Chew, J. W., Cahyadi, A., Hrenya, C. M., Karri, R., & Cocco, R. A. (2015). Review of entrainment correlations in gasâ€“solid fluidization. *Chemical Engineering Journal, 260*, 152-171.

[Karimipour2011] Karimipour, S., & Pugsley, T. (2011). A critical evaluation of literature correlations for predicting bubble size and velocity in gasâ€“solid fluidized beds. Powder Technology, 205(1-3), 1-14.

[Maurer 2014] Maurer, S., Schildhauer, T. J., van Ommen, J.R., Biollaz, S. M., & Wokaun, A. (2014). Scale-up of fluidized beds with vertical internals: Studying the sectoral approach by means of optical probes. Chemical Engineering Journal, 252, 131-140.

[Maurer 2015] Maurer, S., RÃ¼disÃ¼li, M., Teske, S.L., Schildhauer, T.J., van Ommen, J.R., Biollaz, S. M., & Wokaun, A. (2015). Transformation of local bubble rise velocity measurements to global results: Shown by a Monte Carlo simulation of a fluidized bed. International Journal of Multiphase Flow, 70, 89-95.

[Maurer 2016] Maurer, S., Gschwend, D., Wagner, E.C., Schildhauer, T.J., van Ommen, J.R., Biollaz, S.M., & Mudde, R.F. (2016). Correlating bubble size and velocity distribution in bubbling fluidized bed based on X-ray tomography. Chemical Engineering Journal, 298, 17-25.

[RÃ¼disÃ¼li 2012a] RÃ¼disÃ¼li, M., Schildhauer, T.J., Biollaz, S. M., & van Ommen, J.R.

(2012). Evaluation of a sectoral scaling approach for bubbling fluidized beds with vertical internals. Chemical engineering journal, 197, 435-439.

[RuÌˆdisuÌˆli2012b] RuÌˆdisuÌˆli, M., Schildhauer, T.J., Biollaz, S. M., & van Ommen, J.R. (2012). Radial bubble distribution in a fluidized bed with vertical tubes. Industrial & Engineering Chemistry Research, 51(42), 13815-13824.

[RÃ¼disÃ¼li 2012c] RÃ¼disÃ¼li, M., Schildhauer, T. J., Biollaz, S. M., & van Ommen, J.R. (2012). Monte Carlo simulation of the bubble size distribution in a fluidized bed with intrusive probes. International Journal of Multiphase Flow, 44, 1-14.

[Schildhauer2016] Schildhauer T.J. (2016). Methanation for SNG production â€“ Chemical reaction engineering aspects. In Synthetic Natural Gas from Coal, Dry Biomass, and Power-to-Gas Applications, T.J. Schildhauer, S.M.A. Biollaz (Eds.), Wiley & Sons, New York, 77-159

[Schillinger 2015] Schillinger, F., Maurer, S., Wagner, E.C., Biollaz, S.M., van Ommen, J.R., Mudde, R.F., Schildhauer, T.J. (2015). Influence of the internals Orientation on the Hydrodynamics in Gas-Solid Fluidised Beds, Proceedings of the 12th International Conference on Fluidized Bed Technology.

[Schillinger 2017] Schillinger, F., Maurer, S., Wagner, E.C., van Ommen, J.R., Mudde, R.F., Schildhauer, T.J. (2017). Influence of vertical heat exchanger tubes, their arrangement and the column diameter on the hydrodynamics in a gasâ€“solid bubbling fluidized bed. International Journal of Multiphase Flow 97, 46-59 (2017)

[Schillinger 2018] Schillinger, F., Schildhauer, T.J., Maurer, S., Wagner, E., Mudde, R.F., & van Ommen, J.R. (2018). Generation and evaluation of an artificial optical signal based on X-ray measurements for bubble characterization in fluidized beds with vertical internals. International Journal of Multiphase Flow.

[Seemann 2010] Seemann, M. C., Schildhauer, T.J., & Biollaz, S.M.(2010). Fluidized bed methanation of wood-derived producer gas for the production of synthetic natural gas. Industrial & engineering chemistry research, 49(15), 7034-7038.

[vanOmmen2007] van Ommen, J.R., Nijenhuis, J., van den Bleek, C. M., & Coppens, M. O. (2007). Four ways to introduce structure in fluidized bed reactors. *Industrial & Engineering Chemistry Research*, *46*(12), 4236-4244.

Leaching of Potassium from Rock Ilmenite Used As a Bed Material in Fluidized Bed Boiler Performed Under Acidic and Alkaline Conditions

Conference: Fluidization

Year: 2019

Proceeding: Fluidization XVI

Group: General Paper Pool

Session: 5B: Fluidized Bed Applications

Time: Wednesday, May 29, 2019 - 9:42am-9:54am

Authors:

Krymarys, E., Chalmers Univeristy of Technology

Leaching of potassium from rock ilmenite used as a bed material in fluidized bed boiler performed under acidic and alkaline conditions
Ewa Krymarys, Angelica GyllÃ©n, Pavleta Knutsson, Fredrik Lind, Magnus RydÃ©n

Oxygen Carrier Aided Combustion (OCAC) is a new concept for fluidized bed combustion of biomass and waste fuels, where the semi-inert particles typically used as bed material are replaced with chemically active oxygen carrying particles. The oxygen carrier is capable of absorbing oxygen from oxygen-rich zones and release it in zones that are oxygen-depleted, thereby evening out the oxygen fugacity in space and time. The concept has shown promising results and has been implemented on industrially relevant scales from 12 â€“ 75 MW_{th}. Oxygen carriers particles should have high reactivity with fuel and air, be resistant to attrition, have low tendency to agglomerate, be environmentally- benign and have low cost. The most studied oxygen carrier material for OCAC so far has been ilmenite, which is a common mineral ore consisting mainly of titanium and iron oxides. Apart from being a good oxygen carrier ilmenite has been shown to act also as an ash absorber. Previous studies with ilmenite as oxygen carrier in OCAC have shown that there is distinctive elements patterns for particles that has been applied in the combustion process, where iron migrates to the surfaces, titanium enrichment occurs in the core and a calcium-rich double layer covers the formed iron layer. Potassium has been observed to diffuse into the core forming KTi_8O_{16}.

Perpetual absorption of ash on bed material is not possible. Bed agglomeration arises when ash layers are formed around the bed material particles. Potassium compounds are of low melting point, thus especially prone to evaporate at an early stage during the combustion. Evaporation enhances formation of sticky potassium silicates on the surfaces of bed material particles, which pick up on other ash components such as calcium. In both cases whether potassium stays as a layer or migrates to the core, there is an urge to discover the possible ways to leach the potassium out â€“ to avoid possible agglomeration and regain the capability of potassium-capturing.

The present work involves Norwegian rock ilmenite which has been used as bed material during combustion of wood chips in the Chalmers campus 12-MW_{th} biomass-fired Circulating Fluidized Bed (CFB) boiler. The aim was to investigate leaching of potassium from ilmenite samples extracted directly from the boiler during operation. If certain amounts of potassium could be removed from the ilmenite particles under moderate process conditions this would allow for recirculating bed material back to the boiler, significantly reducing the cost and environmental footprint of OCAC. Leached potassium would also represent an additional by-product stream for the process.

Samples of bed material were extracted at different time intervals (after 1 day and 3-4 days of operation) directly from the boiler furnace. Consequently, samples were leached both under acidic (pH=2,5) and alkaline (pH=12) conditions by means of acetic and phosphoric acids both at room (21°C) and elevated temperatures (70, 80°C). The results have been interpreted in terms of optimal conditions for leaching alkali, changes in morphology and elemental content of the bed material and influence on its reactivity for further reuse. Based on the obtained results from a flame atomic absorption spectroscopy (AAS), scanning electron microscopy (SEM) analysis combined with electron dispersive X-ray (EDX) the possible implementation of leaching as an industrially applied step to the OCAC can be discussed.

Modelling of Autothermal Chemical Looping Reforming for Syngas Production

Conference: Fluidization

Year: 2019

Proceeding: Fluidization XVI

Group: General Paper Pool

Session: 5B: Fluidized Bed Applications

Time: Wednesday, May 29, 2019 - 8:42am-8:54am

Authors:

He, Z., Materials and Process Engineering (IMAP), Universite catholique de Louvain (UCL), Louvain-la-Neuve
De Wilde, J., Université Catholique de Louvain (UCL)
Minette, F., Universite catholique de Louvain (UCL)

Hydrogen as well as the syngas (the mixture of CO and H_2) are important industrial raw materials widely used in chemical synthesis, ammonia production and the petroleum industry. The conventional steam methane reforming (SMR) produces about 75% of worldâ€™s total hydrogen but emits about 7 kg of CO_2 per kg H_2 on average, which is responsible for around 3% of the worldwide CO_2 emissions. Since the greenhouse gas emission is becoming more and more problematic, it is necessary to introduce new technologies that are environmentally friendly, low in energy consumption, and high in hydrogen efficiency. Chemical Looping Reforming (CLR) is an emerging technology that potentially reduce the CO_2 emissions and minimize the energy losses in syngas production. CLR shares the same principle as chemical looping combustion (CLC) in which the metal oxygen carrier (OC) is alternatingly oxidized and reduced using a dual-fluidized bed reactor that act as air reactor (AR) and fuel reactor (FR), respectively. The OC particles is oxidized by air in AR, while in FR, the OC is reduced by the methane and syngas while acting as catalyst for methane reforming. Accurate predication of the methane conversion and syngas composition is crucial for the design of industrial scale CLR unit. In this work, a detailed 1-D model coupling fluidized bed hydrodynamics based on Kunii-Levenspiel two-region model with intrinsic reaction kinetics including catalyst deactivation was developed. The back-mixing of solids in BFB is accounted in this model. This model can be used for process optimization and for the prediction of the performance of commercial scale a-CLR unit. An autothermal chemical looping reforming (a-CLR) unit that equivalent to 50 commercial SMR pipes is presented as an example in this paper. The required reactor design to reach a target reactor performance was studied. The effects of the main operating conditions on the performance of both the AR and the FR were analyzed. These results confirm the feasibility of the chemical looping reforming process with reactors of reasonable size for industrial applications.

We highlighted that the main difference between a-CLR and CLC is that, in the a-CLR, the oxygen carrier (OC) is mainly at reduced state in the FR, while it is only slightly oxidized (about 5% Ni is oxidized) in the AR and the autothermal operation can be guaranteed. Conventional well-mixed reactor assumption is not valid due to the low NiO-CH_4 ratio in the a-CLR unit. Thus, 1-D axial dispersion model should be used to model solid phase. The required oxygen to CH_4 ratio to maintain autothermal operating is about 1.14 if heat is sufficiently recovered from the off-gas of both the AR and the FR. The temperature difference between two reactors has less influence on the required oxygen to CH_4 ratio to maintain autothermal operation while the small difference with respect to oxygen to CH_4 ratio is resulted from energy loss of the off-gas of both reactor. The operating temperature of FR and interfacial transfer of the methane from the bubble phase to the emulsion phase limit the conversion of CH_4, while the former also affects the conversion of OC in AR due to higher oxidation rate at higher temperature. With appropriate reactor design, near equilibrium conversion can be achieved.

Production and Stability Assessment of Oxygen Carrier Produced By Sewage Sludge Fluidized Bed Combustion

Conference: Fluidization

Year: 2019

Proceeding: Fluidization XVI

Group: General Paper Pool

Session: 5B: Fluidized Bed Applications

Time: Wednesday, May 29, 2019 - 9:30am-9:42am

Authors:

Coppola, A., University of Naples Federico II
Ruoppolo, G., Italian National Research Council
Urciuolo, M., Istituto di Ricerche sulla Combustione
Chirone, R., Consiglio Nazionale delle Ricerche

CO_2 concentration has increased significantly after the industrial revolution owing to the anthropic emissions caused by the use of fossil fuels. This has been determined by an increase in the greenhouse effect and, consequently, the rising of the temperature of the earthâ€™s surface which is known as â€œglobal warmingâ€. In this scenario, the development of renewable energies, the energy consumption reducing, and the energy efficiency optimization, represents a valid and definitive middle-long term solution to preserve our planet.

Despite many renewable technologies reached a good level of maturity from both technical and economical point of view, these suffer of some limitations which still make our society fossil fuel-dependent. These limitations are represented by the intrinsic intermittent nature of the renewable systems and by the inability to fulfil the total energy demand; this entails that the fossil sources will be predominant for many decades. The Carbon Capture and Storage (CCS) or Utilization (CCU) technologies represent a collection of several processes for an eco-friendly employment of the fossil fuels.

Among the CO_2 capture techniques, the Chemical Looping Combustion (CLC) is one of the most promising technology, which allows to burn fuels without direct contact between fuel and oxygen in the air.

The key aspect for this process is the utilization of oxygen carriers (OC) (typically metal oxides) that undergoes two different process: the first is the fuel oxidation; the second one is the re-oxidation of the OC. These processes can be carried out by using two reactors, typically two interconnected fluidized beds, named respectively Air Reactor (AR) e Fuel Reactor (FR). In the AR the reduced OC, coming from the FR, is regenerated (re-oxidized) by the air. In the FR the fuel is burned with the oxygen supplied by the OC producing a flue gas mainly composed by CO_2 and H_2O.

However, the choice of OC is a complex aspect, because its characteristics must meet well defined physical-chemical properties such as high reactivity, high mechanical strength, and remarkable oxygen transport capacity (1,2). The cost of an OC is a very crucial feature, in particular for the synthetic materials, being the sum of several factors such as the cost of the metal oxide and of the inert support, and the cost of the manufacturing process. This latter is rather low when industrial methods are used and the final cost depends mainly on the price of the raw material.

For this reason, most research is focused on the production of low-cost OC starting from natural ores or reusing waste materials from other activities (3).

As matter of fact, the goal of this work is to develop low-cost OC by combustion of sewage sludges. These sludges, deriving from purification treatment of civil wastewater, contain a large amount of metals, in particular iron, calcium and manganese. The OCs have been prepared by depositing the metals contained in the sludge ash on a support made of high area Î³-Alumina (Al_2O_3), used as bed material during the sewage sludge fluidized bed combustion. Three different combustion tests have been carried out in a 41 mm ID reactor. For each test, different operative conditions were tested, in particular: the fluidization velocity and the particle size, in order to find optimal conditions for the preparation of the OCs. Each test lasted 3 hours, and at predetermined times a small amount of bed material has been sampled, to assess the role played by the duration of the combustion process on the amount of ashes deposited on the alumina particles. These samples have been characterized by TPR (temperature programmed reduction) analysis, carrying out reduction tests with a mixture of 2% H_2/Ar, and a temperature ramp of 10°C/min up to 850°C, to study the reactivity of the carriers and to evaluate the amount of reducible material deposited on the support. The TPR analysis showed that hydrogen consumption increases, increasing the residence time of the samples into the combustor, confirming the progressive accumulation of reducible material (mainly iron oxides) on the support. Moreover, some reduction/oxidation cycles have been performed to evaluate the stability of the OC. In general, material with a higher size (1000 μm) showed better stability during the cycles, on the contrary, smaller particles (400-600μm) exhibited a decay of their reductive activity of about 70% after only 4 cycles.

Moreover, further and deeper studies on the performance of the prepared OCs have been performed using a lab-scale fluidized bed apparatus properly designed for the study of chemical looping processes as close as possible to reality, in terms of cycling of temperatures and of reaction atmospheres. This apparatus, named Twin Beds (TB) (4), consists of two identical bubbling fluidized beds that can be operated separately in batch mode. However, the two reactors are connected each other by a duct, which permits the transfer of the OC from on reactor to the other and vice-versa. The transportation is carried out pneumatically, by the utilization of a suitable valve system placed along the duct and at the outlet of both reactors.

The two reactors were employed as FR and AR respectively, and the OCs have been tested for 10 cycles of oxidation/reduction. The initial amount of OC for the looping tests was ca. 10g. The Oxidation step has been carried out at 850°C at a fixed fluidization velocity of 0.5m/s for 10 min, using methane at 2% in volume as fuel in a stream of nitrogen. The Reduction step has been carried out at the same temperature, fluidization velocity and time of the Oxidation step, but the fluidizing gas was air, in order to regenerate the reduced OC. The progress of reactions was followed by monitoring the CH_4, CO and CO_2 gas profiles at the outlet of the reactors by gas analyzers.

The activity of the OC with methane is, as expected, lower respect to hydrogen. Additionally, these tests confirm that the oxidation capacity decays with the number of the cycles and reaches an asymptotic value after the 6th cycle.

1. AdÃ¡nez J, Abad A, Mendiara T, GayÃ¡n P, de Diego LF, GarcÃa-Labiano F. Chemical looping combustion of solid fuels. Prog Energy Combust Sci. 2018;65:6â€“66.
2. Nandy A, Loha C, Gu S, Sarkar P, Karmakar MK, Chatterjee PK. Present status and overview of Chemical Looping Combustion technology. Renew Sustain Energy Rev [Internet]. 2016;59:597â€“619. Available from:http://dx.doi.org/10.1016/j.rser.2016.01.003
3. Ksepko E. Sewage sludge ash as an alternative low-cost oxygen carrier for chemical looping combustion. J Therm Anal Calorim. 2014;116(3):1395â€“407.
4. Coppola A, Scala F, Gargiulo L, Salatino P. A twin-bed test reactor for characterization of calcium looping sorbents. Powder Technol [Internet]. 2017;316:585â€“91. Available from:http://dx.doi.org/10.1016/j.powtec.2016.11.067

Strategy for Coating of Aerogels Using Spouted Bed Technology

Conference: Fluidization

Year: 2019

Proceeding: Fluidization XVI

Group: General Paper Pool

Session: 5B: Fluidized Bed Applications

Time: Wednesday, May 29, 2019 - 9:18am-9:30am

Authors:

Goslinska, M., Hamburg Univeristy of Technology
Heinrich, S., Hamburg University of Technology

In this work a strategy for the coating of open-pore structured, nanoporous materials with low density is developed. As model material organic aerogel particles are used, which are applied as matrix material for active compounds in food or pharmaceutical products due to their structural properties. These kind of particulate porous solids have to be coated with polymeric films for protective, functional or visual purposes. The coatings can exemplary be used to stabilize the impregnated drug, to protect against the influence of light, oxygen or water penetration during storage or to mask taste and odor. Especially in case of porous aerogels, the exposition of the porous network to water during storage or to gastrointestinal tract fluids during ingestion leads to change or destruction of their structure due to high capillary forces. Therefore, the aerogel particles need to be protected against environmental influence with a secondary coating layer. The functionalization of light-weighting aerogels in µm- to mm-range by film coating is performed using a novel in-house built prismatic spouted bed apparatus. This technique is excellent for coating of particles with dense, thin and uniform layers enabling the production of sustainable particles, which provide new options for food and pharmaceutical applications.

For the coating of aerogels two different types of materials are used: solutions and melts. Both can affect the porous structure of the particles during coating differently and require various process conditions. In case of using solutions, penetration of the pores due to the low-viscosity solvent can take place before its evaporation and thus, lead to collapse of the network. Additionally, evaporation of the solvent and consequently lower concentration of the coating material require longer processing times to get the same film thickness compared to melts. The application of melts requires higher process temperatures, which can destroy the particles due to the thermal stress, but too low temperature causes too fast solidification and formation of solid bridges and agglomerates. Nevertheless, the influence of process parameters like viscosity, bed temperature, droplet size or spray rate on the quality of the aerogel particles and uniformity of the coating are investigated. Therefore, the coated particles are cross-sectioned using focused ion beam to evaluate the interface between particle and coating layer. Based on these investigations, the question, which kind of material at which process conditions is qualified for coating of aerogels to protect the porous network and form shell-core structure, is answered.

During experimental investigations viscosity, contact angle, strength of the solidified films and water vapour permeability of the solidified films made of melts and coating solutions are characterized. The influence of different process parameters during coating, like nozzle air and coating material flows or droplet sizes on the coating layer, is investigated. For this purpose particle growth and coating layer quality are measured. The layer thickness of coated particles is determined by cross-sectioning using focused ion beam (FIB) instrument. The uniformity of the layer is indicated by measurement of specific surface area of treated and untreated particles.

Additionally, coupled CFD (Computational Fluid Dynamics) - DEM (Discrete Element Method) simulations are performed to investigate the stability of the spouting process and optimize the coating parameters. These include determination of the residence time of the particles in the spray zone and its influence on coating layer thickness.

We gratefully acknowledge for the financial support: German Research Foundation (DFG). Project number HE 4526/24-1.

CFD Simulation of Wet Fluidized Beds Using TFM with Variable Particle-Particle Restitution Coefficient

Conference: Fluidization

Year: 2019

Proceeding: Fluidization XVI

Group: General Paper Pool

Session: 5C: Modeling and Simulation

Time: Wednesday, May 29, 2019 - 9:43am-9:55am

Authors:

Xiong, Q., General Motors
Zhong, H., Xi'an Shiyou University

CFD simulation of wet fluidized beds using TFM with variable particle-particle restitution coefficient

Hanbin Zhong[a], Qingang Xiong[b,*], Juntao Zhang[a], Yuqin Zhu[a], Shengrong Liang[a], Xinyu Zhang [a]

[a] School of Chemistry and Chemical Engineering, Xi'an Shiyou University, Xi'an, Shaanxi 710065, China

[b] IT Innovation Center, General Motors, Warren, MI 48092, USA

Abstract

For the wet fluidized beds, the wet restitution coefficients are different with the particle-particle restitution coefficients of dry particles, and affected by the liquid viscosity, liquid layer thickness, and impact velocity. The two-fluid model

(TFM) is widely used in the CFD simulation of fluidized beds due to the relatively low computational cost. However, a constant value for the particle-particle restitution coefficient is normally used to describe the particle-particle interaction in the simulation with TFM, while the previous reports indicated that the predicted hydrodynamics were significantly affected by the value of particle-particle restitution coefficient. Therefore, an advanced TFM with variable particle-particle restitution coefficient was developed to describe the actual wet restitution coefficients in the wet fluidized beds. The elastohydrodynamic model proposed by Davis, et al. (Elastohydrodynamic rebound of spheres from coated surfaces, J. Fluid Mech., 2002, 468, 107-109.) was used to calculate the wet restitution coefficient based on the liquid viscosity, liquid layer thickness, impact velocity, and particle properties. The hydrodynamics of the dry and wet particles were predicted and compared with the experiments. The effects of the liquid viscosity and liquid layer thickness were also investigated.

Keywords:

Two-fluid model; restitution coefficient; wet particle; fluidized bed; CFD

Details on the relationship between the liquid viscosity and liquid layer thickness and the effective wet restitution coefficients

Davis, et al. (Elastohydrodynamic rebound of spheres from coated surfaces, J. Fluid Mech., 2002, 468, 107-109.) have proposed an elastohydrodynamic model to determine the wet restitution coefficient:

$$e_{wet} = \begin{cases} e_{dry}\left(1 - \dfrac{St_c}{St}\right) & St > St_c \\ 0 & St < St_c \end{cases}$$

where e_{dry} and e_{wet} are the dry and wet restitution coefficient, respectively.

The Stokes number is calculated by the following equation:

$$St = \frac{mV_c}{6\pi\mu_l r_p^2}$$

where m and r_p
are the particle mass and radius, respectively. V_c is the impact velocity. μ_l is the liquid viscosity.

The
critical Stokes number St_c is given by:

$$St_c = 0.40 \ln\left(\frac{1}{\varepsilon}\right) - 0.20$$

where ε is
the
dimensionless elasticity parameter calculated as:

$$\varepsilon = \frac{4Y_{eff}\mu_l V_c r_p^{3/2}}{z_0^{5/2}}$$

z_o
is the distance where the viscous forces become significant:

$$z_0 = \frac{2}{3}\delta_l$$

where δ_l
is the liquid layer thickness.

Y_{eff}
is the effective Young's modulus determined by:

$$Y_{eff} = \frac{(1-v_1^2)}{\pi Y_1} + \frac{(1-v_2^2)}{\pi Y_2}$$

where
v_1 and v_2 are Poisson's ration
for the collision particle/plane, Y_1 and Y_2
are the effective Young's modulus for the collision particle/plane.

Euler-Euler Simulation of a Bubbling Fluidized Bed Gasifier: Development of an Intrinsic Particle Conversion Model

Conference: Fluidization

Year: 2019

Proceeding: Fluidization XVI

Group: General Paper Pool

Session: 5C: Modeling and Simulation

Time: Wednesday, May 29, 2019 - 9:07am-9:19am

Authors:

Rößger, P., TU Bergakademie Freiberg
Richter, A., TU Bergakademie Freiberg
Meyer, B., TU Bergakademie Freiberg

The efficiency of fluidized bed gasifiers is strongly linked to the conversion of the solid feedstock. The operating temperature of industrial fluidized bed reactors is limited below ash softening temperature leading to incomplete carbon conversion. To achieve a full carbon conversion, present fluidized bed reactor concepts need to be improved. Computational Fluid Dynamic (CFD) modeling is a time-efficient tool for advanced process development. However, the modeling of solid-gas reactions in fluidized beds is still at the state of current research activities. In the present work an intrinsic conversion model is developed to simulate the carbon conversion in a bubbling fluidized bed gasifier. Char as bed material is fluidized using a mixture of CO_2 and N_2. Due to the reaction of carbon with CO_2 the porosity of the particles increases resulting in a reduction of the particle density, while the particle diameter keeps nearly constant. For that reason the conversion model assumes that the particle size is constant, and the heterogeneous reactions increase the porosity of the particle and decrease its density. The numerical gasifier model combines the new conversion model with the Euler-Euler approach. Numerical results of the shrinking density model are compared against the commonly used shrinking core model for a bubbling fluidized bed gasifier. A parametric study is presented for different process conditions and the influence of the conversion model on the hydrodynamics is discussed.

Evaluation of a Dual-Grid Method for Multiscale TFM and CFD-DEM Simulations of Dense Gas-Solid Flows

Conference: Fluidization

Year: 2019

Proceeding: Fluidization XVI

Group: General Paper Pool

Session: 5C: Modeling and Simulation

Time: Wednesday, May 29, 2019 - 9:31am-9:43am

Authors:

Hirche, D., Technical Universityof Munich
Hinrichsen, O., Technical University of Munich

Simulations of chemical reactors using CFD has become more popular over the last years. Dense gas-solid flow simulations can be achieved through several methods. The most common simulation approaches are the Eulerian-Eulerian, Eulerian-Lagrangian model and the Direct Numerical Simulation (DNS). The Direct Numerical Simulation is the most computationally demanding model as the particle surface is depicted as a boundary during the simulation. Therefore, this model is only suitable for small scale geometries with a small number of particles. With the Eulerian-Lagrangian approach the fluid phase is modeled in a Eulerian way, while the particles are tracked individually. This approach consists of several sub-models, e.g., Multiphase Particle-In-Cell (MP-PIC) [1] or Discrete Element Model [2], often called CFD-DEM. The focus for the Euler-Lagrangian approaches will be on CFD-DEM in this work. This approach is not as computationally demanding as the DNS, but is limited by high numbers of particles. The Eulerian-Eulerian approach is the computationally least demanding method for dense gas-solid simulations compared to the above mentioned models. In this approach the fluid is modeled according to a Eulerian model and the solid phase is regarded as a pseudo fluid and simulated the same way as the fluid. This model is often called Two Fluid Model (TFM) [3]. The particles are not resolved individually, as it is done using CFD-DEM or DNS, but the particles are regarded as a solid phase in a cell with a certain volumetric solid fraction. In order to account for the particulate behavior of the pseudo fluid, i.e. the solid phase, further models have to be considered.

The Kinetic Theory of Granular Flow (KTGF) [4] is widely applied for TFM in dense gas-solid flow simulations. In this theory the particles are modeled as a fluid. Collisions between particles are approximated with statistical approaches. To achieve a high accuracy of the collision forces, a large number of particles in a cell is needed for the calculation of the statistical equations. The finite volume method discretizes the computational domain in three-dimensional cells. To satisfy the above stated conditions of a large number of particles in a cell, the particles have to be either very small compared to the cell size or the grid must be coarsened. Uddin et al. [5] suggest a grid size of 18 times the particle diameter (Geldart B particles) for grid independence. A coarse mesh, however, might not be able to resolve occurring bubbles or in 3D cases the fluid flow accurately, while the particle collisions might be modeled accurately.

Simulations of dense gas-solid flows using the CFD-DEM approach model the fluid as a continuum according to the Eulerian approach, while the solid phase is considered as discrete particles. Each particle is tracked with the equations of Newtonâ€™s Second Law of Motion. The coupling between the fluid and solid phase is done with the calculation of cell averaged forces between the fluid and particles. Important parameters influencing these forces are for example the solid phase fraction and the solid phase velocity. Particles which volumes are overlapping multiple cells are troublesome for the calculation of cell averaged values [6], as for example the solid phase fraction is calculated by dividing the total volume of all particles in a cell by the volume of the cell. Most approaches use the center point of a particle to determine if a particle is within a certain cell. At the worst, i.e., if the center point of a spherical particle is in the corner of a cubic cell, a maximum volume fraction of the particle of 87.5 % is outside of the cell but still accounted for the calculation of the cell-averaged solid volume fraction. To minimize the magnitude of error for this issue, the cell-to-particle volume ratio has to be great. In the literature a value of at least two is suggested [7]. Similar to the TFM, the cell size can be scaled up to overcome this problem with the disadvantage of inaccuracies regarding the fluid flow. Also, simulations of smaller particles are preferable according to this. But with smaller particles the number of total particles often increases as well. A larger amount of particles might result to unfeasible computational times as the particles are simulated with a discrete method.

A dual-grid method for TFM and CFD-DEM simulations was developed to overcome the problem of grid resolution for the fluid and solid phase. With this method the fluid phase and the solid phase are calculated on separate computational grids. The fluid grid consists of a finer grid to resolve the flow structure. The solid grid is a coarser grid. CFD-DEM and TFM simulations require a coarser grid to model the particulate phase with a high accuracy. Parameters which are necessary for the calculation of closure terms, e.g., momentum exchange or cell-averaged volume fraction of the solid phase, are mapped between those grids in a separate inner loop of the time-step until convergence is reached for each transient time-step. TFM and CFD-DEM simulations with the dual-grid method were performed with lab-scale and small pilot-scale geometries found in literature. The grid sizes were varied for the fluid and the solid mesh. All simulation results performed with the dual-grid model are compared to conventional TFM and CFD-DEM solvers. Furthermore, the simulation times for all cases are evaluated to determine the computational efficiency of the dual-grid method compared to the conventional solvers.

References

[1] M. Andrews and P. O'Rourke, Int. J. Multiphase Flow (1996), 22, 379â€“402.

[2] P. A. Cundall and O. D. L. Strack, GÃ©otechnique, (1979), 29, 47â€“65.

[3] Y. Tsuji et al., Powder Technol. (1993), 77, 79â€“87.

[4] J. Ding and D. Gidaspow, AIChE J. (1990), 36, 523â€“538.

[5] M. H. Uddin et al., Particuology (2017), 34, 61-69.

[6] D. Hirche et al., Chem. Eng. J. (in press).

[7] Z. Zhou et al., J. Fluid Mech. (2010), 661, 482-510.

Fluid-Particle Drag in Low-Reynolds-Number Bidisperse Suspensions

Conference: Fluidization

Year: 2019

Proceeding: Fluidization XVI

Group: General Paper Pool

Session: 5C: Modeling and Simulation

Time: Wednesday, May 29, 2019 - 10:19am-10:31am

Authors:

Duan, F., Xi'an Jiaotong University
Chen, X., Xi'an Jiaotong University
Zhou, Q., Xi'an Jiaotong University

A second-order immersed boundary-lattice Boltzmann method (IB-LBM) is used to study the drag force of freely evolving bidisperse suspensions at low Reynolds numbers. The total solid volume fraction of the suspension is varied between 0.1 and 0.4, and the particle size ratio from 1:1.2 to 1:4. The main characteristics of freely evolving suspensions, compared to fix beds that were well reported in the literature, is that the two dispersed phases have different slip velocities relative to the fluid phase. Hence, in freely evolving bidisperse systems, the drag force on each particle phase is normalized by the Stokes drag force based on their corresponding slip velocity. Surprisingly, it is found that the drag force in freely evolving bidisperse suspensions is significantly different from previous results based on computations of systems with static particles. The drag force on small particles increases as the particle size ratio decreases, which is utterly opposite to that obtained from stationary systems. This difference is believed to be caused by the inhomogeneous structures formed by the particles in freely evolving systems, even though the computational domains are very limited. It is also found that the difference between the drag force on each particle phase in dynamic systems and that in stationary systems correlates well with the relative velocity of the two particle phases. Therefore, based on the present simulation results, a new drag force relation is proposed for bidisperse suspensions by including the relative velocity of the two particle phases.

Interactive Realtime Simulation of a Lab-Scale Circulating Fluidized Bed - an Example of Virtual Process Engineering

Conference: Fluidization

Year: 2019

Proceeding: Fluidization XVI

Group: General Paper Pool

Session: 5C: Modeling and Simulation

Time: Wednesday, May 29, 2019 - 8:55am-9:07am

Authors:

Ge, W., Institute of Process Engineering, Chinese Academy of Sciences

Virtual process engineering refers to the research and development of chemical processes using computer simulation a virtual-reality style. That is, the simulations should run at realtime or quasi-realtime speed so that the users can actually operate the simulated system using a virtual reality interface. Such a computational speed is, however, not easily obtained in traditional simulation methods and computer hardware. In this talk, a coarse-grained discrete particle method (CG-DPM) based on the energy minimization multi-scale (EMMS) model, EMMS-DPM, is introduced briefly, together with an efficient parallel computing strategy using CPU(central processing unit)-GPU(graphics processing unit) heterogeneous architecture, which enables interactive simulation of a pilot-scale circulating fluidized bed (CFB). With this capability, hours of physical time may only take days to cover in the simulations and some slow processes, such as particle agglomeration and attrition, deactivation of catalysis and wall wearing, are feasible for direct simulation. This capability also enables detailed dynamic simulation for a network of reactors instead of a single one, suggesting a revolutionary tool for fundamental research, designing, optimization and training in chemical engineering.

Keynote Talk: Modeling of Reacting Flows and industry Applications: Some Examples

Conference: Fluidization

Year: 2019

Proceeding: Fluidization XVI

Group: General Paper Pool

Session: 5C: Modeling and Simulation

Time: Wednesday, May 29, 2019 - 8:30am-8:55am

Authors:

Shen, Y., University of New South Wales

Process design and control plays a significant role in modern industries. Most processes and reactors are very complex, as they usually involve not only multiphase flows but also heat and mass transfers related to chemical reactions and their interactions. The operation must be optimized in order to be competitive and sustainable, particularly under the more and more demanding economic and environmental conditions. This will need continuous innovative research and development. Computer simulation and modelling, supported by experiments, has emerged as an indispensable adjunct to the traditional modes of investigation for design, control and optimization of processes, reactors, and devices. In this presentation, Dr Shen will report his core research on process modelling and design of reacting flows and the applications to a range of complex processes and reactors in conventional and emerging industries. Several examples of industry applications will be used for demonstration, including process metallurgy, coal preparation and utilization. The modelling works are indeed helpful to understand fundamentals and optimize & develop new, cleaner and more efficient technologies with measurable industrial outcomes.

Large-Scale Direct Numerical Simulation for Investigating the Mesoscale Structure in Gas-Solid Flow

Conference: Fluidization

Year: 2019

Proceeding: Fluidization XVI

Group: General Paper Pool

Session: 5C: Modeling and Simulation

Time: Wednesday, May 29, 2019 - 9:19am-9:31am

Authors:

Wang, L., Chinese Academy of sciences
Ge, W., Institute of Process Engineering, Chinese Academy of Sciences

A coupled Lattice Boltzmann Method and Discrete Element Method (LBMâ€“DEM) approach is usually a kind of particle-resolved direct numerical simulation (DNS) algorithms for modeling gas-solid two-phase flows, in which the size of fluid grid is generally one magnitude smaller than particle diameter and force acting on particles directly calculated by integrating both viscos force and pressure gradient force on the particleâ€™s surface. It has been regarded as the most accurate numerical method for simulation of gas-solid flow. However, the main disadvantage is its huge computational cost resulting from small grid size and time step limited by Kolmogorov length and time scales. Only hundreds of particles scale is reported for DNS of gas-solid flow in the latest literature, which is very different from the number of particles in the real gas-solid flows. In order to solve the problem of computational speed and scale, an immersed boundary method in framework of LBM has been adopt to realized the fluid-solid coupling to avoid a stair-step representation of the solid particlesâ€™ surfaces (Wang et al., 2010; Zhou et al., 2011) and the multi graphics processor units (GPUs) parallel computing of LBMâ€“DEM approach has been implemented. Taking advantage of the inherent parallelism of LBM and the attractive Flops/Price ratio of GPU, we have implemented 576 GPUs parallel computing on a Mole-8.5 system and conducted the largest scale DNS of gas-solid suspensions so far, with 1,166,400 solid particles in an area of 11.5cm x 46cm for a two-dimensional system and 129,024 solid particles in a domain of 0.384cm x 1.512cm x 0.384cm for a three-dimensional system (Xiong et al., 2012). The scale of DNS data has been reached the size in traditional computational grid, which implies the really meaningful statistical results from large-scale DNS of gas-solid flows were obtained for the first time. The effects of mesoscale structure on the interaction force between gas and solid phases (Zhou et al., 2014) and the statistical properties of particles (Liu et al., 2017) were explored, which may provide the corresponding constitutive relation and detailed microscopic information for discrete particle simulation and two-fluid model.

Keywords: Lattice Boltzmann method; Discrete element method; Direct numerical simulation; Mesoscale structure; Gas-solid flow

References

1. Limin Wang, Guofeng Zhou, Xiaowei Wang, Qingang Xiong, Wei Ge. 2010. Direct numerical simulation of particle-fluid systems by combining time-driven hard-sphere model and lattice Boltzmann method. Particuology, 8: 379-382.
2. Guofeng Zhou, Limin Wang, Xiaowei Wang, Wei Ge. 2011. Galilean-invariant algorithm coupling immersed moving boundary conditions and Lees-Edwards boundary conditions. Physical Review E 84, 066701.
3. Qingang Xiong, Bo Li, Guofeng Zhou, Xiaojian Fang, Ji Xu, Junwu Wang, Xianfeng He, Xiaowei Wang, Limin Wang, Wei Ge, Jinghai Li. 2012. Large-scale DNS of gasâ€“solid flows on Mole-8.5. Chemical Engineering Science 71: 422â€“430.
4. Guofeng Zhou, Qingang Xiong, Limin Wang, Xiaowei Wang, Xinxin Ren, Wei Ge. 2014. Structure-dependent drag in gas-solid flows studied with direct numerical simulation. Chemical Engineering Science, 116: 9-22.
5. Xiaowen Liu, Limin Wang, Wei Ge. 2017. Mesoâ€“scale statistical properties of gasâ€“solid flowâ€"a direct numerical simulation (DNS) study. AIChE Journal 63:3-14.

Particle-Scale Characterization of Spout Deflection in a Spout Fluidized BED

Conference: Fluidization

Year: 2019

Proceeding: Fluidization XVI

Group: General Paper Pool

Session: 5C: Modeling and Simulation

Time: Wednesday, May 29, 2019 - 9:55am-10:07am

Authors:

Shen, Y., University of New South Wales

Spout deflection is a common instabilityphenomenon in spoutfluidizedbeds. However, spout deflection wasnot well characterizedand quantified in the past. In this work, spout deflection in a flat-bottomed spout fluidizedbed is investigatedat the particle scale by means ofComputer Fluid Dynamic-Discrete Element Method (CFD-DEM) coupling approach. Two types of spout deflections - alternating and non-alternating spout deflections are captured by the CFD-DEM simulationsandqualitatively analyzedin terms ofvoidage distributionsand particle momentum profiles. Spout deflection angle is defined and used to characterizethe intensity of spout deflection quantitatively. Then, the effects of spouting velocity on alternating spout deflection are investigated quantitativelyusing the spout deflection angle. The simulation results indicate that spouting velocity has a limitedeffect on the amplitude and frequency of alternating spout deflection. Thecomparisons between solid flow pattern and spout deflection anglealso show that spout deflection angle is highly relevant to particle distribution non-uniformity at the upper region of the bed. The study offers a cost-effective tool to characterizeand quantify spout deflection in a spout fluidized bed.

Similarity Analysis of Gas Phase Flow Field in the Cyclone Separator

Conference: Fluidization

Year: 2019

Proceeding: Fluidization XVI

Group: General Paper Pool

Session: 5C: Modeling and Simulation

Time: Wednesday, May 29, 2019 - 10:07am-10:19am

Authors:

Yang, Z., Xinjiang Engineering Institute
Cai, X., Xinjiang Engineering Institute
Wei, Y., China University of Petroleum-Beijing at Karamay
Wang, J., Xinjiang Engineering Institute

The gas flow field in a cyclone separator with a volute inlet is numerical simulated based on the Reynolds stress model on the platform of commercial computational fluid dynamics (CFD) software. The effect of the diameter parameter and the flow parameters on the dimensionless velocity were investigated. The results show that the flow fields are similar for different cyclone diameters or different flow parameters, but the dimensionless tangential velocities are different, the flow field is not strictly similar. But, as the diameter and flow parameters are increasing, the flow field develops from the weak similarity to strong similarity, until self-similarity. The mechanics of the self-similarity of flow field is analyzed based on the energy losses and the relationship between Euler number and Reynolds number. The self-similarity of the flow field reflects the contribution of the inertia resistance loss and viscous friction loss to the total energy loss.

Assessing the Validity of Monte Carlo Techniques for Simulating Particulate Flows

Conference: Fluidization

Year: 2019

Proceeding: Fluidization XVI

Group: General Paper Pool

Session: 5D: Modeling and Simulation

Time: Wednesday, May 29, 2019 - 9:55am-10:07am

Authors:

Morris, A., Purdue University
Hong, A., Purdue University

The direct simulation Monte Carlo (DSMC) method is well suited for simulating dilute granular gases, where collisions between particles are uncorrelated and molecular chaos assumptions are valid. In many industrial particulate processes, such as fluidized beds, the system spans dilute to dense flow regimes. In large-scale systems, it is computationally expensive to resolve the motion of individual particles in a deterministic way, as is done in discrete element methods. The direct simulation Monte Carlo method provides a powerful alternative to discrete element modeling because the particles are represented by computational parcels and collisions are computed stochastically based on kinetic theory arguments. Our simulation results show the DSMC method can predict similar phenomena as DEM with an order of magnitude reduction in computational cost. To date, the validity of the DSMC method has been evaluated for dilute particle flows. In dense regions, however, particle arrangement and collisions become correlated, and molecular chaos assumptions inherent to kinetic theory or DSMC methodologies break down. The work presented herein examines the validity of the DSMC method in moderately dense and dense flow regimes. To assess the validity of the DSMC method, simulations of homogeneous cooling systems are first compared to theoretical predictions. Such simulations assess the accuracy of the granular cooling rate as well as the ability to predict hydrodynamic instabilities. Simulations of quasi-1D shear flow are simulated and the resulting solids stress are compared to theory over a range of solids concentrations. We also test the DSMC method for a gas-solid flow in a fluidized bed and compare the simulation results to discrete element and multiphase particle-in-cell simulations.

DEM Simulation on Particle Agglomerates Flow Behaviors during Drying Process in a Bubbling Fluidized Bed

Conference: Fluidization

Year: 2019

Proceeding: Fluidization XVI

Group: General Paper Pool

Session: 5D: Modeling and Simulation

Time: Wednesday, May 29, 2019 - 9:43am-9:55am

Authors:

Tang, T., Harbin Institute of Technology
Wang, B., Harbin Institute of Technology
He, Y., Harbin Institute of Technology

DEM Simulation on Particle Agglomerates Flow
Behaviors during Drying Process in a Bubbling Fluidized Bed

Tianqi
Tang[1], Biao Wang[1], Yurong He[1,*]

[1] School of
Energy Science and Engineering, Harbin Institute of Technology, 150001, Harbin,
China

rong@hit.edu.cn

Dependence on fossil
fuels as primary energy source has caused a serious energy crisis and environmental problems. Renewable sources, such as biomass, are effective resources to provide energy for the human population. Meanwhile, bubbling fluidized beds play an important role in biomass combustion, and biomass is an important clean fuel. Besides,
bubbling fluidized beds have been widely used in drying,
combustion and gasification. Owing

to their unique advantages, bubbling fluidized beds have attracted great attention from researchers. However, most studies have focused on dry particle systems. Cohesive liquids cannot be neglected in relevant industrial processes. For example, wet grinding is employed to reduce dust and energy loss during alumina production.[1]Moreover, the flow characteristics of particles change greatly under the effects of a cohesive liquid between particles[2]. Thus, the flow behaviors of particles in bubbling fluidized beds are important and require further investigation.

The gas-solid two-phase flow behaviors are complex and unique in bubbling fluidized beds. As far, bubbling fluidized beds have attracted more and more attention from researchers. For the experiment, Taghipour et al.[3] have conducted experiment to investigate the hydrodynamics of a two-dimensional bubbling fluidized bed reactor, including solid-phase kinetic energy fluctuation and pressure drop. Radmanesh et al.[4] studied gasification of beech wood particles in a bubbling fluidized bed gasifier, and found that pyrolysis is an important step in the overall gasification model that can determine the distribution of products. With the development of computational fluid dynamics, numerical simulation has become an effective method of further investigating the gas-solid hydrodynamic characteristics in a gas-solid fluidized bed. Since 1993, a modified soft-sphere model has been used, which was improved by Tsuji et al. [5] based on the model proposed by Cundall and Strack[6]. The discrete element method (DEM) has been widely used to investigate complex granular flow[7], and also plays an important role in the analysis of particle flow behaviors in mixers[8], V-blender[9] and rotating drum[10]. Furthermore, DEM was also used to explore particle flow behaviors in bubbling fluidized beds. Gui et al.[11] investigated gas–solid flow behaviors in a bubbling fluidized bed with immersed tubes, and found that particle back-mixing motion and bubble motion seem to bear a strong correlation with each other. Rhodes et al.[12] investigated the mixing characteristics in a bubbling fluidized bed, and pointed that the rate of solids mixing increases with increasing gas velocity. However, most numerical simulation were limited to dry granular systems.

Few investigations have been conducted on wet particle flow behaviors, especially those considering heat and mass transfer behaviors processing. In fact, wet particle systems couldnft be neglected in industrial processing, especially considering heat and mass transfer characteristics. Moreover, heat and mass transfer is important in industrial processing, such as particle drying. In this study, the module of heat and mass transfer has been added. For the heat transfer process, conduction, convection, radiation and liquid evaporation are considered. Besides, the mass transfer between liquid coated on particles and gas are taken into account. Moreover, the effect of gas temperature on wet particle system drying process has been conducted. The distribution of heat transfer coefficient in the bubbling fluidized bed has been investigated. From numerical simulation results, it could be obtained that particles formed agglomerates under the effect of cohesive liquid. With the effect of fluidization gas, agglomerates disappeared with the evaporation of cohesive liquid. The wet particle system converted into dry particle system. With an increases of gas temperature, the rate of drying processing increased gradually. In the bed middle, the heat and mass transfer rates were higher than those near the walls due to higher gas and solid momentum exchange. It could be obtained that the momentum exchange has a strong effect on mass and heat exchange.

Acknowledgement

This work was financially supported by the National Natural Science Foundation of China (Grant No. 91534112).

References

---

[1] Wang, X., Gui, W., Yang, C., Wang, Y. (2011). Wet grindability of an industrial ore and its breakage parameters estimation using population balances. International Journal of Mineral Processing, 98(1-2), 113-117.

[2] Liao, C. C., Hsiau, S. S. (2010). Experimental analysis of dynamic properties in wet sheared granular matter. Powder Technology, 197(3), 222-229.

[3] Taghipour, F.,
Ellis, N., Wong, C. (2005). Experimental and computational study of gas–solid fluidized bed hydrodynamics. Chemical Engineering Science, 60(24), 6857-6867.

[4] Radmanesh, R., Chaouki, J., Guy, C. (2006). Biomass gasification in a bubbling fluidized bed reactor: experiments and modeling. AIChE Journal, 52(12), 4258-4272.

[5] Tsuji, Y., Kawaguchi, T., Tanaka, T. (1993).
Discrete particle simulation of two-dimensional fluidized bed. Powder technology, 77(1), 79-87.

[6]
Cundall, P. A., Strack, O.
D. (1979). A discrete numerical model for granular assemblies. Geotechnique, 29(1), 47-65.

[7]
Guo, Y., Curtis, J. S. (2015). Discrete element
method simulations for complex granular flows. Annual Review of Fluid Mechanics, 47, 21-46.

[8]
Alian, M., Ein-Mozaffari,
F., Upreti, S. R. (2015). Analysis of the mixing of
solid particles in a plowshare mixer via discrete element method (DEM). Powder Technology, 274, 77-87.

[9] Pereira, G. G., Cleary, P. W. (2013).
De-mixing of binary particle mixtures during unloading of a V-blender. Chemical Engineering Science, 94, 93-107.

[10] Arntz, M. M. H. D.,
den Otter, W. K., Briels, W. J., Bussmann,
P. J. T., Beeftink, H. H., Boom, R. M. (2008).
Granular mixing and segregation in a horizontal rotating drum: a simulation study on the impact of rotational speed and fill level. AIChE Journal, 54(12), 3133-3146.

[11] Gui,
N., Fan, J. R., Luo, K. (2008). DEM–LES study of 3-D
bubbling fluidized bed with immersed tubes. Chemical Engineering Science, 63(14), 3654-3663.

[12] Rhodes, M. J., Wang, X. S., Nguyen, M.,

Stewart, P., Liffman, K. (2001). Study of mixing in
gas-fluidized beds using a DEM model. Chemical Engineering Science, 56(8), 2859-2866.

Euler-Euler Model for Charge Transport in Fluidized Beds of Polyethylene Particles

Conference: Fluidization

Year: 2019

Proceeding: Fluidization XVI

Group: General Paper Pool

Session: 5D: Modeling and Simulation

Time: Wednesday, May 29, 2019 - 9:07am-9:19am

Authors:

Ray, M., Iowa State University
Passalacqua, A., Iowa State University
Chowdhury, F., University of Ottawa
Sowinski, A., University of Ottawa
Mehrani, P., University of Ottawa

Abstract

Triboelectric charging encountered in polyethylene production is of deep concern which requires pre-assessment. In this article, a Eulerian model which treats both the solid and gas phases as continuous phases is discussed, which can be used to reasonably predict charge distribution in fluidized beds. The kinetic theory of granular flows has been used to formulate charge diffusion expressions for the monodisperse as well bi-disperse models. The discrete charging model of (Matsusaka, Ghadiri, and Masuda 2000) has been used as the discrete charging model in evaluating the collision integrals. Self-diffusion which assumes non-zero correlation between variance of charge and granular temperature has also been included. For model validation, simulation results were compared to experimental observation in a lab-scale fluidized bed of polyethylene particles. The former predicts that particles less than 450 Î¼m may stick to the wall, while an average layer of 600 Î¼m was seen in the experiments. The model was subsequently modified to simulate bi-disperse particulate flows, based on the work of Jenkins and Mancini (1987). Example results of the same will be presented as well.

Introduction

A particularly interesting example of triboelectrification is represented by charging of polyethylene particles during production in fluidized bed reactors, where electrostatic charging leads to reactor wall sheeting which can clog the reactor inlet and disrupt the manufacturing process. Electrostatic phenomena can lead to safety issues when high electric potentials develop, which sometimes causes sparks and fire. Numerical modelling to simulate the contribution of charge towards the behavior of fluidized systems has been attempted by few previous authors. Of particular note is the work of Rokkam et al. (2013), who used particles of constant charge magnitudes which varied as a function of particle size, predicting charge-driven preferential segregation in a fluidized bed. Their work was based on the particle charge magnitudes observed by Sowinski (2012), (Andrew Sowinski, Mayne, and Mehrani 2012), who had carried out detailed experiments to test the effect of particle size, fluidizing velocity and wall material on bed electrification. Rokkamâ€™s work, dealt with charge transport through only advection, and neglected diffusion and charge generation. (Kolehmainen, Ozel, and Sundaresan 2018) formulated an Euler-Euler model which includes charge generation from the wall, and modeled the charge diffusion coefficient by analogy to the term found for heat transfer by (Hsiau and Hunt 1993). Particle charge and velocity were assumed to be uncorrelated.

In the present work we account for particle charging in the case of monodisperse particles and derive a charge generation and a charge diffusion model, consistent with the kinetic theory model used to describe the granular flow (Jenkins and Savage 1983), where it is assumed that particle velocity and particle charge are correlated. This leads to the formulation of a charge diffusion coefficient and of a charge source term for the charge transport equation. A wall boundary condition to model particle charging due to collisions of particles with a wall was derived considering wall charging model of Matsusaka et al. (2000) and Matsusaka and Masuda (2003). This led to the formulation of an Eulerian charge model consistent with the boundary conditions for slip velocity and pseudo-thermal granular energy of Johnson and Jackson (1987).

Results

The model developed in this work was applied to a two-dimensional computational domain, representing the laboratory-scale fluidized bed employed by Sowinski (2012)where the effect of different particle sizes was investigated. Three different particle sizes of approximate diameters 362, 462 and 550 Î¼m were used in the simulations. Constant fluidization velocity magnitudes 50% higher than the minimum fluidization velocities were assigned. The simulation was run until steady state values of total bed charge and total electric potential were obtained. Like the experiments, adhesion of particles to the fluid bed wall due to charging was predicted for smaller sized particles (< 450 Î¼m) in preference to larger-sized ones. The average bed charge was shown to increase by over 250% in each case with a sharp boundary layer of about 500 Î¼m thickness forming at the wall. The electric potential developed was also found to be below the Paschen curve for air, thereby eliminating the possibility of electric breakdown. The net bed charge was shown to be negative for all particle sizes tested.

References

Hsiau, S. S., and M. L. Hunt. 1993. â€œKinetic Theory Analysis of Flow-Induced Particle Diffusion and Thermal Conduction in Granular Material Flows.â€ *Journal of Heat Transfer* 115 (3): 541â€“48.https://doi.org/10.1115/1.2910720.

Jenkins, J. T., and S. B. Savage. 1983. â€œA Theory for the Rapid Flow of Identical, Smooth, Nearly Elastic, Spherical Particles.â€ *Journal of Fluid Mechanics* 130 (May): 187.https://doi.org/10.1017/S0022112083001044.

Johnson, P. C., and R. Jackson. 1987. â€œFrictionalâ€“collisional Constitutive Relations for Granular Materials, with Application to Plane Shearing.â€ *Journal of Fluid Mechanics* 176 (March): 67â€“93.https://doi.org/10.1017/S0022112087000570.

Kolehmainen, Jari, Ali Ozel, and Sankaran Sundaresan. 2018. â€œEulerian Modelling of Gasâ€“solid Flows with Triboelectric Charging.â€ *Journal of Fluid Mechanics* 848 (August): 340â€“69.https://doi.org/10.1017/jfm.2018.361.

Matsusaka, Shuji, Mojtaba Ghadiri, and Hiroaki Masuda. 2000. â€œElectrification of an Elastic Sphere by Repeated Impacts on a Metal Plate.â€ *Journal of Physics D: Applied Physics* 33: 2311â€“19.

Matsusaka, Shuji, and Hiroaki Masuda. 2003. â€œElectrostatics of Particles.â€ *Advanced Powder Technology* 14 (2): 143â€“66.https://doi.org/10.1163/156855203763593958.

Rokkam, R.G., A. Sowinski, R.O. Fox, P. Mehrani, and M.E. Muhle. 2013. â€œComputational and Experimental Study of Electrostatics in Gasâ€“solid Polymerization Fluidized Beds.â€ *Chemical Engineering Science*92 (April): 146â€“56. https://doi.org/10.1016/j.ces.2013.01.023.

Sowinski, A. 2012. â€œThe Study of Reactor Wall Fouling in Gas-Solid Fluidized Beds Caused by Electrostatic Charge Generation.â€ Thesis dissertation, University of Ottawa.

Sowinski, Andrew, Antonio Mayne, and Poupak Mehrani. 2012. â€œEffect of Fluidizing Particle Size on Electrostatic Charge Generation and Reactor Wall Fouling in Gasâ€“solid Fluidized Beds.â€ *Chemical Engineering Science* 71 (March): 552â€“63.https://doi.org/10.1016/j.ces.2011.11.031.

Keynote Talk: Generating Macroscopic Quantities of Particle-Fluid Flows: Applications to Systems with Coarse or Cohesive Particles

Conference: Fluidization

Year: 2019

Proceeding: Fluidization XVI

Group: General Paper Pool

Session: 5D: Modeling and Simulation

Time: Wednesday, May 29, 2019 - 8:30am-8:55am

Authors:

Hou, Q., Monash University
Wu, Y., Monash University
Zhou, Z., Monash University
Curtis, J. S., University of California Davis
Yu, A., Monash University

There are continuum and discrete approaches for describing granular flows. A continuum approach relies on local quantities which can be derived through an averaging method based on a discrete approach. But the selection of an averaging domain and the validity of local quantities for constitutive relations are not well established, particularly for transient particle-fluid flows. Here, we demonstrate that invariant local quantities such as stress and solid volume fraction can be generated on a proper averaging domain in a fluidized and a moving bed with coarse or cohesive particles. Furthermore, the relation between solid pressure and solid volume fraction is established and compared to the ones in the literature. For coarse particle systems the relation agrees qualitatively well but is different to all the existing monotonic ones in the literature. It shows a bifurcation at a high solid volume fraction, essentially linked to the variation of short and enduring contacts among particles with flow state and solid volume fraction. For cohesive particle systems, it is found that the solid pressure is underestimated by the classical kinetic theory, even when the solid fraction is smaller than 0.49. Accordingly, a modified equation considering particle contacts and particle fluctuating velocity is proposed, which agrees well with the averaged relation. These findings should be well considered in continuum modelling of systems with coarse/cohesive particles.

Lbm-DEM Analysis of Inertial Migration of Neutrally Buoyant Ã‚Â Particle Suspensions in Planar Poiseuille Flow

Conference: Fluidization

Year: 2019

Proceeding: Fluidization XVI

Group: General Paper Pool

Session: 5D: Modeling and Simulation

Time: Wednesday, May 29, 2019 - 10:07am-10:19am

Authors:

Wu, C., University of Surrey
Liu, W., University of Surrey

We establish a hybrid numerical framework for modelling solid-liquid multiphase flow based on a single-relaxation-time lattice Boltzmann method and the discrete element method implemented with the Hertz contact theory. The numerical framework is then employed to systematically study the effect of particle concentration on the inertial migration of neutrally buoyant particle suspensions in a planar Poiseuille flow. The results show that the influence of particle concentration on the degree of inertial migration primarily depends on the characteristic channel Reynolds number Re_0, which is defined as the Reynolds number for the pure channel flow with the same condition but no particles. For relatively low Re_0(sayRe_0<20), evident migration behaviour is only observed at a very low particle concentration (<1%), although the migration behaviour can still be observed at a high concentration (â‰¥10%) when Re_0>20.Furthermore, we discover a unique feature regarding the influence of the solid fraction on the flow field: the normalized Reynolds number (the actual Reynolds number in presence of particles divided by Re_0) can be well regressed by an exponential law. Finally, we introduce an empirical focusing number to characterise the degree of inertial migration, and find that when the focusing number is above a critical value, a fully inertial migration occurs; when it is below another critical value, particles are laterally unfocused.

Multiscale Simulation of Different-Sized Methanol-to-Olefins Fluidized Bed Reactors with Consideration of Coke Distribution

Conference: Fluidization

Year: 2019

Proceeding: Fluidization XVI

Group: General Paper Pool

Session: 5D: Modeling and Simulation

Time: Wednesday, May 29, 2019 - 8:55am-9:07am

Authors:

Lu, B., Dalian National Laboratory for Clean Energy
Li, H., Dalian National Laboratory for Clean Energy and National Engineering Laboratory for MTO, Dalian Institute of Chemical Physics, Chinese Academy of Sciences
Ye, M., Dalian Institute of Chemical Physics, Chinese Academy of Sciences
Wang, W., University of Chinese Academy of Sciences

Multiscale
simulation of different-sized methanol-to-olefins fluidized bed reactors with consideration of coke distribution

Bona
Lu[1]*, Hua Li[2], Mao Ye[2], Wei Wang[1]

[1] State
Key Laboratory of Multiphase Complex Systems, Institute of Process Engineering, Chinese Academy of Sciences, Beijing 100190, China.*Corresponding author: bnlu@ipe.ac.cn

[2] National
Engineering Laboratory for MTO, Dalian Institute of Chemical Physics, Chinese Academy of Sciences, Dalian 116023, China

1. Introduction

The methanol to

olefins (MTO) process is an important route to produce light olefins and has attracted much attention in recent years. In China, more than 10 industrial units using the dimethyl ether or methanol to olefins (DMTO) technique have been successfully operated for producing light olefins. The main unit of DMTO process follows the design of modern fluid catalytic cracking process, but was modified due to the unique features of MTO reactions. For example, as the ethylene selectivity is closely related to the coke deposition on SAPO-34 catalyst, a densely fluidized bed is hence preferred to achieve the desired coke content. Understanding of hydrodynamics and related reaction features can help us to further optimize the process. Computational Fluid Dynamics (CFD) is a powerful tool to help understand the complex hydrodynamics and reactions. However, reactor scale-up is always involved with variation of flow structures and reaction behaviors, accurate simulations of a series of MTO reactors ranging from the lab scale to industrial scale pose bigger challenges to CFD modeling.

2. Methods

Fig.1 presents geometries of four different-sized reactors and boundary settings. Two-fluid model (TFM) with kinetic theory for granular flow and a seven lumped reaction network are employed with FluentÒ15 as the solver. EMMS/bubbling drag model[1] is employed for the micro-, pilot- and demo-scale reactors operating under low gas velocity, and its newly developed two-step version[2] which was found suitable for coarse-grid simulation is used for the simulation of an industrial reactor. A reactor model developed in our previous study[3] is used to provide initial distribution of the coke content. Both gas and solid phases are treated as mixtures, nine species (CO_2, CH_4, C_2H_4, C_3H_6, C_3H_8, C_4H_8, C_5H_{10}, H_2O and methanol) for the gas mixture, and two species (coke and catalyst) for solid phase. The catalyst particles belong to Geldart type A particles (d_p=97mm and r_p=1500 kg/m^3). More details can refer to our previous studies [4-5].

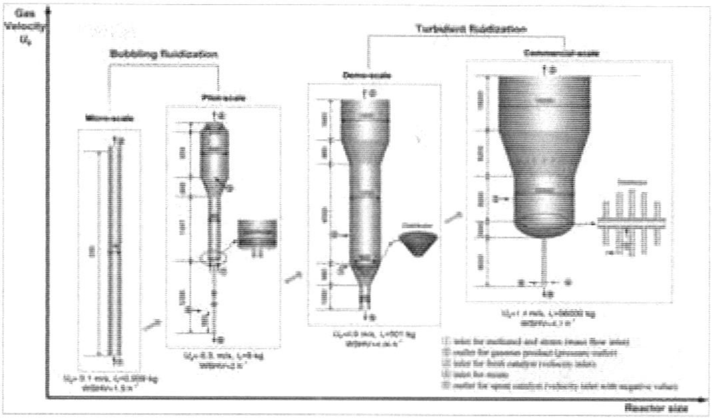

Figure
1. Schematic diagram of four different-sized DMTO reactors

3. Results and discussion

As shown in Fig. 2a, a relatively uniform flow structure is found in the micro-scale fluidized bed. As the gas velocity increases, flow structures evolve to have a heterogeneous distribution in larger DMTO reactors, such as typical bubbling fluidization in pilot-scale reactor, very diffuse bed surface in demo-scale reactor, particle clusters in the commercial reactor. The predicted pressure drops of reaction zones for four reactors agree well with the experiment, showing a reasonable choice of drag models. The predicted mass fractions of products for both micro-scale and pilot-scale reactors are in good agreement with experiment, but large discrepancy is found between simulation and measurement for both demo-scale and commercial reactors (particularly ethylene is over-predicted).

The experiment found that there exists a wide distribution of coke content in large MTO reactors, but the above TFM simulations predict uniform coke distribution. As the TFM simulation averages different coke contents in a computational cell, thus fails to predict ethylene which is sensitive to the coke content. We introduced the population balance model (PBM) to describe the evolvement of coke content and combined it with TFM. The catalysts with different coke contents are denoted by a group of solid species. The simulation of 2D demo-scale reactor shows the ethylene prediction is greatly improved (Fig. 2b).

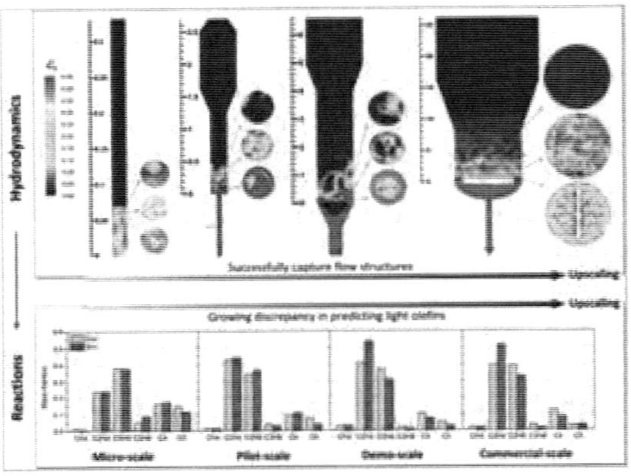

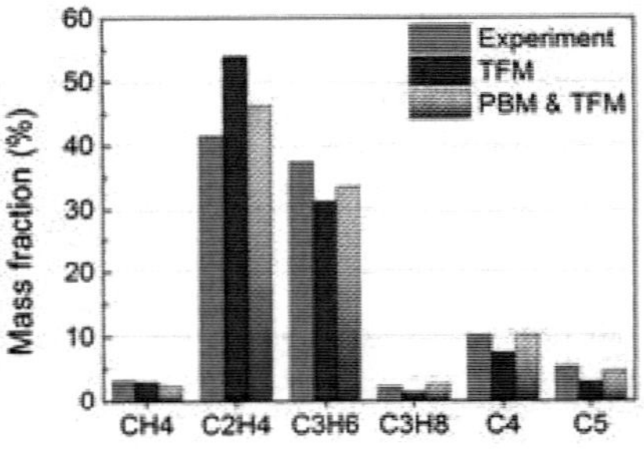

(b)

centration of
of gaseous products; (b) mass
ning PBM and TFM.

MTO reactors, but
tration. The possible
ting PBM into TFM enables
ng the prediction in

B. Lu, J. Zhang, H. Luo, W. Wang, H. Li, M. Ye, Z. Liu, J. Li. Chem. Eng. Sci. 2017, 171: 244-255.

[5]
J. Zhang, B. Lu, F. Chen, H. Li, M. Ye, W. Wei. Chem. Eng. Sci. 2018, 189:212-220

Simulation of Proppant Transport in Fractures with DNS-Derived Drag Correlations

Conference: Fluidization

Year: 2019

Proceeding: Fluidization XVI

Group: General Paper Pool

Session: 5D: Modeling and Simulation

Time: Wednesday, May 29, 2019 - 9:31am-9:43am

Authors:

Yin, X., Colorado School of Mines
Li, X., Colorado School of Mines

Multiphase flow simulations of proppant transport, a particulate flow process encountered in hydraulic fracturing of oil and gas reservoirs, need problem-relevant drag correlations. In this study, DNS was employed to study the influences of several dimensionless numbers, namely the Reynolds number of cross flows, the Archimedes number, the density of particles relative to that of the carrying fluid, particle volume fraction, and the ratio of fracture width over particle size on the settling velocity. DNS results show that fracture width significantly impedes the settling velocity. Cross flow and particle-fluid density ratio (provided that the Archimedes number is held as a constant) on the other hand do not have significant effects.

The aim of DNS is not only to understand the influence of the dimensionless numbers, but also to obtain data for developing the needed drag correlation. A drag correlation developed from DNS data using quadratic polynomials and interpolations was incorporated into MFIX. MFIX simulations of particle transport in narrow fractures predicted slower formation of particle sediment layers at the bottom of the fracture compared to default drag laws, because particle-fluid drag is enhanced by the width of fractures. Simulations using the DNS-derived drag correlation can better match experimentally determined sediment layer thickness, compared to those using other drag laws.

Simulation of Spatial Trajectory and Regulation of Chemical Pesticide Droplets

Conference: Fluidization

Year: 2019

Proceeding: Fluidization XVI

Group: General Paper Pool

Session: 5D: Modeling and Simulation

Time: Wednesday, May 29, 2019 - 9:19am-9:31am

Authors:

Xue, S., Dalian University of Technology
Ma, X., Dalian University of Technology
Lan, Z., Dalian University of Technology
Song, X., Dalian University of Technology
Zhao, J., Dalian University of Technology

The use of chemical pesticides is extremely indispensable for pest and disease control in crop protection and agricultural production. However, due to the fact that most of the actual operations are based on human long-term experiences without enough theoretical guidance, particle drift and evaporation drift always occur under the action of current after droplets are released from nozzles. The effectiveness of the pesticide operation is still significantly limited by the spray technology and management. According to the statistics provided by professional organizations, the pesticides utilization efficiency is only 20-30% and spray drift can pose a considerable threat to bystanders, residents, livestock, even terrestrial and aquatic ecosystem. In this study, in order to explore the evolution laws of droplets motion and predict the particle drift ratio influenced by multiple factors in the current, a computational fluid dynamics (CFD) model is used to track the spatial trajectory of the droplets cloud through varying surface tension (0.035N/m-0.1N/m), viscosity (0.001Pa·s-0.1Pa·s), diameter and velocity distribution, wind velocity (1-5m/s), spray height (0.5m-1.0m). Multiple regression analysis and multi-objective optimization are integrated to obtain the optimal spray parameter combination and ensure adequate coverage on intended canopy, so as to lower the spatial loss of chemical pesticides. It shows that increasing surface tension and viscosity of pesticide solution can make nozzle produce more larger droplets with better performance of resisting entrainment, which can have a reducing effect on drift ratio. Spray height can cause greater impact on deposition velocity of droplets falling on canopy leaves than deposition diameter. And impacting droplets with higher velocity are prone to jump off the canopy target, also causing loss of chemical active components. Simultaneously, the optimal spray application is recommended with the wind velocity of 1-3m/s. This study demonstrates that CFD model can be used to evaluate the spray application performance in field operation. More importantly, it can provide a beneficial concept to regulate the current spray application technique and design an optimal scheme with lower drift ratio and higher coverage on intended target.

Characterization of Multi-Scale Structures in Gas-Solids Circulating Fluidized Beds By Recurrence Quantification Analysis

Conference: Fluidization

Year: 2019

Proceeding: Fluidization XVI

Group: General Paper Pool

Session: 6A: Fundamentals of Fluidization

Time: Wednesday, May 29, 2019 - 11:38am-11:50am

Authors:

Wang, C., China University of Petroleum
Lan, X., China University of Petroleum
Jinsen, G., State Key Laboratory of Heavy Oil Processing

Circulating fluidized bed (CFB) is one of the most complicated hydrodynamic systems. Typical particle-fluid two-phase flow patterns of CFBs exhibit nonlinear and nonequilibrium dynamic characteristics with multiscale flow structures. Power spectrum and recurrence quantification analysis can be used as an effective approach to characterize multi-scale flow behavior in CFB systems. Pressure fluctuations were measured in a circulating fluidized bed of 0.15 m in diameter and 18 m in height under a wide operating conditions with superficial gas velocity from 7 to 9 m/s and solids circulation rates from 100-800 kg/m^{2}s. The measured signals were analyzed using power spectrum and recurrence quantification analysis. The results shows that the complex dynamics of CFB system can be presented as macro, meso and micro structures. It was found that a minimum in wide band energy with an increase in the velocity from 7 to 9 m/s corresponds to the transition between macro structures and finer structures. Based on the PS and RQA methods, three different structures can be determined with frequency range of 0â^'5 Hz for macro, 5â^'50 Hz for meso, and 50â^'200 Hz for micro structures. The results of RQA showed that the micro structures of the CFB systems have the least periodicity.

Development of a Heterogeneous Drag Model for Gas-Solid Flows Based on Particle-Resolved Direct Numerical Simulations

Conference: Fluidization

Year: 2019

Proceeding: Fluidization XVI

Group: General Paper Pool

Session: 6A: Fundamentals of Fluidization

Time: Wednesday, May 29, 2019 - 11:14am-11:26am

Authors:

Song, N., Xi'an Jiaotong University
Zhou, Q., Xi'an Jiaotong University
Chen, X., Xi'an Jiaotong University
Jiang, M., Xi'an Jiaotong University

The conventional drag models commonly used in gas-solid flow are mainly developed based on the system of random and homogeneous distribution of particles. Since the existence of particle clusters and bubbles in fluidized beds have a pronounced influence on the interaction between fluid and particles, the effect of heterogeneity on drag force in fixed beds at low Reynolds numbers is studied with particle-resolved direct numerical simulation (PR-DNS). Two kinds of inhomogeneous structures are studied in this paper: (i) the linear spatial distribution of solid volume fraction, (ii) the simple stepwise heterogeneous structure formed with dilute phase and dense phase. Our PR-DNS results show that though the drag coefficient agrees well with the BVK (Beetstra et al. AIChE J. 2007) model at dense or dilute phase with a certain gradient of volume fraction, obvious deviation occurs in the latter configuration, especially near the interface between dense and dilute phase.

Further focus on the drag force at the stepwise heterogeneous structures, the PR-DNS results suggest that: (i) the drag coefficient decrease if the gas flow is perpendicular to the gradient of the volume fraction while increase if gas flow is parallel to the gradient; (ii) the deviations between the PR-DNS results and the homogenous drag model increase with the increasing of the filtered grid size at the same gradient of the volume fraction; (iii) the drag curves versus solid concentration vary with gradient of the volume fraction at the same filtered grid size.

Based on these results, a heterogamous drag coefficient model considering the solid volume fraction, particle Reynolds numbers, filtered grid size, flow direction, and the gradient of volume fraction is developed. Note that, no consistency rule could be observed when the filter grid size is one particle diameter or below, which indicates the proposed drag model no longer applies at this situation.

Furthermore, the proposed new drag model is validated with a serial of PR-DNS of gas-solid flows with obvious clusters. Our proposed drag model could obtain not only the well-known drag reduction but also the local drag increasing due to the local heterogeneity. The further validation of the new drag model in large scale simulations using CFD-DEM or TFM will be our future work.

Experimental and Numerical Investigations into the Charge Distribution and Particle Velocity Characteristics in the Freeboard of a Bubbling Fluidized Bed

Conference: Fluidization

Year: 2019

Proceeding: Fluidization XVI

Group: General Paper Pool

Session: 6A: Fundamentals of Fluidization

Time: Wednesday, May 29, 2019 - 11:26am-11:38am

Authors:

Zhang, W., North China Electric Power University
Qi, B., North China Electric Power University
Wang, X., North China Electric Power University
Yan, Y., University of Kent

Due to the intensive and frequent collisions and interactions between particles with each other and with the wall, tribo-charging is inevitable in the fluidization process. The electrostatic charge on the particles can influence the particle flow characteristics in the fluidized bed, resulting in problems such as agglomeration, wall fouling and even electric discharge. It is therefore necessary to investigate the electrostatic characteristics and the interplay between electrostatics and hydrodynamics in the fluidized bed. Several methods have been proposed to characterize the electrostatics in the fluidized bed such as Faraday cups and electrostatic probes. The parameters of local charge levels, charge spatial distributions and particle charge density are determined in these methods. In addition, particle velocity characteristics are also determined using several methods, including optical probes, particle image velocimetry and radioactive particle tracking. However, none of the techniques is capable of measuring both the charge distribution and the particle velocity characteristics. More work is required to achieve an in-depth understanding of the relationship between the electrostatics and the particle flow dynamics in the fluidized bed. In this study, experimental and numerical investigations into the charge distribution and particle flow characteristics are conducted. Wire-mesh electrostatic sensor arrays are installed in the freeboard of a bubbling fluidized bed. The charged particles in the sensing volume of the sensor arrays generate induced charge on the electrodes. The induced charge on the electrodes is measured and used to reconstruct the cross-sectional charge distribution in the freeboard. Meanwhile, the axial and lateral velocities of particles are estimated by cross-correlating the signals from the corresponding electrodes. To study the electrostatic charging effect, numerical simulations of the bubbling fluidized bed are conducted using coupled computational fluid dynamics (CFD) and discrete element method (DEM). Charge on particles is computed from an established triboelectric model. The particle behavior in the freeboard region is numerically tracked using the DEM method. With the available experimental and numerical results on charge distribution and particle velocity characteristics, the electrostatic properties in the freeboard region and their relationship with the particle hydrodynamics under different superficial air velocities are thoroughly investigated.

Key Sub-Grid Quantities Affecting the Filtered Drag Force

Conference: Fluidization

Year: 2019

Proceeding: Fluidization XVI

Group: General Paper Pool

Session: 6A: Fundamentals of Fluidization

Time: Wednesday, May 29, 2019 - 10:50am-11:02am

Authors:

Zhou, Q., Xi'an Jiaotong University
Jiang, M., Xi'an Jiaotong University
Chen, X., Xi'an Jiaotong University

Abstract

Based on numerous correlative analyses of fully resolved simulations, the filtered drag force is found to correlate well two sub-grid quantities: the drift velocity and the scalar variance of solid volume fraction, with the drift velocity as the primary marker. We refer to the drift velocity as the gas drift velocity since it is defined as the difference between the filtered gas velocity seen by the particle phase and the resolved filtered gas velocity and a drift velocity for the solid phase can be defined similarly. In present study, the dependence of the filtered drag force on sub-grid quantities are analyzed by employing the second order Taylor polynomial approximation to the microscopic drag coefficient. Through theoretical analyses of the coefficients of the new appearing sub-grid quantities in this approximation, it is found that, other than the gas drift velocity and the scalar variance of solid volume fraction, the solid drift velocity and a third-order moment defined as the covariance of squared solid volume fraction and the fluctuation of slip velocity, are significant for an accurate estimation of the filtered drag force at low Reynolds numbers. The data from fully resolved simulations support this finding and also reveal that the contributions of these four quantities vary at different solid volume fractions.

Temperature Influence on Fluid Dynamics at the Transition from Bubbling to Turbulent Fluidization for Geldart's Group B Particles

Conference: Fluidization

Year: 2019

Proceeding: Fluidization XVI

Group: General Paper Pool

Session: 6A: Fundamentals of Fluidization

Time: Wednesday, May 29, 2019 - 11:02am-11:14am

Authors:

Wytrwat, T., Hamburg University of Technology
Hartge, E. U., Hamburg University of Technology
Yazdanpanah, M. M., TOTAL
Heinrich, S., Hamburg University of Technology

For upscaling of fluidized bed processes, knowledge about fluid dynamic behavior is indispensable. Especially, Chemical Looping Combustion (CLC), a modern process for CO_2 capture, shows high complexity in its fluid dynamics because it is designed as a system of different interconnected fluidized bed reactors. An oxygen carrier is circulating between two main reactors, called the air reactor and the fuel reactor. To guarantee a complete combustion of biomass or char particles in the fuel reactor, the residence time of these particles must be high enough and reaction should take place in the dense phase of the fluidized bed. Thus, fuel reactors are often designed as bubbling or turbulent fluidized beds. Fluid dynamics of the turbulent regime and its transition from the bubbling state are widely investigated in literature for Geldartâ€™s group A particles, but not for group B particles. Besides, most studies focus on ambient conditions, however combustion processes are operated at high temperatures.

Generally, oxygen carriers used for CLC are in the range of group B according to Geldartâ€™s classification. Thus, a thorough investigation of the turbulent fluidized bed regime for this kind of particles, starting at superficial gas velocities in the bubbling state and ending at fast fluidization is the aim of this work. Temperature influences on fluid dynamics are investigated in a range between ambient conditions and temperatures of 800Â°C.

Three fluidized bed plants in pilot scale are used for this investigation. Two of them are heated electrically to reach the temperatures mentioned above. The first one has a diameter of 100 mm and the second one a diameter of 150 mm with heights of 17 m and 9 m, respectively. The third plant, having a diameter of 400 mm and a height of 17 m, is operated at ambient conditions.

In addition, the measurements are compared to results carried out in two laboratory scale plants with diameters of 50 mm and 100 mm.

A superficial gas velocity range between 0.2 m/s and 3 m/s is investigated. Aspect ratios (static bed height divided by bed diameter) are varied between 1 and 4. Two quartz sand fractions of different sizes (Sauter mean diameter: 188 Î¼m and 348 Î¼m) and ilmenite (Sauter mean diameter: 146 Î¼m) are used as bed material.

Pressure signals are recorded and analyzed related to their standard deviation and frequency spectra to get information about the transition from bubbling to turbulent fluidization and thus the flow structure. As local online measurement technique, capacitance probes are used for the determination of local solids concentrations and local velocities of bubbles/voids in the dense zone of the bed.

Static bed height and plant diameter are found to have a significant influence on the transition velocity from bubbling to turbulent fluidization for the range investigated. Higher static bed heights lead to higher transition velocities. With increasing temperature of the fluidizing gas, its viscosity and density change. Due to this change, in contrast to group A particles according to Geldartâ€™s classification, for group B particles an increase in the transition velocity with increasing temperature can be observed. Finally, a correlation for the determination of the transition velocity from bubbling to turbulent fluidization is introduced by taking into account the static bed height, bed diameter and temperature.

Effects of Feed Jets on Pressure Distribution in FCC Riser

Conference: Fluidization

Year: 2019

Proceeding: Fluidization XVI

Group: General Paper Pool

Session: 6B: Fluidized Bed Applications

Time: Wednesday, May 29, 2019 - 11:50am-12:02pm

Authors:

Yan, Z., China University of Petroleum, Beijing
Fan, Y., China University of Petroleum, Beijing
Xu, C., China University of Petroleum, Beijing
Lu, C., China University of Petroleum-Beijing

Fluid Catalytic Cracking (FCC) is an important process for upgrading heavy oil in modern refining industry. In recent years, riser reactor is widely used to realize a fast cracking reaction. In the feed injection zone of FCC riser, atomized raw oil is injected into the riser from the atomizing nozzles to contact with the catalyst particles. Cracking reactions occur immediately during the mixing between oil and catalysts. Therefore, the dynamic mixing of feed jets and catalysts in this zone is very important. In most commercial FCC systems, the nozzles are pointed upward at 30°~40° relative to the riser axis. However, a large number of studies on the gas-solid flow and reaction have indicated that there are some disadvantages in the traditional feed injection zone. To solve these problems, many efforts have been done by researchers. According to the theoretical analysis and 3D-simulation results, a better jet-catalysts mixing was found when feed jets were downward. By this way, the disadvantages of secondary flow are expected to be minimized. Whatâ€™s more, the hydrodynamic characteristics in different types of jet influence zones are need to be further investigated, especially by experiments. Therefore, in this paper, effects of different feed jets on the pressure distributions in FCC riser were analyzed by large scale cold model experiments.

To study the pressure distributions under the influence of upward and downward feed injections, a large scale cold-riser model was established based on the characteristics and operating conditions of FCC reactor-regenerator system. The riser section was 0.186 m in inner diameter and 11 m in height. Four feed nozzles were equipped at a height of 4.5 m above the gas distributor. Both upward and downward oriented nozzles were mounted in the experiment. The angles between the nozzles and the riser axis were both set to 30° no matter the nozzles were mounted upward and downward to make a comparison.

Pressure sensors were used to measure the pressure distributions and pressure drop in the jet influence zone. Generally, comparing with the upward injection, the pressure drop in the downward feed injection zone is a little larger. Within the operating conditions, the pressure drops caused by the downward jets are about 1%~10% higher than that of the upward jets. When the velocity of jet gas increases, the pressure drop in the upward and downward feed injection zones both increases. However, it is more significant when nozzles are mounted upward. Under lower jet gas velocity condition (U_j=41.8 m/s), the pressure drop of downward injection zone is 9.88% higher than that of the upward injection zone. When the jet velocity is higher (U_j=78.5 m/s), the pressure drop is nearly the same for the two cases. Furthermore, composition of the pressure drop was analyzed. In the feed injection zone, the overall pressure drop can be divided into three parts, i.e., the static pressure drop, the accelerated pressure drop and the frictional pressure drop. The static pressure drop and accelerated pressure drop are easy to be obtained by measuring the solid holdup and particle velocity. Then the frictional pressure drop in the jet influence zone is calculated by knowing the overall pressure drop. Characteristics of frictional pressure drop under the influence of upward and downward injections are presented. Based on the experiment results, correlations between frictional pressure drop and operating parameters are obtained.

Enhancement of Particle Segregation By Air-Staging in a Circulating Fluidized Bed

Conference: Fluidization

Year: 2019

Proceeding: Fluidization XVI

Group: General Paper Pool

Session: 6B: Fluidized Bed Applications

Time: Wednesday, May 29, 2019 - 11:14am-11:26am

Authors:

Jiang, D., University of Chinese Academy of Sciences
Zhu, Z., University of Chinese Academy of Sciences
Zhang, H., Institute of Engineering Thermophysics, Chinese Academy of Sciences

Coal gasification is endothermic and slow reaction between gas and solid. It requires high solid holdup and uniform temperature profile in a circulating fluidized gasifier. Air-staging means the total air to the riser is split into the primary and secondary introduced at different location. It affects the flow structure and thermal balance in a gasifier, and it is a promising method to meet the requirements of coal gasification in a CFB. The regulation effect of the secondary air on solid circulating flow rates, axial profiles of solid holdup and particle size distribution was studied in a small circulating fluidized bed test rig.

The CFB test rig consists of a riser with an internal diameter of 0.15 m and a height of 3 m, a cyclone, a loop seal. Three layers of the secondary air nozzles locate at 0.3 m, 0.46 m and 0.7 m, respectively, above the air distributor. The secondary air is introduced from layers in different combination modes. The ratio of secondary air to total air, the distribution mode and the distribution ratio between layers are considered the key factors in the experiments. To adjust solid circulating flow rates in the standpipe, aeration rate in the loop seal was also studied. For fixed secondary air flow rate, the secondary air introduced from several layers together results in an especially strengthened particle segregation phenomenon. Compared with the experiment without air-staging, solid holdup decreases in the dense phase zone and increases in the dilute phase. Axial profile of solid holdup is more uniform. The distribution ratio also has effect on flow structure. The results can be used for design of CFB gasifier and optimizing operating parameter to promote gasification reaction.

Fluidization Characteristics of Silicon Carbide in Fluidized Bed Under Microwave Irradiation

Conference: Fluidization

Year: 2019

Proceeding: Fluidization XVI

Group: General Paper Pool

Session: 6B: Fluidized Bed Applications

Time: Wednesday, May 29, 2019 - 10:50am-11:02am

Authors:

Zhang, Y., University of Minnesota
Xiong, Q., General Motors

Fluidization characteristics of silicon carbide in fluidized bed under microwave irradiation

Cunfeng Ke [a], Qingang Xiong [b], Yaning Zhang [a], Yunlei Cui [a],

Liqing Li [a], Bingxi Li [a]

[a] *School of Energy Science and Engineering, Harbin Institute of Technology, Harbin 150001, China*

[b] *IT Innovation Center, General Motors, Warren, MI 48092, USA*

Abstract Microwave-assisted pyrolysis of biomass is increasingly hot in recent years due to the many advantages of microwave irradiation over electric heating, i.e., fast, efficient, etc. However, this promising technology suffers from low temperatures in the reaction unit due to the poor microwave absorbability of biomass. In this case, microwave absorbent is needed. The main objective of this study is to investigate the fluidization characteristics of silicon carbide (SiC) which can be used as an efficient microwave absorbent in fluidized bed under microwave irradiation. SiC volume fractions and temperature fields at different fluidization velocities (0.3, 0.4, 0.5, 0.6, and 0.7 m/s), microwave powers (500, 600, 700, 800 and 900 W), initial volume fractions (0.20, 0.25, 0.30, 0.35, 0.40), and particle diameters (0.3, 0.4, 0.5, 0.6, 0.7 mm) are detailed. The results show that initial increases in the fluidization velocity benefit the fluidization whereas a column appears in the bed when the fluidization velocity is further increased, and the critical fluidization velocity is 0.6 m/s. Increases in microwave power accelerate the heating while the heat cannot be transferred to the airflow in time. The microwave power of 500 W shows better SiC distribution and better temperature distribution. Increases in initial volume fraction result in better particle distribution and better heat transfer performance, and the initial volume fraction should be lower than the highest value (0.40). Increases in particle diameter deteriorate the fluidization, and the best particle diameter is 0.3 mm.

Keywords: fluidization characteristics; silicon carbide; microwave irradiation; numerical simulation

Gaseous Emissions during Oxy-Fuel Combustion of Sewage Sludge in a Circulating Fluidized Bed

Conference: Fluidization

Year: 2019

Proceeding: Fluidization XVI

Group: General Paper Pool

Session: 6B: Fluidized Bed Applications

Time: Wednesday, May 29, 2019 - 11:26am-11:38am

Authors:

Luckos, A., University of the Witwatersrand
Kosowska-Golachowska, M., Czestochowa University of Technology
Kijo-Kleczkowska, A., Czestochowa University of Technology
Musial, T., Czestochowa University of Technology
Wolski, K., Czestochowa University of Technology
Sroda, K., Czestochowa University of Technology

Increasing production of sewage sludge in Poland has created some environmental problems related to its disposal. These problems have resulted in recent changes in legislation. In the light of the Minister of Economy Regulation of 16 July 2015, regarding the criteria and procedures for releasing wastes for landfilling, the thermal disposal of sewage sludge is important because of its calorific value, which is greater than 6 MJ/kg, and problems associated with its use. Consequently, a restrictive legislation was introduced on the 1st January 2016 for sewage sludge storage in Poland. Sewage sludge incineration is an attractive option because it significantly reduces the volume of the waste material, minimizes odor, and thermally destroys the organic and toxic components of the off pads. Currently, 11 plants in the country use sewage sludge as fuel thus this technology must be further developed in Poland.

Furthermore, greenhouse gases emissions from coal-fired systems, particularly carbon dioxide,become an important issue in the European Union countries. Oxy-fuel combustion is one of the promising technologies for power generation with CO_2 capture. This technology can also significantly reduce emissions of NO_x and improve the thermal efficiency of the combustion process through the reduction of flue gas volume by about 80%. In the oxy-fuel combustion, fuel is burnt in a mixture of pure oxygen and recycled flue gas. Because nitrogen is eliminated from the oxidizing gas, the flue gas leaving the combustion chamber is highly enriched in CO_2 which means that the combustion process takes place in an O_2/CO_2 environment. Partial recycling of flue gas helps to control the flame temperature in the combustion chamber.

Results of sewage sludge oxy-fuel combustion tests carried out in a 12-kW circulating fluidized-bed (CFB) bench-scale combustor are presented in this paper. The electrically heated rectangular column, 680×75×35 mm, is the main component of the CFB test unit. The front of the column is made of transparent quartz window through which the combustion process can be directly observed. Silica sand (particles smaller than 400 Î¼m) constitutes the inert bed. The tested material consisted of sewage sludge samples in the form of spherical granules. They originated from large-scale commercial sewage treatment plants operating in Poland. Granulated sludge samples were burnt in air (base case) and mixtures of O_2/CO_2 at 850Â°C. Concentrations of emitted pollutants were measured using Gasmet DX-400 FTIR analyzer.

The main objective of this study was to investigate pollutant emissions from the combustion process of sewage sludge under oxy-fuel conditions. The influence of the inlet O2/CO2 ratio on emissions of NO, NO_2, N_2O, SO_2, CO and HCl is reported and discussed in this paper. Compared with combustion in air, combustion in a mixture of 21% O_2 and 79% N_2 caused a reduction in NO emission, a slight increase in N_2O and SO_2 emissions and a dramatic increase in CO emission. An increase in the O_2 inlet concentration resulted in higher emissions of NO and SO_2 and lower emissions of N_2O and CO.

Swing Adsorption Reactor Cluster (SARC) in Post Combustion CO2 Capture: Multistage Fluidization Effect on Heat Transfer and Reactor Performance

Conference: Fluidization

Year: 2019

Proceeding: Fluidization XVI

Group: General Paper Pool

Session: 6B: Fluidized Bed Applications

Time: Wednesday, May 29, 2019 - 11:38am-11:50am

Authors:

Zaabout, A., SINTEF Industry
Dhoke, C., NTNU
Cloete, S., SINTEF Industry
Seo, H., Korean Research Institute for Chemical Technology, Daejeon, South Korea
Park, Y. K., Korea Research Institute of Chemical Technology
Amini, S., SINTEF Industry

1. Introduction

The SARC concept is an adsorption-based technology that aims at minimizing the energy penalty of post combustion CO_2 capture. It consists of a cluster of multistage fluidized bed reactors running simultaneously and alternating sorbent carbonation and regeneration steps to ensure steady state processing of a flue gas stream. Heat is transferred from carbonation to regeneration using a heat pump that can work very efficiently if the temperature difference between the two stages is minimized by applying a partial vacuum in the regeneration step.

A multistage fluidized bed reactor configuration was proposed for the SARC reactor to simultaneously achieve high heat transfer and maximize CO_2 uptake by the sorbent [1]. Heat transfer tubing, containing the heat pump working fluid, transfers heat between the reactors in the cluster. Competitive CO_2 energy penalty could be achieved using the SARC concept applied to a coal-fired power plant .[2]

2. Methods and results

The multistage fluidized bed reactor results in a low solid back mixing, helping achieving a better performance in terms of CO_2 capture as compared to single stage fluidized bed [1]. A multistage fluidized bed reactor with 10 cm ID was therefore designed and being constructed, with different internal/mesh dividing the reactor in four separate stages. Additionally, the reactor will be embedded with heat transfer surfaces in each stage, wherein heat pump working fluid will be used for heating and cooling the sorbent. The performance of Polyethyleneimine (PEI) sorbent in terms of CO_2 working capacity will be studied in this multistage reactor with different internals segregating each stage. Moreover, the effect of multistage fluidized bed on heat transfer rate at various stages will be evaluated and compared.

The results will indicate the gain in CO_2 uptake during carbonation step by using multistage fluidized bed because of the reduction in solid back mixing. Further, the optimal type of partition between the four reactor stages will be identified. Partitions should have large enough openings to allow individual particles to pass through (to avoid blockage), but the openings should still be small enough to prevent large-scale back mixing flow patterns from developing. Finally, the effect of multistage fluidized bed on heat transfer rate at each stage will set a basis for the demonstration of the complete SARC cycle.

References

1. Zaabout, A., et al., *Thermodynamic assessment of the swing adsorption reactor cluster (SARC) concept for post-combustion CO_2 capture.* International Journal of Greenhouse Gas Control, 2017. **60**: p. 74-92.
2. Cloete, S., et al., *The effect of sorbent regeneration enthalpy on the performance of the novel Swing Adsorption Reactor Cluster (SARC) for post-combustion CO_2 capture.* Chemical Engineering Journal, 2018.

Testing Different Sorbents for Sorption-Enhanced Methanation in a Dual Fluidized Bed System

Conference: Fluidization

Year: 2019

Proceeding: Fluidization XVI

Group: General Paper Pool

Session: 6B: Fluidized Bed Applications

Time: Wednesday, May 29, 2019 - 11:02am-11:14am

Authors:

Coppola, A., University of Naples Federico II
Salatino, P., Università degli Studi di Napoli Federico II
Scala, F., Istituto di Ricerche sulla Combustione, Consiglio Nazionale delle Ricerche
Massa, F., Università degli studi di Napoli Federico II

The increasingly growing debate on climate change is pushing our society toward the development of innovative technologies for limiting the emission of greenhouse gases deriving from the exploitation of fossil resources. On the one side, Carbon Capture and Storage (CCS) and Carbon Capture and Utilization (CCU) technologies are intended to exploit sustainably fossil fuels, and represent the transition solution toward the complete conversion of our society to renewable energy (RE). On the other hand, the scientific community is working hard to find solutions to the typical problems of REs, which are mainly represented by their intrinsic intermittent nature. In particular, the concept of energy storage is a key feature for the development of a fully renewable energy systems. Currently, energy storage techniques are mostly based on the accumulation of sensible or latent heat with different materials, either liquid or solid. Unfortunately, these techniques suffer in general of a low storage energy density, and, in some cases, problems related to the chemical stability of the storage medium. For these reasons, in the last years the concept of chemical storage is gaining an increasing interest as a valid energy accumulation system: the idea is based on the utilization of solar radiation to produce fuels, named solar fuels (SF). Among them, hydrogen can be considered as the simplest SF or as the precursor of other SFs, because it can be produced by photochemical or thermochemical electrolysis of water (Smestad and Steinfeld 2012).

In this framework, methanation can assume an important role because it can be strictly linked to the concepts of CCU, solar fuels and chemical storage. In particular, catalytic methanation with hydrogen of either CO (Eq.1) or CO_2 (Eq.2) occurs exothermically following the reactions:

$$3H_2 + CO = CH_4 + H_2O \text{ -206kJ mol-1 (@298K) (1)}$$

$$4H_2 + CO_2 = CH_4 + 2H_2O \text{ -164kJ mol-1 (@298K) (2)}$$

Thus, methane could represent the final product for the storage of solar energy in the framework of the Power-to-Gas (PtG) technology (Götz et al. 2016), initially converted into hydrogen by water splitting with photochemical or thermochemical processes. Specifically, CO_2 methanation could be considered as a process for the utilization of CO_2 coming from combustion of fossil fuels, while CO methanation could be the final conversion step of syngas produced by coal or biomass gasification.

Methane, contrary to hydrogen, is already extensively used in the civil and industrial sectors, as well as in the automotive industry, and as a consequence has a well-developed distribution and storage infrastructure in many countries, and finally benefits from a relatively large public acceptance.

The commercially available methanation plants consist of a cascade of fixed bed reactors, operated at high pressure, and equipped with gas recycling and intermediate cooling, for the management of the heat of reaction (Rönsch et al. 2016). However, to reduce investment and operational efforts, there is large scope in finding new solutions to improve process temperature control, enhancement of efficiency and flexibility.

Some research groups (Borgschulte et al. 2013; Walspurger et al. 2014) recently proposed and demonstrated, in a lab-scale fixed bed apparatus, the concept of Sorption-Enhanced Methanation (SEM), where the steam generated during methanation is continuously captured by means of a suitable regenerable sorbent (Eq. 3):

$$\text{Sorbent} + H_2O = \text{Sorbent}(H_2O) \text{ (3)}$$

In this way, the equilibrium of reactions (1) or (2) can be shifted toward higher methane yields, allowing to lower the operating pressures.

In this work an innovative configuration system for the methanation process based on the technology of dual interconnected fluidized beds is presented, coupling the concepts of SEM and of chemical looping.

The idea is to employ one fluidized bed, named *Methanator/Hydrator*, for the catalytic methanation reaction with inherent capture of steam by a selective sorbent at low or atmospheric pressure, in order to drive the equilibrium towards product formation; while the regeneration of the hydrated sorbent takes place in a second fluidized bed, named*Dehydrator*, operated at higher temperature. In particular, this study was focused on the evaluation of the suitability of two different sorbents to capture and release water at different temperatures and reaction environments relevant for catalytic methanation. The two sorbents used are CaO, derived from natural limestone, and a commercial Zeolite 3A. The CaO sorbent was obtained by fluidized bed calcination at 850 °C of an Italian limestone named Massicci.

The tests were carried out in a lab-scale apparatus called Twin Beds (TB) (Coppola et al. 2017) consisting of two identical bubbling beds operated batch-wise and connected by a rapid solid transfer line. Transfer of sorbent is accomplished pneumatically by using a valve system arranged along the duct and at the outlet of the reactors. The two reactors were employed as hydrator and dehydrator respectively.

Each test consisted of 10 complete cycles, with a fluidization velocity fixed at 0.5m/s, while the time of each hydration or dehydration step was fixed at 10 min.

The main operating conditions were: hydration with 10% steam (balance air) and dehydration in nitrogen or in air. An initial sensitivity analysis on temperature was carried out fixing the hydration temperature at 250 or 300°C and varying the dehydration temperature at 350, 400 and 450°C in air.

Generally, the results for CaO showed, for all conditions investigated, that the H_2O capture capacity (Î³), expressed as the mass of captured water per mass of initial sorbent, increases after the first cycle, with a maximum around the 2nd-3rd cycle, and successively it tends to decrease with the number of cycles, reaching an asymptotic value after approximately the 5th-6th cycle. The increase of the dehydration temperature entails an increase of Î³. In fact, for the tests with hydration at 300Â°C the asymptotic values of Î³ were 0.04, 0.13 and 0.15 g/g for a dehydration temperature of 350, 400 and 450 Â°C, respectively. This behavior could be determined by a different change of the sorbent microstructure (pore structure), with a consequent variation of the sorbent capacity. A positive effect on Î³ was also registered with the rising of the hydration temperature.

As opposed to CaO, for the Zeolite 3A no decay effect of Î³ with the number of cycles was observed. However, the hydration temperature has an opposite effect on the Zeolite as for CaO: the higher the hydration temperature, the lower the sorbent capacity. On the contrary, the dehydration temperature seems to have no remarkable effect on the zeolite performance. In general, the value of Î³ fluctuates from 0.02 up to 0.05 g/g which are on average lower with respect to CaO.

Borgschulte, Andreas et al. 2013. â€œSorption Enhanced CO2methanation.â€ *Physical Chemistry Chemical Physics* 15(24): 9620â€“25.

Coppola, Antonio, Fabrizio Scala, Liberato Gargiulo, and Piero Salatino. 2017. â€œA Twin-Bed

Test Reactor for Characterization of Calcium Looping Sorbents.â€ *Powder Technology* 316: 585â€“91.http://dx.doi.org/10.1016/j.powtec.2016.11.067.

GÃ¶tz, Manuel et al. 2016. â€œRenewable Power-to-Gas: A Technological and Economic Review.â€ *Renewable Energy* 85: 1371â€“90.

RÃ¶nsch, Stefan et al. 2016. â€œReview on Methanation - From Fundamentals to Current Projects.â€ *Fuel* 166: 276â€“96.

Smestad, Greg P., and Aldo Steinfeld. 2012. â€œReview: Photochemical and Thermochemical Production of Solar Fuels from H2O and CO2 Using Metal Oxide Catalysts.â€ *Industrial and Engineering Chemistry Research* 51: 11828â€“40.

Walspurger, StÃ©phane et al. 2014. â€œSorption Enhanced Methanation for Substitute Natural Gas Production: Experimental Results and Thermodynamic Considerations.â€ *Chemical Engineering Journal* 242: 379â€“86.

A CFD-DEM-IBM Method for Cartesian Grid Simulation of Gas-Solid Flow in Complex Geometries

Conference: Fluidization

Year: 2019

Proceeding: Fluidization XVI

Group: General Paper Pool

Session: 6C: Modeling and Simulation

Time: Wednesday, May 29, 2019 - 10:50am-11:02am

Authors:

Wang, J., State Key Laboratory of Multiphase Complex Systems, Institute of Process Engineering, Chinese Academy of Sciences
Zhao, P., State Key Laboratory of Multiphase Complex Systems, Institute of Process Engineering, Chinese Academy of Sciences

The use of CFD-DEM method to accurately simulate gas-solid flows in complex geometries is challenging, mainly due to the complexity related to the use of unstructured computational grids. To this end, the gas-solid flows in complex geometries were simulated using Cartesian grids in present study, where the interactions between gas phase and complex geometries were firstly modelled using the immersed boundary method (IBM) implemented in OpenFOAM, it was found that the computational efficiency is quite low. To improve the efficiency of the CFD-DEM-IBM solver, a new IBM method was then proposed by removing the neighboring immersed boundary cells from the interpolated extended stencil in the reconstruction of the velocity field near the wall, and proposing a new zero-gradient boundary condition to replace the original Neumann boundary condition for reconstructing the pressure field.

Single-phase flow past a stationary cylinder, single-phase pipe flow, gas-solid flow in a cylindrical fluidized bed and fluidized bed with immersed tubes were simulated with four different IBM imposition methods to assess the accuracy and efficiency of CFD-DEM-IBM solver. It was shown that the results of CFD-DEM-IBM simulations, which are more accurate than those of using unstructured grids, agree well with the reported experimental, analytical and/or numerical results available in literature, and the computational efficiency of newly proposed CFD-DEM-IBM solver is one or two order of magnitudes faster than that of the original IBM in OpenFOAM.

A Comprehensive Model for the Interface-Based Crystal Particle Autoselection Via Membrane Crystallization

Conference: Fluidization

Year: 2019

Proceeding: Fluidization XVI

Group: General Paper Pool

Session: 6C: Modeling and Simulation

Time: Wednesday, May 29, 2019 - 11:38am-11:50am

Authors:

Jiang, X., Dalian University of Technology
Li, J., Dalian University of Technology
He, G., Dalian University of Technology
Li, G., Dalian University of Technology

A
comprehensive model for interface-based crystal particle autoselection via membrane crystallization:

Jin Li, Guannan Li, Xiaobin Jiang*,
Gaohong He

Corresponding author E-mail: xbjiang@dlut.edu.cn

State Key Laboratory
of Fine Chemicals, R&D Center of Membrane Science and Technology, School of Chemical Engineering, Dalian University of Technology, Dalian, Liaoning 116024, China

Accurate
crystallization control is an essential aspect for chemical engineering and separation process. With the fast development of membrane combined separation technology, the unique features of membrane crystallization (MCr) in simultaneous crystal and solution recovery with low

energy consumption were realized (PNAS
2017, 114, 6936; Adv Mater, 2016, 28, 610; AIChE J. 2016, 62, 829;AIChE J. 2017,63, 2187).

While, the physical control mechanism of the nucleation and particle motions (adhere or detach) via MCr is still not clear, which limits the operation design and industrial application of MCr. In addition, the membrane induced crystallization should be distinguished from the conventional membrane scaling and uncotrolling separation in both conceptual aspect and engineering aspect (show in Fig. 1).

In this work, we endeavored to develop a model combined the particle motion behavior with classic nucleation theory and establish an experiment equipment to directly observe the simulated particle motion. For the first time, beyond a separation technology, the novel Interfacial crystal particle auto-selection function of advanced membrane crystallization (MCr) was revealed and verified in theoretical and experimental aspect. As illustrated in Fig. 1, with the definition of particle detachment criterion K and the three particle motion modes (temporary adhere, detachment and perpetual adhere), the three different function regions of MCr (nucleation detection, scaling control and crystallization regulation) was systematically demonstrated via the directly observed results.

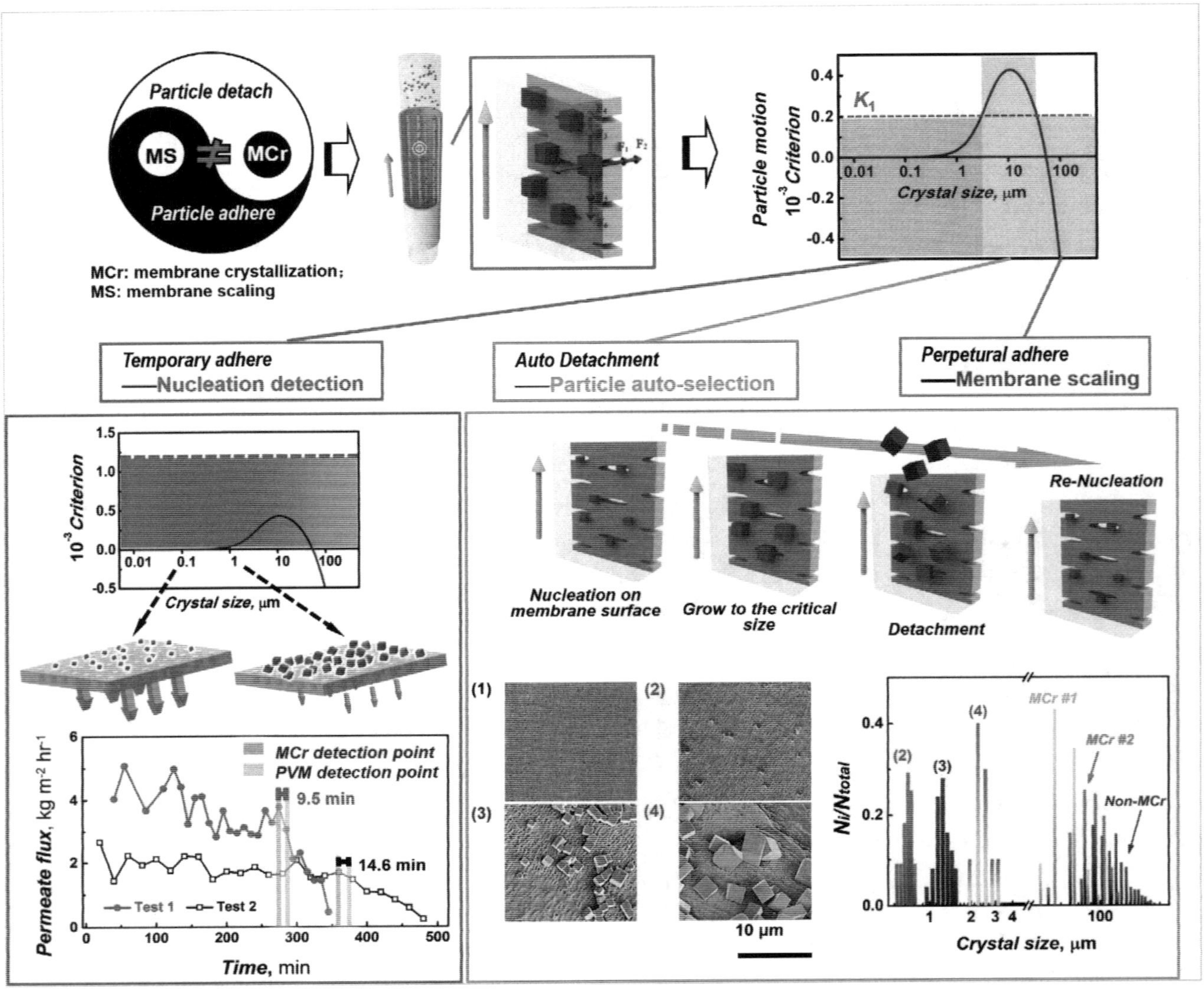

Fig.
1 Schematic diagram of the key functions illustrated via the developed MCr model

The exciting findings in this paper can uncover the optimization strategy for membrane scaling adjustment (which is one of the foundational problems in membrane separation engineering: J. Membrane Sci. 2016, 499, 201; Environ. Sci. Technol. 2015, 49, 4235) and the nucleation controlling mechanism (which are the general concerns in crystallization engineering: Science, 2013, 341, 855; AIChE J. 2016, 62(9), 3505; AIChE J. 2014, 60 (1), 275).

The advantages of MCr on accurate crystal seed auto-selection based on the natural membrane interfacial characteristics, which also shed light on the potential applications of the uniform size distribution, morphology modified particles manufacture and in situ nucleation detection, etc. Thus, the results in this paper are expected to be rapidly and widely distributed, discussed, and built upon in multiple disciplines, such as chemical engineering, physical, pharmaceutical, bioengineering, etc.

Acknowledgments

We acknowledge financial contribution from National Natural Science Foundation of China (Grant No. 21527812, 21676043, U1663223, 21606035), Changjiang Scholars Program (T2012049), the Fundamental Research Funds for the Central Universities (DUT16TD19, DUT17ZD203).

CFD-DEM Simulations of Heat Transfer inside Fluidized Beds at Elevated Temperatures

Conference: Fluidization

Year: 2019

Proceeding: Fluidization XVI

Group: General Paper Pool

Session: 6C: Modeling and Simulation

Time: Wednesday, May 29, 2019 - 11:26am-11:38am

Authors:

Mihajlovic, M., Eindhoven University of Technology
Roghair, I., Eindhoven University of Technology
Van Sint Annaland, M., Eindhoven University of Technologhy

For many different applications the good heat transfer rates in gas-solid fluidized beds is key to the overall performance. However, the formation of densified zones near confining walls or submerged surfaces, or the formation of clusters of particles, or even agglomerates, significantly affect the fluidization behaviour and the heat transfer inside the fluidized bed and the heat exchange with the walls (a.o. Patil et al., 2015; Li et al., 2016). In addition, the operating temperature influences the particle collisional properties and inter-particle forces, altering the fluidization behaviour and consequently the heat transfer processes.

Inter-particle forces are known to cause detrimental effects in industrial drying processes for example for the production of polyolefins, where the formation of particle agglomerates causes reduced particle mixing, which affects the solids circulation patterns and may cause hot spot formation. Recent experiments have shown that at higher temperatures inter-particle forces become more pronounced, indicating a temperature dependency of the Hamaker constant (Campos Velarde et al., 2016). Moreover, also the particle collisional properties are known to strongly depend on the particle temperature.

In this work, the effects of the inter-particle forces and particle properties on the heat transfer processes inside a gas-solid fluidized bed are investigated, employing an in-house developed Discrete Particle Model (DPM), which is an Euler-Lagrange model with a discrete description of the solids phase and a continuous description of the gas phase. Particle-particle collisions are dealt with deterministically, using a soft-sphere collision model, which allows multiple simultaneous contacts between several pairs of particles. The motion of each individual particle is tracked and described with Newtonâ€™s second law of motion. Contrary to commonly used studies, where the particle restitution coefficients are set to fixed constants for all particles, in our DPM we let the restitution coefficients depend on the particle impact velocity and particle temperature using the correlations by Thomton and Ning (1998) and Dong et al. (2014). The van der Waals forces were described using the equations by Hamaker (1937) and Israelachvili (1992). The gas-particle heat transfer coefficient was evaluated with Gunnâ€™s correlation. Special attention was paid to the numerical implementation of the inter-particle forces.

The extended DPM code was used to investigate in detail the effects of the inter-particle forces and collisional properties on the minimum fluidization velocity, the bed porosity at minimum fluidization, and cluster size and the formation of dead zones at different fluidization conditions. In this work we extend this study to characterise the influence of inter-particle forces and change in restitution coefficients on the heat transfer processes in the fluidized bed.

REFERENCES

PATIL, A.V., PETERS, E.A.J.F., KUIPERS, J.A.M., â€œComputational study of particle temperature in a bubbling spout fluidized bed with hot gas injectionâ€, *Powder Technology* 284:475-485, 2015

LI, Z., VAN SINT ANNALAND, M., KUIPERS, J.A.M., DEEN, N.G., â€œCFD-DEM modeling of fluidized beds with heat production: Influence of the particle size distribution and heat sourceâ € *Fluidization XV*, 2016.

VELARDE, I.C, GRIM, R., GALLUCCI, F., VAN SINT ANNALAND, M., â€œInfluence of Temperature on Minimum Fluidization Properties of Gas-Solid Fluidized Bedsâ€, *Powder Technology*, 2016 Submitted.

DELIMONT, JACOB M. "Experimental Investigation of Temperature Effects on Microparticle Sand Rebound Characteristics at Gas Turbine Representative Conditions." (Doctoral dissertation) 2014

THOMTON, C., NING, Z.M. â€œA theoretical model for the stick bounce behaviour of adhesive, elastic-plastic spheresâ€, *Powder Technology* 99, 154â€"162., 1998,

DONG, M., XIE, J., BAI, L., AND LI, S., â€œAn Experimental Investigation on the Influence of Temperature on the Normal Impact of Fine Particles with a Plane Surfaceâ€ *Energies* 1:2079-2094, 2014

HAMAKER, H. C. â€œThe Londonâ€"van der Waals attraction between spherical particlesâ € *Physica* 4(10), 1058â€"1072., 1937

ISRAELACHVILI, J.N., â€œIntermolecular and surface forcesâ€*Academic Press*, London, 1992

MIHAJLOVIC, M. ROGHAIR, I. VAN SINT ANNALAND, M. â€œHigh temperature fluidization - influence of inter-particle forces on fluidization behaviourâ€ In J. E. Olsen, & S. T. Johansen (Eds.), Progress in Applied CFD â€“ CFD2017 (pp. 107-114). Oslo: SINTEF Academic Press, 2017

Characterization of Drag Force on Agglomerates in Fluidized Beds

Conference: Fluidization

Year: 2019

Proceeding: Fluidization XVI

Group: General Paper Pool

Session: 6C: Modeling and Simulation

Time: Wednesday, May 29, 2019 - 11:02am-11:14am

Authors:

Tang, Y., Eindhoven University of Technology
Cox, J., Eindhoven University of Technology
Kuerten, H., Eindhoven University of Technology

Agglomeration of bed materials is an important phenomenon in several applications that use fluidized bed technologies such as granulation, compacting, drying, combustion and gasification, etc. Agglomerates complicate the flow structure, thus influence the hydrodynamics as well as mass and heat transfer in fluidized beds. For instant, fluidization of Geldart C/A particles is expected to be particularly difficult due to agglomeration behaviour by cohesive forces existing between particles, which become more and more prominent as the particle size decreases. On the other hand, in fluidization drag force is one of the dominant particle-fluid interactions to control particle movements, circulation and concentration distribution. It is well understood that formation of agglomerates (similar to particle clusters) significantly reduces the drag force comparing to single particles. Therefore, understanding the mechanisms underlying particle agglomeration and their effect on the hydrodynamics (mainly gas-solid drag force) is crucial for the design and control of the desired process performance. Unfortunately, currently available drag correlations that are applied in modelling of fluidization do not truly take into account the agglomeration phenomenon.

Computationally, direct numerical simulations (DNS) is proven effective to study in detail the gas-solid interactions in multiphase flows. Such simulations employ no assumptions for solving the fluid motions, and resolve the flow field from the boundary layer of the particle surface. The hydrodynamic forces acting on the solid particles are directly computed in DNS from the fluid flow via enforcing a no-slip boundary condition at particle surface.

In this work, a detailed understanding and quantification of agglomeration behaviour of fine particles as well as drag resistance of the formed agglomerates under fluidization conditions is provide using DNS. The sedimentation of hundreds of primary particles of same diameters (orders of micrometers) in a fluidizing fluid is considered under different flow conditions (Reynolds numbers upto 200). The particles experience gravity, drag as well as van der Waals forces. During the sedimentation process, the particles can stick together and form agglomerates, which can then also be broken up. The obtained agglomerate morphologies are characterized by various structural parameters such as sphericity, a convex hull-based porosity and the radius of gyration. The time-dependent growth of agglomerates size is simultaneously tracked. The formed agglomerates with different morphologies are then analysed with respect to the drag force in comparison with the drag on single primary particles. Finally, by comparing all the simulation results to the existing drag correlations for non-spherical particles, a new drag correlation is formulated. This new correlation is a correction to the Schiller-Naumann [2] correlation for spheres but includes the dependency on the agglomerates morphologies. The improved drag correlation can be used to improve the accuracy of prediction of CFD-DEM and TFM models for fluidized beds with agglomeration phenomenon.

Design and Simulation of a New Micro Fluidized Bed Reactor By Computational Fluid Dynamics

Conference: Fluidization

Year: 2019

Proceeding: Fluidization XVI

Group: General Paper Pool

Session: 6C: Modeling and Simulation

Time: Wednesday, May 29, 2019 - 11:50am-12:02pm

Authors:

Latifi, M., Ecole Polytechnique Montreal
Mokhtari, M., Ecole Polytechnique de Montreal
Berruti, F., Western University / Institute for Chemicals and Fuels from Alternative Resources (ICFAR)
Briens, C., Western University / Institute for Chemicals and Fuels from Alternative Resources (ICFAR)
Chaouki, J., Ecole Polytechnique Montreal

New applications are emerging for fluidized bed reactors such as baking and thermal cracking of ores, chemical vapor deposition, waste valorization, fluidization of nanoparticles, and catalytic reactions for feedstocks that are unconventional with a low grade of the desired elements etc.

Such reactors are usually endothermic and should operate at a high temperature with a desired gas residence time. Micro fluidized bed reactors are promising reactors to save time and cost of experimental investigations when especially designed to mimic an operation at commercial scale and to screen operating conditions, feedstock formulations, and bed material physicochemical properties in terms of feed conversion, product selectivity and gas-solid interactions, e.g. agglomeration and attrition.

Authors have designed and developed a micro fluidized bed reactor where both gas and solid phases operate in batch mode, and an induction heating mechanism supplies the required heat with a high heating rate to stay at the setpoint temperature, i.e. up to 1200 °C. A centrifugal impeller takes in the outlet gas from the bed and returns it to the bed though internal tubes immersed into the reactor, i.e. gas sparger. The tubes are also heating elements; that is, the generated heat on the external surface of the tubes by the induction heating is transferred to the bed.

Computational fluid dynamics was employed to optimize dimensions of the reactor and the gas circulation system so that the impeller rotates at a speed below 5000 rpm enabling fluidization of Geldartâ€™s group B particles with negligible wall effect.

The sliding mesh technique was used for numerical simulation of the impeller rotation. Prediction of the turbulence in the system was done by using the K-omega turbulent model and the hydrodynamics of the fluidized bed reactor was simulated by the Eulerian multiphase approach. The SIMPLE algorithm was used for pressure-velocity coupling and the second order upwind was utilized to obtain precise results. The results show that the gas flow rate is proportional to the impeller diameter, however, the shape of the blades does not have a considerable effect. Based on the numerical results, the design of the impeller and the reactor was optimized.

Numerical Investigation of Electrostatic Charging on the Hydrodynamics of Gas-Solid Fluidized Beds

Conference: Fluidization

Year: 2019

Proceeding: Fluidization XVI

Group: General Paper Pool

Session: 6C: Modeling and Simulation

Time: Wednesday, May 29, 2019 - 11:14am-11:26am

Authors:

Kordnejad, M., TUe
Roghair, I., Eindhoven University of Technology
Van Sint Annaland, M., Eindhoven University of Technologhy

Numerical investigation of electrostatic charging on the hydrodynamics of gas-solid fluidized beds

1. Introduction

Fluidized beds are used commonly in industry, for example for hydrocracking, polymer production, drying of minerals, foods and pharmaceuticals, and many more applications. A fluidized bed consists of a bed of particles suspended in a gas flow from below, termed the emulsion phase, which behaves as a fluid. Through the emulsion phase, gas bubbles form at the injector, and rise through the bed. These bubbles are responsible for the creation of solids circulation throughout the equipment, which results in very good heat and mass transfer characteristics. The minimum fluidization velocity is a key parameter, which determines many other hydrodynamic parameters and is central to the design of fluidized bed reactors.

In gas-solid fluidized beds, the emulsion phase is usually dense, and many collisions occur between particles and between particles and the walls (Hendrickson 2006; Sowinski, Miller, and Mehrani 2010). This causes particles to gain electrostatic charge, a phenomenon called triboelectrification or contact electrification. These charges cause additional inter-particle and particle-wall forces which may result in detrimental phenomena, such as wall sheeting and particle agglomeration.

How triboelectrification affects the minimum fluidization velocity, and thereby the hydrodynamic behavior, is however not fully understood. Carrying out experiments to investigate the effect of electrostatics on the hydrodynamics of gas-solid fluidized beds has proven rather difficult, especially related with the difficulties associated with the determination of the charge distribution over the particles. Without detailed information on the charge distribution, it will be very difficult to draw quantitative conclusions. Therefore, in this work a modelling study has been carried out to elucidate the effects of triboelectrification on the hydrodynamic behavior of gas-solid fluidized beds, particularly focusing on the minimum fluidization velocity, the solids circulation patterns and bubble behavior in beds with electrostatically charged particles.

2. Model Description

In this work an Euler-Lagrange model, referred to as CFD-DEM or the Discrete Particle Model (DPM), has been used to study the hydrodynamics of a fluidized bed with charged particles. The DPM allows for a fine-grained investigation on the particle-level, but is restricted to relatively small system sizes because of the required calculation times. Basically, every particle with its own individual properties such as mass, velocity, and charge, is tracked individually using Newtonâ€™s law of motion, while the gas flow is solved on the Eulerian (background) mesh. The inter-phase momentum coupling is performed by a mapping term in the momentum balance, corresponding to the drag force term in the equations for the particle motion.

The Cundall and Strack linear spring/dashpot soft-sphere model is used to describe the contact collision force of particles (Cundall and Strack 1979). The electrostatic interaction between particles is calculated by Coulombâ€™s law, assuming that the charge is distributed uniformly over the particle surfaces and that the calculation can be based on point charges. The electrostatic force consists of particle-particle and particle-wall interactions.

$F_{electrostatic} = F^{e}_{p-p} + F^{e}_{p-w}$ 1

It is assumed that the system wall has no charge and the interaction between particles and the conducting wall is due to the induction of charge at the wall (image charge).

$F^{e}_{p-p} = q_i q_j\ n_{ij}/4$Ï€Îµ$_0 r_{ij}^2$ r_{ij}â‰¤ r_{cut} 2

$F^{e}_{p-w} = q_i q_i\ n_{ij}/4$Ï€Îµ$_0 (2r_{ij})^2 r_{ij}$â‰¤ r_{cut} 3

where and represent the charge of particle *i* and *j*, the distance between them, the vacuum permittivity (equal to) and the unit vector.

The consideration of all the electrostatic interactions between all particle combinations quickly becomes computationally prohibitive. Considering only the most significant interactions (i.e. nearby particles) can be an effective way of reducing the required computation time. In this work a neighbor list, or Verlet list, has been used, storing for each particle a list of nearest-neighboring particles (Cal et al. 1967), for which the electrostatic interactions have been calculated.

3. Results

- **Minimum fluidization and minimum bubbling velocity**

To determine the minimum fluidization velocity, the same pressure-drop procedure has been used as is standard in experimental works. The simulations are started with a bed consisting of Geldart B type particles (either charged and uncharged, depending on the case), which is subjected to a slowly linearly increasing superficial velocity from nihil till a velocity where bubbles have started to appear, and subsequently, the gas flow rate is linearly decreased again. During the simulation, the pressure drop over the bed is logged. Our simulations have shown that in charged systems the appearance of bubbles is delayed, while the system starts to fluidize earlier. Helped by the electrostatic forces, the bed starts to fluidize and expand already at lower superficial velocities, since the particles repel each other. Moreover, more energy is needed to push particles into densified zones and form bubbles in the bed. Thus, the minimum fluidization velocity decreases and the minimum bubbling velocity increases with the degree of electrification of the particles.

- **Bubble injection**

To better understand the effect of electrostatic forces on the hydrodynamics, a single bubble was injected into the bed. The rise velocity and bubble shape of a system with uncharged and charged particles were compared. Results show that in a system of uncharged particles, the bubble is clearly visible, whereas in a system with charged particles, the bubble experiences more excessive particle raining from the roof of the bubble and the bubble starts to disappear in the middle of the bed. Due to the electrical repulsion between the particles, the bubble start to fill with particles already as soon as it starts to rise through the emulsion phase, making a detailed analysis of the bubble rise velocity impossible. The extent of particle raining has been investigated as a function of the particle charge for different particle properties (size and density).

- **Circulation patterns**

By comparing the solids circulation patterns of charged and uncharged systems, it is clear that a bed with charged particles exhibits more homogeneity and due to the image charge of the walls the particles show a tendency to stick to the system walls. Calculations have shown that even 10% charge of the maximum charge can already cause serious wall sheeting in a gas-solid fluidized bed consisting of polymer particles. Due to smaller size of the bubbles in charged fluidized beds, the solids circulation is decreased significantly. Also this effect is quantified as a function of the particle charge for different particle systems.

4. Conclusions

The DPM was extended to model gas-solid fluidized beds in the presence of mono-charged particles. It was found that that the minimum fluidization velocity decreases, while the minimum bubbling velocity increases significantly in fluidized beds with charged particles. Moreover, a charged system shows more homogeneous fluidization with smaller bubbles suffering from heavy particle raining, causing significantly reduced solids circulation in the bed. These effects are quantified as a function of the particle charge for different particle systems. Future work will focus on the extent of fouling in polymer fluidized beds and its effect on the bed-to-wall heat transfer rates.

References

Cal, P. Y. S. I., R. E. Evi, Properties Lennard, and Jones Molecules. 1967. â€œComputer Experiments on Classical Fluids. I. Thermodynamical Properties of Lennardâ€“Jones Molecules.â€ *Physical Review* 159(1):98.

Cundall, P. A. and O. D. L. Strack. 1979. â€œA Discrete Numerical Model for Granular Assemblies.â€ *GÃ©otechnique* 29(1):47â€“65. Retrieved (http://www.icevirtuallibrary.com/doi/10.1680/geot.1979.29.1.47).

Hendrickson, Gregory. 2006. â€œElectrostatics and Gas Phase Fluidized Bed Polymerization Reactor Wall Sheeting.â€ *Chemical Engineering Science* 61(4):1041â€“64.

Sowinski, Andrew, Leigh Miller, and Poupak Mehrani. 2010. â€œInvestigation of Electrostatic Charge Distribution in Gas-Solid Fluidized Beds.â€ *Chemical Engineering Science* 65(9):2771â€“81.

A New Particle Cluster Diameter Model for Fluidized Bed

Conference: Fluidization

Year: 2019

Proceeding: Fluidization XVI

Group: General Paper Pool

Session: 6D: Modeling and Simulation

Time: Wednesday, May 29, 2019 - 11:26am-11:38am

Authors:

Li, F., Chinese Academy of Sciences
Wang, X., Chinese Academy of Sciences
Wang, W., Institute of Process Engineering, Chinese Academy of Sciences

Particle cluster is the characteristic structure in gas-solid fluidization. It is of great importance in hydrodynamic modeling and simulation of fluidized bed. Generally, there are two main factors to describe a particle cluster: the cluster density denoting the solid volume fraction inside a cluster and the cluster diameter denoting the scale of a cluster. Recent reports show that the cluster density can be correlated with average solid volume fraction and operating condition of fluidized bed. However, it is still very hard to determine the cluster diameter via experimental method, due to its irregular shape and transient characteristics. The present work makes use of the recently developed cluster density model and the energy minimization multi-scale (EMMS) model to analyze the fluidization system. And the particle cluster diameter is finally obtained through this analysis and energy minimization restrictions. This new cluster diameter model has been used in the EMMS/DP model to deduce a corresponding drag force model, and is finally applied in some riser simulations. The simulation results agree well with experimental results under different operating conditions, proving the high accuracy in riser simulation of the new cluster diameter model.

Evaluation of Drag Models for Simulating Gas-Solid Fluidization of Geldart A and B Particles

Conference: Fluidization

Year: 2019

Proceeding: Fluidization XVI

Group: General Paper Pool

Session: 6D: Modeling and Simulation

Time: Wednesday, May 29, 2019 - 11:38am-11:50am

Authors:

Zhao, J., Harbin Institute of Technology
Liu, G., Harbin Institute of Technology
Wu, Y., Harbin Institute of Technology
Lu, H., Harbin Institute of Technology
Wang, S., Harbin Institute of Technology
Zhai, Z., Harbin Institute of Technology

In the gas-solid fluidized bed reactors, two-fluid model based on the kinetic theory of granular flow exhibits a significant dependence on drag models, especially for Geldart A and B particles. Many drag models are available in the literature, which have been reported to work for different systems. In this work, the three-dimensional simulations by two-fluid model for gas-solid fluidized beds, including bubbling fluidized bed (BFB) and circulating fluidized bed (CFB), were established. The evaluation of different drag models for Geldart A and B particles was carried out. By using the different single particle drag coefficient C_d, the calculation for five drag models (Gidaspow, Symlal, HKL, BVK and De felice) were performed. Comparison with the available experiment data, the most accurate drag model for predicting mesoscale structures (i.e., bubbles and clusters) has been acquired. The results of this work provide a guideline for choosing appropriate drag models for simulating Geldart A and B particles in the gas-solid system and develop the more accurate drag models by changing the single particle drag coefficient C_d. In the further study, we will explore the modification of drag model due to the effect of mesoscale structures and the drag model in the liquid-solid sytem.

Experimental Study and DEM Modelling of the Contact Behaviour of Cylindrical Particles

Conference: Fluidization

Year: 2019

Proceeding: Fluidization XVI

Group: General Paper Pool

Session: 6D: Modeling and Simulation

Time: Wednesday, May 29, 2019 - 11:14am-11:26am

Authors:

Grohn, P., Technische Universität Kaiserslautern
Weis, D., Technische Universität Kaiserslautern
Antonyuk, S., Technische Universität Kaiserslautern
Heinrich, S., Hamburg University of Technology

In the food, chemical and pharmaceutical industry, cylindrical particles are often used and further processed in various production steps e.g. fluidization, drying and coating. To control and optimize these processes, it is essential to know and describe the particle kinematics and dynamics. An important tool to obtain these data is numerical simulations with the computational fluid dynamics (CFD) coupled with the discrete element method (DEM). In DEM, the interactions are calculated for each single particle based on models describing physical properties of particles and their mechanical behavior during slow, fast and repeated stressing [1-3]. For cylindrical particles, there is a lack of understanding of how particle shape influences the particle contact behavior in different scenarios. Especially for the exact calculation of the particle interactions in DEM, it is essential to describe the particle contact behaviour with a compatible contact model.

In this work, the influence of loading direction (contact geometry) and deformation behaviour (from elastic to plastic) on cylindrical particles during compression and shearing is investigated. The compression tests are performed with a Texture AnalyserÂ®. For the measurement of particle-wall sliding friction, a self-developed setup at the University of Kaiserslautern is used. During the measurement of tangential forces, the particles are fixed to a flat holder and moved over a defined wall surface at different normal loads. In addition, experiments in a fluid bed rotor granulator equipped with a Magnetic Particle Tracking (MPT) system [4] are conducted in order to visualize and quantify the real particle kinetics and dynamics.

The intention of these investigations is the experimental calibration of contact models, which can be applied for numerical DEM studies of fluidization processes with cylindrical particles.

References

[1] Antonyuk S, Palis S and Heinrich S 2011 *Powder Technology* **206**88â€“98

[2] Antonyuk S, Heinrich S, Tomas J, Deen N G, van Buijtenen M S and Kuipers J A M 2010 *Granular Matter* **12** 15â€“47

[3] Oschmann T, Hold J and Kruggel-Emden H 2014 *Powder Technology* **258** 304â€“23

[4] Neuwirth, J., Antonyuk, S., Heinrich, S., Jacob, M.: CFD-DEM study and direct measurement of the granular flow in a rotor granulator, Chemical Engineering Science 86 (2013), 151-163.

Numerical Simulation of Flow Characteristics of Two-Component Particles in Magnetically Controlled Gas-Solid Bubbling Bed

Conference: Fluidization

Year: 2019

Proceeding: Fluidization XVI

Group: General Paper Pool

Session: 6D: Modeling and Simulation

Time: Wednesday, May 29, 2019 - 10:50am-11:02am

Authors:

Wang, B., Harbin Institute of Technology
Wang, T., Harbin Institute of Technology
He, Y., Harbin Institute of Technology

Numerical
simulation of flow characteristics of two-component particles in magnetically controlled gas-solid bubbling bed

Biao Wang,
Tianyu Wang, Yurong He *

[1] School of Energy Science and
Engineering, Harbin Institute of Technology, 150001, Harbin, China

rong@hit.edu.cn

The gas-solid
fluidized bed under the action magnetic field refers to adding magnetic field to ordinary gas-solid fluidized bed as the external energy field, and the solid particles in the flow field are magnetic particles. A magnetic field is added outside the reactor, and the magnetic field lines pass through the interior of the bed layer, attracting the particles of magnetic materials to be regularly

distributed in the bed layer. In the fluidization process, the bubbles are evenly distributed, thus improving the fluidization quality[1].
The magnetic fluidized bed has outstanding advantages, such as low vibration, low noise, recyclable use and higher mass and heat transfer rate. It has high development value and broad application prospect[2].
It is widely used in many industrial processes, such as filtration, dust removal, separation of materials, fluidization of viscous particles and fluidization of particles under microgravity.

So far, domestic
and foreign scholars have conducted some researches on the flow characteristics of particles in magnetic fluidized beds. For example, in the area of enzymatic catalysis, Bahar[3] et
al. fixed the reaction enzyme on magnetic particles and performed the catalytic reaction in the magnetic fluidized bed. Jovanovic[4] et
al. conducted experiments and theoretical investigations on the magnetic-assisted liquid-solid fluidized bed under the normal and microgravity conditions, and proved that the gradient magnetic field can make the fluidized bed under microgravity normal fluidization. Wang[5] et
al. studied the mechanism of flue gas desulfurization in magnetic fluidized bed, and improved the utilization rate of desulfurizer. Magnetic-fluidized beds have been applied in the fields of chemical catalysis and material separation and transportation. However, the practical application of magnetic fluidized bed is lacking, and most researches are focused on theoretical research.

Fluidized beds
are widely used to separate materials in the fields of industrial production, metallurgy, chemical industry, cleaning and pharmacy. However, it is particularly difficult to separate materials by fluidized bed without external physical fields due to similar particle density and size. Therefore, it is necessary to study how to improve the separation degree of two-component particles by adding energy field. Mohanta[6] et
al. applied the magnetic fluidized bed to the field of coal washing and made the particle flow more stable in the bed.

In this paper,
MFIX is used as the research tool, and a two-component system composing of ferromagnetic particles and non-magnetic particles is studied. The discrete element model is applied to investigate the two-component particles flow behaviors in a two-dimensional gradient magnetic field with different gradient. By comparing the simulation results in the bubbling bed with magnetic field and without magnetic field, it could be found that the ferromagnetic particles

form chains under the action of magnetic field. The separation degree increases with the increase of magnetic field gradient.

Acknowledgement

This work was financially supported by
the National Natural Science Foundation of China (Grant No. 91534112).

References

[1] Bin Zhang.
Numerical simulation and experimental study of gas-solid two-phase flow in magnetic fluidized bed [D]. Nanjing: Southeast University, 2004: 1-4.

[2] Ping Zeng,
Tao Zhou, Guanqun Chen, Zhiqiang Ge. Research and
application of magnetic fluidized bed [J]. Chemical progress, 2006, 25(4): 371-373.

[3] Bahar T,
Celebi S S. Performance of immobilized glucoamylase in a magnetically stabilized fluidized bed reactor (MSFBR)[J]. Enzyme and Microbial Technology, 2000, 26(1): 28-33.

[4] Jovanovic G
N, Sornchamni T, Atwater J E, et al. Magnetically assisted liquid¨Csolid fluidization in normal and microgravity conditions: experiment and theory[J]. Powder Technology, 2004, 148(2-3): 80-91.

[5] Zhixiao
Wang, Yunfeng Zhang. Study on mechanism of enhanced
flue gas desulfurization by magnetic fluidized bed [J]. Proceedings of the Chinese Society for Electrical Engineering, 2005, 25(14): 68-52.

[6] Mohanta S,
Rao C S, Daram A B, Chakraborty S, Meikap B C. Air dense medium fluidized bed for dry beneficiation of coal: technological challenges for future[J].
Particulate Science and Technology, 2013, 31(1): 16-27.

The MP-PIC Method for CFD-Simulation of Fluidized Beds - Comparison of Two Different Implementations

Conference: Fluidization

Year: 2019

Proceeding: Fluidization XVI

Group: General Paper Pool

Session: 6D: Modeling and Simulation

Time: Wednesday, May 29, 2019 - 11:02am-11:14am

Authors:

Dymala, T., Hamburg University of Technology (TUHH)
Wytrwat, T., Hamburg University of Technology
Hartge, E. U., Hamburg University of Technology
Heinrich, S., Hamburg University of Technology

Fluidized beds are very common in the industry, for example in combustion or gasification processes. Due to the large dimensions, the high number of particles and the reaction kinetics of industrial-scale combustors the simulation is commonly too computationally expensive to resolve every particle interaction. To allow the simulation of industrial plants with reasonable computational time the multi-phase â€“ particle in cell (MP-PIC) method is applied. This method results from a modification of the Euler-Lagrange approach, where the fluid is described by the Navier-Stokes equations on a Eulerian grid in combination with Lagrangian particles. To reduce the computational costs a defined number of particles with the same properties is represented by so-called parcels. This facilitates the consideration of various effects like particle size distributions as well as shrinkage effects and reduces the computational costs compared to the discrete particle model (DPM).

In this study two different implementation of the MP-PIC method in Barracuda VRÂ® and OpenFOAMÂ® are compared with experimental results of a lab-scale fluidized bed reactor to validate the applicability of the MP-PIC method. The fluidization velocities are set in the range from 0.2 â€“ 1.3 m/s with a reactor diameter of 0.1 m. The bed height of quartz sand with a sauter mean diameter of 220 Âµm is 0.1 m. For the comparison between simulation results and experimental data video recordings as well as frequency analysis are used. To evaluate the simulation results radial and axial profiles of the particle volume fraction are compared to literature.

Preliminary results indicated that both implementations of the MP-PIC method could predict the fluidization behavior at lower fluidization velocities with reasonable agreement, while the simulations fail to predict the impact of higher fluidization velocities. This is presumably due to the application of the homogeneous Gidaspow drag model, which neglects the reduced drag force by the formation of clusters and therefore overestimates the momentum exchange between the fluid and the particle phase. To allow for more accurate simulations a drag model based on the promising energy minimization multi-scale (EMMS) theory will be applied.

We gratefully acknowledge for the ï¬nancial support the German Research Foundation (DFG) (Germany). Project number HE 4526/21-1.

A Novel Method for Gas to Particulate Mass Transfer Measurements in Fluidized Beds

Conference: Fluidization

Year: 2019

Proceeding: Fluidization XVI

Group: General Paper Pool

Session: 7A: Fundamentals of Fluidization

Time: Wednesday, May 29, 2019 - 2:31pm-2:43pm

Authors:

Köhler, A., Chalmers University of Technology
Soria Verdugo, A., University Carlos III of Madrid
Pallarès, D., Chalmers University of Technology
Johnsson, F., Chalmers University of Technology

Mass transfer between the gas and the particulate phase is a key process in fluidized bed applications, in which the particulate phase can be to a greater or lesser extent active: while for example chemical looping combustion applies solely active particles, the particles in combustion and gasification are inert or less active. Despite the mass transfer between the gas and particulate phase being a decisive phenomenon for the modelling and design of fluidized bed reactors, the difficulty to accurately measure this phenomenon has resulted in little experimental data available in literature. Experimental studies providing such data have typically been carried out based on a mass balance closure by means of gas measurements, which implies uncertainties related to the gas measurements (with respect to for example accuracy of the gas analyzer and reproducibility, convolution and possible post-reaction of the gas sample).

This paper presents a novel method to measure the mass transfer between the gas and particulate phase in laboratory fluidized beds, which is based on the bed resting on a high-precision scale (±0.01 g). The scale can monitor the mass gain of the bed, initially consisting of dry silica gel particles, while fluidized by humified air. From the sampled signal of mass gained by the bed due to moisture adsorption by the silica gel particles, mass transfer coefficients can be calculated with different spatial resolution and compared for different operational parameters. This represents a novel, accurate and non-intrusive method to measure mass transfer in fluidized beds, which makes it possible to determine mass transfer from the gas phase to beds of active particles (e.g. oxygen carrier particles in a fuel reactor of a CLC unit) and to active particles (fuel particles) dispersed in an inert bed.

Application of the method proposed in this work shows that with an increase in fluidization velocity, the mass transfer between the gas and particulate increases, reaches a maximum and then decrease. This is due to the change in fluidization behavior with increased velocity where the bed enters different bubble regimes, which will be discussed in the paper. Further, the mass transfer is reduced with increased bed height.

CFD-DEM Simulation of a Fluidized Bed of Particles with Inter-Particle Cohesive Force

Conference: Fluidization

Year: 2019

Proceeding: Fluidization XVI

Group: General Paper Pool

Session: 7A: Fundamentals of Fluidization

Time: Wednesday, May 29, 2019 - 1:55pm-2:07pm

Authors:

Liu, D., Southeast University
Zhao, Z., Southeast University
Ma, J., Southeast University
Chen, X., Southeast University

Gas-solid fluidized beds with the cohesive particles are applied widely in industry, while agglomeration and poor fluidization could happen as a result of inter-particle cohesiveness. Different methods for incorporating cohesive force with CFD-DEM models are discussed. In this paper, a hysteresis contact model is coupled with CFD-DEM model. The fluidization characteristics of the cohesive particles under different conditions in a 2D fluidized bed are simulated. Generally, it is found that with increasing inter-particle cohesiveness, the particles begin to be agglomerated and the bubbles become gas channels. By increasing the cohesiveness, particle velocity decreases generally and the internal recirculation of cohesive particles is restrained. The fluctuation of particle velocity decreases generally and the probability density of particle velocity around zero increases. When the cohesiveness is increased further, the internal recirculation of particles is restrained and the gas channels become stable. Also, the fluidization of cohesive particles under different conditions in a 2D fluidized bed with immersed tubes is simulated. The immersed tubes play the frameworks promoting the formation of gas channels.

Characterization of Granular Electrostatics Generation

Conference: Fluidization

Year: 2019

Proceeding: Fluidization XVI

Group: General Paper Pool

Session: 7A: Fundamentals of Fluidization

Time: Wednesday, May 29, 2019 - 2:43pm-2:55pm

Authors:

Yao, J., China University of Petroleum-Beijing

Characterization of granular electrostatics generation

Jinzhui Li[1], Jun Yao[1], Yanlin Zhao[1]*, Chi-Hwa Wang[2]

[1] Beijing Key Laboratory of Process Fluid Filtration and Separation, College of Mechanical and Transportation Engineering, China University of Petroleum-Beijing, Beijing 102249, Peopleâ €™s Republic of China

[2] Department of Chemical and Biomolecular Engineering, National University of Singapore, 4 Engineering Drive 4, 117585, Singapore

* Corresponding author, Email address: ylzhao@cup.edu.cn (Y. Zhao)

ABSTRACT

Granule electrostatics is widely found in most granule engineering areas. However, the working mechanism has never been fully understood due to its sensitivity to working conditions, such as granule geometrical shape (including length-ratio and front-facing angle), contact area, granule material, relative humidity, sliding velocity, sliding orientation and normal stress. In this work, single granule was designed to slide along a metal plate and the electrostatics charges generated were measured via an electrometer connected with a computer. The working factors above have been considered for two kinds of granules made of PVC or coal. Regarding the geometrical factors, it is found that granules with larger length-ratio tend to produce more electrostatics, the slender or sharper granules tend to have higher capability of electrostatics generation and triangular granules generate more electrostatics than trapezoidal granules. In addition, electrostatics increase with decreasing granule front-facing angle and charges tend to gather at the granule tip. Regarding the granule material, it is found that coal granule tends to generate more electrostatics in comparison with PVC granule. The sequence of the relative work function for the three materials used in this work from high to low is given by the following sequence: PVC, stainless steel and coal. Regarding surrounding factors, it is found that electrostatics increases with granule contact area and such trend increases with decreasing relative humidity. Electrostatics generated increases with decreasing relative humidity and sliding velocity. Granule sliding orientation does affect electrostatics. In this work, granules sliding with smaller front-facing angle as well as shorter side tend to generate more electrostatics. Electrostatics increases with normal stress.

Experimental and Numerical Investigation of Coherent Motion in Slugging Gas-Solid Fluidization

Conference: Fluidization

Year: 2019

Proceeding: Fluidization XVI

Group: General Paper Pool

Session: 7A: Fundamentals of Fluidization

Time: Wednesday, May 29, 2019 - 2:07pm-2:19pm

Authors:

Lungu, M., Zhejiang University
Wang, H., Zhejiang University
Wang, J., Zhejiang University
Yang, Y., Zhejiang University
Chen, F., Zhejiang University
Siame, J., Copperbelt University

Gas-solid flows are characterized by a multitude of coherent flow structures displaying varying spatial-temporal behavior. The novel high speed particle image velocimetry, HSPIV measurement method coupled with advanced signal processing protocols are utilized to identify and characterize these structures. High speed pressure fluctuations from experiments and TFM simulations are also processed to reinforce the findings from the particle velocity fluctuations. Spectral analysis of particle velocity and pressure fluctuations reveal large scale oscillations at low frequencies. The classical Kolmogorov -5/3 power law is obeyed at low frequencies of the particle velocity fluctuations. The time-frequency properties of the bed dynamics are studied using the wavelet coherent analysis and the Hilbert Huang transform. The time localized coherent structures are identified using the wavelet coherence analysis while the multi-resolution of the particle fluctuations is realized with the Hilbert-Huang transform.

Experimental and Numerical Studies of the Fluidization and Electrostatic Characteristics of Non-Spherical Particles in a Pseudo-2D Fluidized Bed

Conference: Fluidization

Year: 2019

Proceeding: Fluidization XVI

Group: General Paper Pool

Session: 7A: Fundamentals of Fluidization

Time: Wednesday, May 29, 2019 - 3:19pm-3:31pm

Authors:

Qi, B., North China Electric Power University
Yan, Y., University of Kent
Zhang, W., North China Electric Power University
Wang, X., North China Electric Power University

Electrification is a common phenomenon in a gas-solid fluidized bed that involves intensive collisions and interactions between particles with each other and particles with the wall. There have been extensive studies of the mechanism of triboelectrification in particle systems using experimental and numerical methods. However, these studies focus mostly on spherical particles and investigations into non-spherical particle are scarce. It is challenging to quantify directly the electrostatic characteristics of non-spherical particles in a fluidized bed because of the high number and complex behaviors of particles in the bed. In addition, the charge transferred between the particles is difficult to measure accurately. In this study, by conducting experiments with single grains in a sliding process, the charges transferred between different materials are quantified. In the sliding experiments a range of 3D printed regular non-spherical particles with different shapes and surface areas are used. The effects of sphericity, material properties, ambient temperature and relative humidity on the charge generation of non-spherical particles are investigated. In addition, a new dynamic model for the electrification of non-spherical particles is established using the above parameters. In order to verify the established model, the fluidization of non-spherical particles was carried out in a pseudo-2D fluidized bed. Rice grains and mung beans were used as test material. To evaluate the hydrodynamic and electrostatic characteristics in the fluidized bed, digital imaging and electrostatic sensing techniques are applied. Moreover, computational fluid dynamics and discrete element modeling (CFD-DEM) method, which includes the new electrification model, is used to simulate the fluidization process. Non-spherical particles are represented by the multi-sphere model in the DEM method. The simulation results show more detailed information about charge characteristics and particle distribution in the whole bed. Besides, the relationship between the electrostatics and the hydrodynamics of particles is investigated. A close agreement between the experimental results and numerical simulation data on the characteristics of electrostatics and hydrodynamics is observed. The results confirm that the new dynamic model for the electrification of non-spherical particles is capable of simulating the electrostatic phenomena in the fluidized bed with a maximum error of 4%.

Experiments and Numerical Simulations of Hydrodynamics in Gas-Liquid-Solid Min-/Micro-Fluidized Beds

Conference: Fluidization

Year: 2019

Proceeding: Fluidization XVI

Group: General Paper Pool

Session: 7A: Fundamentals of Fluidization

Time: Wednesday, May 29, 2019 - 2:19pm-2:31pm

Authors:

Liu, M. Sr., State Key Laboratory of Chemical Engineering (Tianjin University)

In this presentation, recent researches on the experiments and numerical simulations of hydrodynamics in the gas-liquid-solid min-/micro-fluidized beds were reviewed. The main contents include experiments minimum fluidization velocity, single bubble behavior and flow regimes in the gas-liquid-solid mini-fluidized beds with single gas orifice, meso-scale modeling in the gas-liquid-solid mini-fluidized beds with multi gas orifices, bed expansion and multi-bubble behavior in the gas-liquid-solid micro-fluidized beds and VOF-DEM numerical simulations of hydrodynamics in the gas-liquid-solid mini-fluidized beds with single gas orifice.

Instabilities in an Underflow Standpipe Operating in a Bubble-Upflow Regime

Conference: Fluidization

Year: 2019

Proceeding: Fluidization XVI

Group: General Paper Pool

Session: 7A: Fundamentals of Fluidization

Time: Wednesday, May 29, 2019 - 3:07pm-3:19pm

Authors:

Karri, R., Particulate Solid Research, Inc.
Du, B., ExxonMobil
Findlay, J., PSRI
Harandi, M., ExxonMobil Research and Engineering

An extensive testing was conducted to understand the operation of an underflow, bottom-constrained vertical 25-cm-diameter standpipe operating in a bubble upflow regime with Geldart Group A particles. Standpipe operation was studied over a range of solids mass fluxes, amount of external aerations at the bottom of the standpipe with slide valve controlling the catalyst circulation rates. The effects of both aeration amount and catalyst flux on standpipe operation in a bubble upflow regime were investigated. Stable operation was demonstrated over a wide range of catalyst fluxes up to 60 kg/s-m^2 and an imposed superficial gas velocity of 0.04 to 0.15 m/s in the standpipe. However, increasing flow instabilities and catalyst flow problems were observed as the fines content less than 44 microns decreased below 7 to 8% by weight in the circulating catalyst.

Helium tracer gas testing indicated that some degree of gas bypassing is occurring in the 6 m tall standpipe, even at the 8% fines content. Research at PSRI[1] has shown that the pressure ratio across a fluidized bed greater than 1.05 was expected to promote gas bypassing in deep beds. The pressure ratio across the standpipe operating in this test unit was approximately 1.38 and gas bypassing would occur, which was accounted for the root cause of instabilities in standpipe operation. The study also showed that the higher fines contents in the catalyst helps reduce the tendency for gas bypassing[2] to occur for a given bed height.

[1] Karri, S. B. R., A. Issangya and T. M. Knowlton, Gas bypassing in deep fluidized beds,

in *Fluidization XI*, U. Arena, R. Chirone, M. Miccio and P. Salatino, eds., 515 â€“ 521, 2004.

[2] Issangya A., Karri, S. B. R. and T. M. Knowlton, Why Gas bypassing occurs in deep fluidized beds of Geldart Group A particles and how to prevent it, 10*th Int. Conf. on Multiphase Flow in Industrial Plant, Tropea (VV), Italy*, , XI-XXXI, 2006.

Keynote Talk: A Definition of Micro Fluidized Bed in Terms of in-Bed Gas Back-Mixing

Conference: Fluidization

Year: 2019

Proceeding: Fluidization XVI

Group: General Paper Pool

Session: 7A: Fundamentals of Fluidization

Time: Wednesday, May 29, 2019 - 1:30pm-1:55pm

Authors:

Xu, G., Shenyang University of Chemical Technology
Geng, S., Institute of Processing Engineering
Yue, J., Institute of Process Engineering
Wang, F., Institute of Process Engineering
Han, Z., Shenyang University of Chemical Technology
Hu, D., Shenyang University of Chemical Technology
Zeng, X., Institute of Processing Engineering
Yu, J., Institute of Processing Engineering
Liu, X., Institute of Processing Engineering

Micro fluidized bed (MFB) has received good attention in recent years because of its many traits different from traditional large fluidized beds and potential specific applications. This work systematically studied the gas back-mixing characteristics in MFB through experiments, simulation and verification by application. It revealed the fact that the gas flow in MFBs can maximally approach the plug flow and in turn for the first time defined what is â€œmicro fluidized bedâ€ in terms of gas back-mixing. We found that the residence time distribution (RTD) of gas is subject to an unique quantitative correlation between $E(t)_h$ and Ïƒ$_t^2$, the peak and variance of RTD for MFB, respectively. The critical values of $E(t)_h$ and Ïƒ$_t^2$ ensuring the actual gas residence time to deviate within 10% from the ideal plug flow are identified to define "micro fluidized bed" as the bed with gas flow maximally approaching the gas plug flow. Computational Fluid Dynamics (CFD) simulation was further performed to verity the experimental observations and the gas flow patterns in the bed. Measurement of a gas-solid reaction was also finally carried out using MFB to show the influence of operating conditions and the necessary to ensure the conditions guaranteeing the maximal approach of the in-bed gas flow to gas plug flow.

Nonequilibrium Features in a Bubbling Fluidized Bed

Conference: Fluidization

Year: 2019

Proceeding: Fluidization XVI

Group: General Paper Pool

Session: 7A: Fundamentals of Fluidization

Time: Wednesday, May 29, 2019 - 2:55pm-3:07pm

Authors:

Wang, H., Institute of Process Engineering, Chinese Academy of Sciences
Chen, Y., Institute of Process Engineering, Chinese Academy of Sciences
Wang, W., University of Chinese Academy of Sciences

The spatio-temporal evolution of the meso-scale structures, such as bubbles or clusters in a gas-solid fluidized bed, has been a hot area of research in recent decades [1, 2]. These dynamical heterogeneous structures can be characterized by nonequilibrium features. In the present work, the scaleâ€"dependent nonequilibrium features are investigated experimentally in a pseudo 2D bubbling fluidized bed. Velocities of individual particles are measured by using a particle tracking velocimetry (PTV) method, and void fractions are obtained with the Voronoi tessellation. A bimodal shape of probability density function (PDF) for particle vertical velocity is found in not only time-averaged but also time-varying statistics, which is caused by the transition between the dense and dilute phases and breaks the local-equilibrium assumption in continuum modeling of fluidized beds. The results of time-varying radial distribution function and voidage distribution also confirm this finding. Moreover, the analysis of voidage, particle velocity, granular temperature and turbulent kinetic energy of particles shows that there is no scale-independent plateau over the interface, and it seems hard to find a scale-independent plateau to separate the micro- and meso-scales of fluidized beds. Such scale-dependent nonequilibrium properties require sub-grid meso-scale modeling for continuum or coarse-graining methods of gas-fluidized systems [2, 3].

References:

[1] Li J, Zhang J, Ge W, Liu X. Multi-scale methodology for complex systems. Chem Eng Sci. 2004;59(8-9):1687-1700.

[2] Wang W, Chen Y. Mesoscale modeling: beyond local equilibrium assumption for multiphase flow. Adv Chem Eng. 2015;47:193-277.

[3] Tian Y, Geng J, Wang W. Structure-dependent analysis of energy dissipation in gas-solid flows: Beyond nonequilibrium thermodynamics. Chem Eng Sci. 2017;171:271-281.

Agglomeration in Fluidized Bed Gasification of Biomass

Conference: Fluidization

Year: 2019

Proceeding: Fluidization XVI

Group: General Paper Pool

Session: 7B: Fluidized Bed Applications

Time: Wednesday, May 29, 2019 - 2:31pm-2:43pm

Authors:

Zhao, L., Technical University of Denmark
Lin, W., Technical University of Denmark
Wu, H., Technical University of Denmark
Dam-Johansen, K., Technical University of Denmark (DTU)

Fluidized beds (FB) are widely applied to gasification of biomass due to its advantages of high fuel flexibility and conversion. However, bed agglomeration induced by the high content of alkali compounds can have strong impact on the operation of a FB gasifier and may lead to defluidization and unscheduled plant shutdown in severe cases. Agglomeration at high temperature, such as combustion and gasification, is mainly attributed to the presence of molten phase, i.e. the low melting point alkali containing compounds originated from biomass. During gasification, the alkali species in biomass can be transformed into gaseous phase, such as KCl and KOH vapor, and condensed phase, such as K_2SO_4 aerosols and K-silicates particles, depending on the local conditions. The interactions of the alkali species with bed materials (mainly SiO_2) may result in the formation of molten phase, causing agglomeration. However, the agglomeration tendencies and mechanisms during biomass gasification with different gasification agents are still not fully understood.

In this work, the agglomeration behaviors of different types of biomass during gasification are studied in a lab-scale fluidized bed reactor. The impact of the composition of biomass, e.g. types of biomass, and the operational conditions, such as temperature and gasification agents (steam and CO_2), are studied. The different agglomeration mechanisms are revealed based on SEM-EDX analyses of the agglomerate samples and chemical equilibrium calculations.

Energy Storage By Sand in Fluidized BED

Conference: Fluidization

Year: 2019

Proceeding: Fluidization XVI

Group: General Paper Pool

Session: 7B: Fluidized Bed Applications

Time: Wednesday, May 29, 2019 - 3:19pm-3:31pm

Authors:

Liu, Z., IHI Corporation
Zhang, H., Tsinghua University
Chen, Q., Tsinghua University
Ishikawa, A., IHI Corporation

Renewable energy such as wind or solar energy is becoming more and more important. However, electricity generation by renewable energy is usually unsteady and cannot match the demand of electricity consumption, especially, in regions where wind or solar power covers a high percentage of power generation. Parts of wind or solar power have to be abandoned sometimes to maintain stability of the electricity grid. Therefore, it appears reasonable to store electricity when it is in excess. There are several approaches to store electricity, such as direct storage in secondary batteries or storage in form of heat, mechanical energy or chemical energy. It is believed that transformation of electricity into heat is the cheapest method for large scale storage (for example, 100MWh).

Energy storage by sand in fluidized bed (referred to as â€œESFBâ€ below) is proposed in this paper. ESFB is one kind of thermal energy storage. ESFB is used to store heat generated by electricity in heating mode and to release heat when necessary in heat-release mode.

Heating mode is executed, when more electricity is generated than needed. Electrical heaters are set in wind box and wall of the riser. Cold particles are fed from a bubbling fluidized bed of low temperature, heated in the riser and then stored in a hopper. The hot air used for particle transportation is ducted from the cyclone to the low temperature fluidized bed in order to recover heat from the air.

The heat-release mode is executed, when heat (or electric power) is needed. The hot particles are fed from the high-temperature hopper to the riser in order to heat air supplied from the bottom. The hot air is then exported to working equipment, which can be a power generator or a furnace.

Advantages of the ESFB are low cost of the storage material, high energy density and system flexibility. The ESFB is cheap, because that the material used to store thermal energy is cheap sand. It has a high energy density, because the temperature can be as high as 1000â„ƒ. It provides high flexibility both, in heating mode and heat-release mode, because load and temperature can be adjusted by changing feeding rate of sand and inlet velocity of air into the riser.

Numerical simulations are performed for the electric heater in the riser, because that it is an important component of the ESFB system. Equivalent heat transfer coefficient, pressure drop, erosion, residence time distribution and segregation of sand particles are assessed for several design parameters of the heater. Optimization of the heater is outlined.

Evaluation of the Fluid-Dynamically Down-Scaled Model of a ~200 MWth CFB Boiler

Conference: Fluidization

Year: 2019

Proceeding: Fluidization XVI

Group: General Paper Pool

Session: 7B: Fluidized Bed Applications

Time: Wednesday, May 29, 2019 - 2:43pm-2:55pm

Authors:

Pallarès, D., Chalmers University of Technology
Djerf, T., Chalmers University of Technology
Johnsson, F., Chalmers University of Technology
Ylä-Outinen, V., Valmet Technologies Oy

Experimental campaigns in commercial CFB boilers are often limited to some extent by safety demands, narrow operational range imposed by requirements on safe and cost-efficient operation, limited access to in-situ measurements and lack of experimental methods due to harsh in-furnace environment. The use of fluid-dynamical scaling laws allows the design of lab-scale experimental setups which can operate under ambient conditions, while still resemble mixing and fluid-dynamics under hot conditions and allowing for a large span in operational condition, detailed measurements and visual observations of the flow.

This work presents first operating and measurement experience from a 1/13 fluid-dynamically down-scaled model of a 200 MW_{th} CFB boiler, which is operated according to Glicksmanâ€™s simplified set of scaling parameters for fluid-dynamical scaling. Having this, the measured vertical profile of solids concentration is validated by comparing with corresponding data from the full-scale boiler. The external solids circulation rate is directly measured in the cold flow model and values obtained for the industrial cases resembled are studied. Finally, the size distribution of the bed material in different parts of the unit is also studied.

Gas Velocity Distributions in Conical Spouted Beds with High Density Particles

Conference: Fluidization

Year: 2019

Proceeding: Fluidization XVI

Group: General Paper Pool

Session: 7B: Fluidized Bed Applications

Time: Wednesday, May 29, 2019 - 3:07pm-3:19pm

Authors:

Guler, N., Middle East Technical University
Koksal, M., Hacettepe University
Kulah, G., Middle East Technical University

Conical spouted beds have wide range of applications in many industrial processes such as pyrolysis, drying, coating, etc. due to their unique solids circulation characteristics. Coating of nuclear fuel element to be used in next-generation high temperature gas cooled reactor is an especially emerging topic. During this coating process, the density of the nuclear particles (uranium kernels) decreases from 10000 kg/m3 to 2500 kg/m3 where the average density of the particles is around 6000 kg/m3. While most of the studies in literature generally comprise spouted beds operating with low density particles, studies conducted in spouted beds operating with heavy particles (r_p >2500 kg/m3) are limited and these studies are mainly focused on the determination of the minimum spouting velocity, time- averaged and dynamic pressure drops, fountain height, particle velocity and solids hold-up (Zhou, 2008, Rojas, 2010, Olazar, 1992, 1996, 2001). However, it is important to have knowledge about the local gas flow structure to characterize the hydrodynamics, for successful design, scale up and manufacture of the spouted beds. However, to our knowledge, the only study of the gas velocity has been conducted in spouted beds operating with heavy particles (r_p >2500 kg/m3) was presented by Zhou et al., even investigation of gas velocity has been conducted in spouted beds operating with light particles can be found in a limited number of researches. The flow pattern of the gas can shed light on gas mixing phenomenon, which is a crucial aspect in reactor design The gas dispersion within the spouted beds is a complex subject that is not fully enlightened due to the difficulties of determining the local gas velocity distribution. To properly model the gas dispersion in spouted beds, exact flow structure must be investigated. The gas velocity profiles can also be used to validate the results of the computational fluid dynamics (CFD) studies. Therefore, the main objective of this study was to investigate the gas velocity distributions in the conical spouted beds with both low and high density particles.

Two different methods have been generally used to determine the gas velocity profiles in spouted beds (Mathur and Epstein, 1974). In the first method, the gas velocity is calculated from the static pressure drop change in the axial direction measured at different axial heights on the bed wall. For the application of this method, many assumptions are made such as the annulus zone is assumed to be in loosely packed bed regime (Throley et al., 1959) and the radial component of the gas in the spouted bed is neglected. This method ignores the flow of gas between the annulus and spout zones. All these assumptions cause errors in the gas velocity predictions (Mathur and Epstein, 1974). In the second method, the local gas velocity is measured with a pitot tube or a Prandtl tube.
Although pitot tube is a robust and easy to use device, the use of data obtained by pitot tube in multiphase systems is very complex. In the previous studies, although physically the gas velocity was significantly different in the spout and annulus regions, no significant difference was observed in measurements (Becker, 1961, Mamuro and Hattori, 1968 and Van Velzen et al., 1972). In addition, the volumetric flow rate of the gas at each axial position calculated by utilizing the cross-sectional average of the measured gas velocity profiles were found to be much higher than the volumetric flow rate of the gas entering to the system. Mamuro and Hattori commented that since the pitot tube is a device designed for use in single phase systems, the measurements performed in the spout region could be assumed to be correct due to high voidage in this region. However, the gas velocities measured in the annulus region were incorrect and needed adjustment. Therefore, the first task in this study was the calibration of the pitot tube for determination of gas velocities in the annulus section. For this purpose, experiments were conducted in a 78 mm ID cylindrical system where particles were kept at almost loosely packed condition to simulate the conditions in the annulus section of the system. The results showed that the local gas velocities calculated using the calibration equation of the pitot tube supplied by the manufacturer resulted in

more than 1300% higher values than the actual local gas velocities which verified the inaccurate results observed in previous studies reported in the literature. Instead of using this calibration equation, a new calibration equation was developed to be used for the measurements in the annulus section of the spouted bed for each type of particles (glass bead, Al_2O_3 and ZrO_2) used in this study. Once the calibration equations were determined, the gas velocity measurements were conducted with all three particles (d_p=1 mm for all, r_{glass}=2460, r_{Al2O3}=3700, r_{ZrO2}=6050 kg/m3) in conical spouted beds having two different cone angles (g = 30° and 60°) and cylindrical diameter (D_c) of 150 mm. The local gas velocity measurements were performed at several radial and axial positions. To investigate the effects of the bed size, experiments were also conducted in a large scale conical spouted bed (g = 66°, D_c = 250 mm) with glass beads and zirconia particles. The interstitial gas velocity profiles in both small and large scale systems showed that the gas velocity was at its maximum in spout center and significantly decreased in the annulus region. The amount of gas flowing through the spout and annulus zone was also calculated using the radial gas velocity profiles. For both size beds, the results showed that the amount of gas passing through the spout decreased with increasing axial height for all conditions. In the small scale conical spouted bed, the fraction of the total gas passing through the annulus was relatively smaller compared to that in the large scale bed for all particles tested. In addition, the amount of gas passing through the spout decreased with increasing density of the particle.

References

Becker, H.A., An investigation of laws governing the spouting of coarse particles. Chem. Eng. Sci. 13, 245, 1961.

D.L. Rojas, Preliminary results for the investigation of hydrodynamic scaling relationships in shallow spouted beds, Seperation Sci. Tech. 45, 1928–1934, 2010.

J. Zhou, Characterizing and Modeling the Hydrodynamics of Shallow Spouted Beds, Ph.D Thesis, University of Tennessee, Knoxville, 2008.

Mamuro, T., and Hattori, H., Flow pattern of fluid in spouted beds. J. Chem. Eng. Jap. 1, 1, 1968.

Mathur, K., Epstein, N., Dynamics of spouted
beds. Advan. Chem. Eng. 9, 11-191, 1974.

M. Olazar, M.J.
San Jose, A.T. Aguayo, J.M. Arandes, J. Bilbao,
Stable operation conditions for gas–solid contact regimes in conical spouted
beds, Ind. Eng. Chem. Res. 31, 1784–1792, 1992.

M. Olazar, M.J.
San Jose, M.A. Izquierdo, A. Ortiz de Salazar, J.
Bilbao, Effect of operating conditions on solid velocity in the spout, annulus
and fountain of spouted beds, Chem. Eng. Sci. 56, 3585–3594, 2001.

M. Olazar, M.J. San Jose, E. Cepeda, R. Ortiz de Latierro, J. Bilbao, Hydrodynamics of fine solids
in conical spouted beds, Fluidization VIII, 196–201, 1996.

Thorley, B., Saunby,
J.B., Mathur, K., Osberg,
G.L., An Analysis of Air and Solid Flow İn A Spouted Wheat Bed, The
Canadian Journal of Chemical Engineering, p. 184-192, 1959.

Van Velzen, D., Flamm, H.J., Langerkamp, H., Gas
flows in spouting beds. Dragon Proj. Rep. No. 785. Euratom, Ispra, Italy, 1972.

Keynote Talk: Fluidization Engineering in Chemical Looping Combustion/Gasification

Conference: Fluidization

Year: 2019

Proceeding: Fluidization XVI

Group: General Paper Pool

Session: 7B: Fluidized Bed Applications

Time: Wednesday, May 29, 2019 - 1:30pm-1:55pm

Authors:

Guo, Q., Ningxia University

Chemical looping, decomposing a reaction into multiple sub-reactions in separated reactors, offers a clean and efficient way to convert fuels or to conduct redox reactions. Chemical looping process refers to many key issues of fluidization engineering, such as the preparation of oxygen carrier particles and their fluidization characteristics in chemical looping reactors. In this paper, issues of fluidization engineering in chemical looping combustion/gasification were discussed. Firstly, the design, formulation and manufacture of oxygen carrier particles were introduced. It was pointed out that composite oxygen carrier with catalytic function or with oxygen decoupling was the promising candidate materials. Large-scale preparation of these promising oxygen carriers is of great significance for chemical looping process. Due to the alternate redox reactions of oxygen carrier in separated reactors, the inter-connected fluidization bed was an optimize choice of chemical looping reactors. The bubble and fast fluidization characteristics of oxygen carrier particles in difference reaction atmospheres at high temperature should be explored. Based on previous findings, it indicated that the matching of reaction rate and oxygen carrier capacity and the non-touching of two reaction atmospheres between two inter-connected fluidization bed is the vital factor for guaranteeing the steady operation of reactors. The syngas, hydrogen, and methane can be produced depended on the different oxygen carriers and various reactors configuration combination. In recent years, we propose to yield high-valued chemicals by chemical looping process using biomass or coal, a new field for fluidization engineering.

Movement and Mixing of Biomass or Coal Particles in a Fluidized Bed Combustor Under High Temperatures

Conference: Fluidization

Year: 2019

Proceeding: Fluidization XVI

Group: General Paper Pool

Session: 7B: Fluidized Bed Applications

Time: Wednesday, May 29, 2019 - 2:55pm-3:07pm

Authors:

Yang, Z., Southeast University
Duan, L., Southeast University
Li, L., Southeast University
Liu, D., Southeast University

The combustion characteristics of a single fuel particle in fluidized beds are widely studied. However, in most simulation of the single particle combustion in fluidized beds, the particle is always assumed to be staying in the emulsion phase, which is quite different from the real motion characteristics. Monitoring the motion characteristics of fuel particles at different positions in the furnace is of great significance for understanding the conversion process of fuel in the fluidized bed. In this work, the motion characteristics of a single fuel particle (coal or biomass) at different locations (the emulsion phase and the splash zone) were investigated in a 2D visualized fluidized bed combustor. The movement during its real combustion process, including devolatilization and char combustion, was recorded with a high-resolution video camera. It is found that the motion characteristics of different fuel particles are significantly different. Residence time in the emulsion phase of coal char particles was longer than that of biomass char particles. In the devolatilization process, the particles stay most of the time in the emulsion phase. As the char combustion continued, the particle density decreased and the char appeared more in the splash zone or floated on the surface of the bed. With increasing the fluidizing velocity, the time share of the char particle in splash zone increased. The residence time of the particles in the emulsion phase was increased with the particle size improved.

Oxy-Fuel Combustion of Solid Recovered Fuels in the Fluidized Bed Calciner of a 1 MWth Calcium Looping Unit

Conference: Fluidization

Year: 2019

Proceeding: Fluidization XVI

Group: General Paper Pool

Session: 7B: Fluidized Bed Applications

Time: Wednesday, May 29, 2019 - 2:07pm-2:19pm

Authors:

Haaf, M., Technische Universität Darmstadt
Hilz, J., Technische Universität Darmstadt
Peters, J., Technische Universität Darmstadt
Unger, A., SUEZ Deutschland GmbH
Ströhle, J., Technische Universität Darmstadt
Epple, B., Technische Universität Darmstadt

Carbon capture and storage (CCS) technologies represent an effective mean for climate change mitigation. Furthermore, the utilization of waste derived fuels, such as municipal solid waste (MSW) or solid recovered fuel (SRF) in CCS-equipped power or industrial systems might lead to the removal of CO_2 from the atmosphere, due to the biogenic fuel fractions. The calcium looping (CaL) process is a promising approach for CO_2 capture from various power or industrial systems. It is a post-combustion CO_2 capture technology using natural limestone based sorbents. The CO_2 in the flue gas stream is absorbed by calcium oxide (CaO) according to the reversible carbonation reaction in the carbonator. The formed calcium carbonate ($CaCO_3$) is transferred to the oxy-fired calciner to be regenerated subsequently. Whereas the carbonator operates at 650 °C, the temperature in the calciner needs to be raised to approximately 900 °C, in order to achieve a complete regeneration of the solid stream. The required heat is supplied by oxy-fuel combustion of additional fuel in the calciner. Within the past years, the CaL process has been tested thoroughly during thousands of operational hours in numerous test rigs worldwide. However, in most experimental investigations coal is the dominant fuel for the heat supply in the calciner.

Within the framework of this study, experimental results achieved during two consecutive test campaigns, each two weeks long, conducted at the 1 MW_{th} CaL Unit at Technische Universität Darmstadt are presented. During the test periods, two different types of SRF were used for the heat supply in the circulating fluidized bed (CFB) calciner. The two fuel candidates are derived from a German waste treatment facility, and are typically used in cement plants and dedicated combustion facilities. They show different properties such as maximal particle size ($d_{max,SRF\ I}$ < 30 mm, $d_{max,SRF\ II}$ <50 mm), lower heating value ($LHV_{SRF\ I}$ > 16 MJ/kg, $LHV_{SRF\ II}$ > 21 MJ/kg) and volatile content (SRF I: 60-70 m/m$\%_{ar}$, SRF II: 70-80 m/m$\%_{ar}$). The flue gas to be decarbonized in the CFB carbonator was based on the combustion of lignite in a combustion chamber on side. The main objective of these campaigns was the demonstration the steady state CO_2 capture by means of CaL technology while firing SRF in the calciner. Special focus has been given to assess the influence of the chlorine present in the fuel on the overall process performance. A continuously FTIR measurement at the calciner off gas and frequent solid stream sampling at the inlet and the outlet of the calciner cover the chlorine mass balance. In addition to the type of fuel for the calciner, the influence of several crucial CaL process parameters such as the operation temperatures of carbonator and calciner, the global solid circulation rate and the calcination conditions is discussed.

This work gives a deep insight to the experimental results achieved. In addition to pressure and temperature profiles of both CFB reactors, the particle size distribution of the circulating solid streams are presented. It was proven, that stable operation of the coupled CFB unit with CO_2 capture rates over 80 % in the carbonator are feasible while firing SRF in the calciner. The total CO_2 capture rate of the system accumulates to more than 90 %. Furthermore, the combustion of SRF and the calcination of the sorbent under air and oxy-fuel atmosphere are compared and assessed.

Release and Migration Behavior of Nitrogen in Chemical-Looping Gasification Process Based on Iron-Based Oxygen Carrier

Conference: Fluidization

Year: 2019

Proceeding: Fluidization XVI

Group: General Paper Pool

Session: 7B: Fluidized Bed Applications

Time: Wednesday, May 29, 2019 - 2:19pm-2:31pm

Authors:

Li, Y., Ningxia University
Guo, Q., Ningxia University

Release and migration behavior of nitrogen in chemical-looping gasification process based on iron-based oxygen carrier

LI Yankun, GUO Qingjie*, MA Jingjing, CHANG Guozhang, HU Xiude

(Ningxia University£¬State Key Laboratory of High-efficiency Coal Utilization and Green Chemical Engineering, Yinchuan,750021,China)

Abstract: The combined thermo gravimetric-infrared-mass spectrometry (TG-FTIR-MS) analysis was used to conduct chemical-looping gasification experiments of coal under Ar atmosphere, and the mass change and gas composition during the reaction were monitored in real time. The surface element analysis of the solid phase was carried out by X-ray photoelectron spectroscopy, when the change of nitrogen functional groups in the solid phase at different stages of chemical-looping gasification was investigated. The results show that the oxygen carrier has an effect on the release of nitrogen-containing gas in different stages of the chemical-looping gasification process. During the pyrolysis stage, the oxygen carrier promotes the generation of free radicals, which accelerates the release of nitrogen-containing gas in a pyrolysis stage. At high temperature, the oxygen carrier promotes the conversion of NH_3 to HCN; The addition of the oxygen carrier during the gasification stage reduces the degree of graphitization of the semi-coke and increases the release rate of the nitrogen-containing gas. For the nitrogen functional groups in the solid phase, the oxygen carrier promotes the decomposition and conversion of pyrrole nitrogen in the pyrolysis stage. At high temperature, the protonated pyridine is converted to pyrrole nitrogen and pyridyl nitrogen due to the presence of the oxygen carrier. During the pyrolysis process carried out at a high temperature, the protonated pyridine convert to pyridinium nitrogen, During the gasification process, the presence of water vapor increases the concentration of hydrogen radicals at high temperature, and the protonated pyridine is converted into pyrrole nitrogen which is more easily decomposed and converted.

Key words: Coal; chemical-looping gasification; iron-based oxygen carrier; nitrogen-containing gas; nitrogen functional groups; TG-MS-FTIR

The Impact of Multifunctional Additives on NOX Emission and Bed Agglomeration in Fluidized Bed Combustion of Biomass

Conference: Fluidization

Year: 2019

Proceeding: Fluidization XVI

Group: General Paper Pool

Session: 7B: Fluidized Bed Applications

Time: Wednesday, May 29, 2019 - 1:55pm-2:07pm

Authors:

Ulusoy, B., Sino-Danish Centre for Education and Research
Anicic, B., Technical University of Denmark
Wu, H., Technical University of Denmark
Lin, W., Technical University of Denmark
Dam-Johansen, K., Technical University of Denmark (DTU)

NO_X emission and bed agglomeration are two major concerns in fluidized bed combustion of biomass. While the NO_X emission is determined by a series of competing formation and reduction reactions, bed agglomeration is mainly determined by the release and transformation of inorganic elements, especially potassium, during combustion. A variety of additives exists with the separate purpose of minimizing NO_X emission or bed agglomeration. However, these additives are often multifunctional thereby affecting several ongoing processes. Therefore, the purpose of this study was to investigate the impact of multifunctional additives on the NO_X emission and bed agglomeration behavior during biomass combustion.

Straw and sunflower husk combustion experiments were performed in a fluidized bed reactor with and without the use of additives. The additives included compounds such as kaolin, lime, ammonium sulphate, coal fly ash, clay, magnesium carbonate etc. Different air staging conditions and additive addition methods (mixing with fuel and batch addition to the bed) were additionally investigated. The NO_X emission was measured using a series of online gas analyzers, while defluidization was observed from the bed pressure drop and temperature. The bed samples were further examined using SEM-EDS and XRD in order to obtain better insight into the agglomeration mechanism.

The results indicate that the conversion of fuel-N to NO increased slightly when pre-mixing the fuels with lime and magnesium carbonate, while a decrease was observed when adding ammonium sulphate due to the facilitation of thermal deNO_X reactions. In contrast to this, kaolin, coal fly ash, and clay had little effect on the NO_X emission from straw combustion, while it increased NO_X emission from sunflower husk combustion. In terms of agglomeration behavior, the results show that kaolin, lime, and magnesium carbonate prevented defluidzation during straw combustion, while ammonium sulphate, coal fly ash, and clay prolonged the defluidization time to some degree. This was explained by the role of additives either in capturing the gaseous potassium species from combustion (kaolin) or reducing the stickiness of formed melts (lime and magnesium carbonate) and thereby the defluidization tendency. Additionally, a comparison between the two addition methods indicates that pre-mixing the additive with the fuel is more efficient in terms of NO_X and agglomeration tendency reduction, most likely due to the better mixing between the fuel and additive.

Detailed CFD-DEM Simulation of Biomass Gasification in a Fluidized Bed Reactor

Conference: Fluidization

Year: 2019

Proceeding: Fluidization XVI

Group: General Paper Pool

Session: 7C: Modeling and Simulation

Time: Wednesday, May 29, 2019 - 2:43pm-2:55pm

Authors:

Ostermeier, P., Techinal University of Munich
Fischer, F., Techinal University of Munich
Fendt, S., Techinal University of Munich
DeYoung, S., Techinal University of Munich
Spliethoff, H., Techinal University of Munich

Gas-solid fluidized beds play an important role in many industrial operations. Nevertheless, there is still a lack of knowledge of the processes inside the bed which impedes proper designing and upscaling of fluidized bed operations. In this work, the biomass gasification process with steam in a fluidized bed reactor is investigated with a detailed CFD-DEM approach.

The numerical model in the Eulerian-Lagrangian framework treats the gas phase as a continuum and describes the particle interactions with the discrete element method (DEM). The non-spherical shape of the particles is accounted for in the momentum exchange calculation with the gas phase. The considered systems consists of steam (ideal gas, temperature 800 °C) as the fluidization gas, entering the bottom region of the complex three dimensional reactor geometry through inclined nozzles in an inner duct at the center axis. The wet wood pellets (sphere equivalent diameter 6.1 mm, particle density 1160 kg/m^3, sphericity factor 0.86) are fed into the reactor in the freeboard region together with a nitrogen purge gas stream (ideal gas, 80 °C).

Two different operation modes are investigated. For the start-up procedure, the reactor is filled with 2.565 kg of sand particles (50-450 µm, particle density 2210 kg/m^3, sphericity factor 0.86, coarse-graining 1.5e5 parcels). The inert material is allowed to settle for three seconds and is fluidized for another three seconds until quasi steady-state is achieved. Afterwards, the wet wood pellets are added for 35 seconds at a rate of one pellet every two seconds and undergo the processes of heating, drying, pyrolysis and coke conversion. From the simulation results and literature correlations, the average time intervals for pyrolysis and coke conversion, as well as the respective particle properties (temperature, diameter, density) are estimated. With the knowledge of the average composition of reactive material in the reactor, variable operating points can be simulated.

The second case investigated is after approximately 50 hours of operating time. Additionally to the start-up procedure, 1.5 kg of inert coke residue (100-4000 μm, particle density 400 kg/m^3, sphericity factor 0.86, coarse-graining 5e4 parcels) is added with the sand particles. Again, the inert material is allowed to settle for three seconds and is fluidized for another three seconds. Subsequently, the average reactive material obtained from the start-up simulations is patched into the bed region and wet wood pellets are fed to the freeboard region.

The following phenomena are implemented in the numerical model: gas-solid momentum exchange, solids collisional behavior, turbulence, heat and mass transfer, particle shrinkage and change in material properties, pyrolysis, and homogeneous and heterogeneous chemical reactions. Here, the focus is on the influence of properly modeling turbulence and heat transfer.

Simulation results are analyzed both qualitatively and quantitatively. The particle flow fields and the mixing between sand, wood, and residue are investigated for laminar and turbulent flow and for different filling of the reactor. Furthermore, the heating and conversion of the wood pellets is examined for different heat transfer models. Bed pressure drop and product gas composition are compared to experimental data to verify the numerical model. For this purpose, literature correlations for the wood pellet conversion are considered as well.

The results show that the average heating rate for the wood pellets is similar with and without turbulence model. However, the pellets in the simulation with turbulence exhibit a more repeatable heating and thus conversion behavior (smaller spread in the temperatures and properties of the different pellets) than in the laminar case. On the other hand, the choice of the heat transfer model to calculate the heat exchange between the different phases has a higher influence than the turbulence model. The user-defined approach including the Gunn-correlation (recommended for dense fluidized beds) results in faster heating of the particles than the correlation according to Ranz and Marshall (the default model in ANSYS Fluent). Finally, the ranges for the bed pressure drop, the gas composition, and the conversion time obtained with the numerical simulation agree well with experimental data and literature correlations. This indicates that the proposed model can make a significant contribution towards understanding and improving the internal processes in fluidized bed reactors for biomass gasification and combustion.

Drying of Cohesive Powders in a Pilot Plant Scale Vibrated Fluidized Bed

Conference: Fluidization

Year: 2019

Proceeding: Fluidization XVI

Group: General Paper Pool

Session: 7C: Modeling and Simulation

Time: Wednesday, May 29, 2019 - 1:55pm-2:07pm

Authors:

Lehmann, S. E., Hamburg University of Technology
Hartge, E. U., Hamburg University of Technology
Jongsma, A., Tetra Pak CPS
Innings, F., Tetra Pak Processing Systems
Heinrich, S., Hamburg University of Technology
deLeeuw, I. M., Tetra Pak CPS

Hindrance of fluidization of fine particles in form of agglomeration, channeling or complete de-fluidization is caused by cohesive forces. Mechanical vibration may be applied to fluidized bed equipment in processing of fine or sticky powders to improve or enable fluidization by counter acting cohesive effects. Vibrated fluidized beds are widely applied in industrial scale in the pharmaceutical and food industries.

However, almost all research in the field of vibrating fluidized beds is conducted with lab-scale apparatuses. Within the current project a pilot scale unit with a cross section of 250 x 500 mm^2 and a total height of 3 m was built. The experimental equipment allows comprehensive parameter studies of hydrodynamics and drying kinetics.

The drying kinetics of cohesive powders has been investigated via parameter studies with regard to different drying air conditions and vibration states. Temperature and humidity of drying air at inlet and outlet were measured online as well as the vibration parameters. Furthermore, powder samples were taken during operation for further analysis. Additionally, visual observations can be made through one transparent wall of the apparatus.

Heat and mass transfer as well as drying curves were calculated to evaluate the influence of vibration on drying kinetics. Based on the experimental results a flow sheet simulation model for vibrated fluidized bed dryers was developed and validated.

Eulerian Modeling of Monodisperse Gas-Particle Flow with Electrostatic Forces

Conference: Fluidization

Year: 2019

Proceeding: Fluidization XVI

Group: General Paper Pool

Session: 7C: Modeling and Simulation

Time: Wednesday, May 29, 2019 - 2:31pm-2:43pm

Authors:

Montilla, C., Université de Toulouse, CNRS-Toulouse
Nasro-Allah, Y., Université de Toulouse, CNRS-Toulouse
Ansart, R., Université de Toulouse, CNRS-Toulouse
Simonin, O., Université de Toulouse, CNRS-Toulouse
Fox, R. O., Iowa State University

Eulerian modeling of monodisperse gas–particle flow with electrostatic forces

Carlos Montilla[1] , Youssef Nasro-Allah[1] , Renaud Ansart[1] , R. O. Fox[2] and O. Simonin[3]
[1] Laboratoire de Génie Chimique, Université de Toulouse, CNRS–Toulouse. France.

[2] Department of Chemical and Biological Engineering, Iowa State University, Ames, IA, USA.

[3] Institut de Mécanique des Fluides de Toulouse, Université de Toulouse, CNRS–Toulouse. France.

carlos.montillaestrella@ensiacet.fr

Keywords: gas–particle flows, kinetic theory, electrostatic force, CFD simulation

Abstract

Due to triboelectric charging, the solid phase in gas–particle flows can become electrically charged, which induces an electrical interaction between all the particles in the system. Because this force decays very rapidly, many current models neglect the contribution of the electrostatic interaction between the particles. Nevertheless, the impact that this force can have in many industrial configurations is well documented. In this work an Eulerian–Eulerian model for gas–particle flow is proposed in order to take into consideration the electrostatic interaction between the particles. We use the kinetic theory of granular flows to derive the transport equation for the electrical charge for dense gas–particle systems. In order to close the collision integrals in the Boltzmann–Enskog equation, we employ the Grad's 13-moments theory to approximate the non-equilibrium probability distribution. The model is validated using two different problems from the literature: a fully periodic box, for which an analytical solution can be found for some special cases; and a periodic channel. The Eulerian–Eulerian simulation results are compared against DEM simulations for the same cases.

Introduction

Nowadays, gas–particle flows play an extremely important role in many industrial technologies. Fluidized beds, cyclonic separators and the transport of air pollutants are just a few examples of this type of flow. In some configurations the particles collide with other solid materials (either another particle or a solid boundary). During these particle–particle or particle–wall interactions, an electron can move from one surface to the other inducing an electrical charge on the particles. This effect is called triboelectrification (Matsusaka, S. & Masuda, H. 2003). The electrically charged particles can now interact with other charged particles via the Lorentz force. Because the particle velocity is very small compared to the speed of light, the magnetic contribution can be dropped, and only the electrostatic term is relevant.

Currently, many Eulerian-Eulerian gas–particle models neglect the effect of this force, because the electrostatic interaction decays like the inverse square law of the distance between two particles. Nevertheless, it has been known for many decades that this type of interaction can modify the dynamics of the solid phase (Miller C. & Logwinuk, A. 1951). In many cases, the presence of electrical charges is undesirable. For example, they can reduce the efficiency of many industrial reactors, or even increase safety hazards due to the rise in the electric potential (Hendrickson,

G. 2006).

Eulerian modeling of the electrostatic force

Recently some attempts have been made to include the electrostatic force into an Eulerian–Eulerian model. One of the first attempts was made by Rokkam, R. G. et al. (2013). These authors added an extra force term to the solid momentum equation:

." src="https://www.aiche.org/sites/default/files/aiche-proceedings/conferences/..." class="documentimage"> (1)

where F_e is the electrostatic force, q_p is the particle charge, and α_p the solid volume fraction. The electrical potential φ is governed by

â^‡ â‹…(Ïµm â^‡Ï•) = -p-p Ïµo
" src="https://www.aiche.org/sites/default/files/aiche-proceedings/conferences/..." class="documentimage"> (2)

where ϵ_m is the mixture permittivity and ϵ_o the vacuum permittivity. This model was applied to a fluidized-bed reactor, successfully capturing the bed height and solid volume fraction distribution, as well as the segregation of the solid phases and the wall-adhesion phenomenon. The model, however, assumes that the solid phases have a constant electrical charge, limiting its application to cases where q_p is known a priori.

More recently, a more complete approach was proposed by Kolehmainen, J. et al. (2018). These authors derived the charge transport equation using the kinetic theory:

â^‚- Î±pQp + â^‚-- Î±pUiQp =
â^‚t Vp â^‚xi Vp ( ) -â^‚-(Ïƒ E)+ â^‚-- (Îº +Îº *) â^‚Qp- â^‚xi p i â^‚xi q q â^‚xi
" src="https://www.aiche.org/sites/default/files/aiche-proceedings/conferences/..." class="documentimage"> (3)

where V_p is the particle volume, U the fluid velocity, σ_p the triboelectric conductivity, κ_p the diffusion coefficient and κ_p^* the self-diffusion coefficient. Assuming an uncorrelated Maxwellian probability distribution for the velocity and the electrical charge, they found an expression for the diffusion coefficient (κ_p). However, this assumption significantly underestimates the magnitude of diffusion, and therefore the self-diffusion coefficient (κ_p^*) was added a posteriori in analogy to heat conduction.

In our work, we propose a closure for the collisional and kinetic dispersion terms in the mean charge balance (3) derived in the framework of the kinetic theory of granular medium (Jenkins, J. T. & Richman, M. W. 1985). In particular, we show that the closure assumption for the collisional contribution ($\kappa_q \nabla Q_p$) can be derived without presuming an uncorrelated Maxwellian probability distribution for the particle electrical charge. In addition, we derive a closure for the dispersion term due to transport by the random velocity ($\kappa_q^* \nabla Q_p$) from the transport equation governing the electrical charge due to the fluctuating particle velocity.

Numerical simulations

In order to test the proposed model, we conduct the two numerical experiments described by Kolehmainen, J. et al. (2018). First, we solve a fully periodic box with a known initial charge distribution. Making some simplifying assumptions, an analytical solution can be derived for this configuration. The second problem we address is a periodic dense channel with conducting walls. Due to the complexity of this problem, the exact solution cannot be found, and therefore we solve it using CFD simulations. Our results for both problems are compared against the DEM simulations from Kolehmainen, J. et al. (2018).

The CFD simulations are conducted using NETPUNE_CFD, a state-of-the-art code developed by EDF (Électricité de France). This software is capable of doing transient multiphase flow simulation using an Eulerian n-fluid approach for each phase. The governing equations are solved using a cell-centered finite-volume method. Furthermore, NEPTUNE_CFD can handle structured and unstructured meshes, and can also take advantage of parallel architectures using MPI protocol (Hamidouche et al. 2018).

References

Hamidouche, Z., Masi, E., Fede, P., Ansart, R., Neau, H., Hemati, M. & Simonin, O. Numerical simulation of multiphase reactive flows. Advances in Chemical Engineering. Vol. 52, 51–124 (2018)

Hendrickson, G. Electrostatics and gas phase fluidized bed polymerization reactor wall sheeting. Chemical Engineering Science. Vol 61, 1041–1064 (2006)

Jenkins, J. T. & Richman, M. W. Grad's 13-moment system for a dense gas of inelastic spheres. Archive for Rational Mechanics and Analysis. Vol 87, 355–377 (1985)

Kolehmainen, J., Ozel, A. & Sundaresan, S. Eulerian modelling of gas–solid flows with triboelectric charging. Journal of Fluid Mechanics. Vol 848, 340–369 (2018)

Matsusaka, S. & Masuda, H. Electrostatics of particles. Advanced Powder Technology. Vol 14, 143–166 (2003)

Miller C. & Logwinuk, A. Fluidization studies of solid particles. Industrial & Engineering Chemistry. Vol 43, 1220–1226 (1951)

Rokkam, R. G., Sowinski, A., Fox, R. O., Mehrani, P. & Muhle, M. E. Computational and experimental study of electrostatics in gas-solid polymerization fluidized beds. Chemical Engineering Science. Vol 92, 146–156 (2013)

Keynote Talk: CFD Simulation of Combustible Solid Waste Pyrolysis in a Fluidized Bed Reactor

Conference: Fluidization

Year: 2019

Proceeding: Fluidization XVI

Group: General Paper Pool

Session: 7C: Modeling and Simulation

Time: Wednesday, May 29, 2019 - 1:30pm-1:55pm

Authors:

Xiong, Q., General Motors
Zhang, Y., University of Minnesota

CFD simulation of combustible solid waste pyrolysis in a fluidized bed reactor

Kuan Ding [a, b], Qingang Xiong [c, e], Zhaoping Zhong [a], Daoxu Zhong [d], Yaning Zhang [e]

[a] *Key Laboratory of Energy Thermal Conversion and Control of Ministry of Education, School of Energy and Environment, Southeast University, Nanjing, 210096, China*

[b] *Center for Biorefining and Department of Bioproducts and Biosystems Engineering, University of Minnesota, 1390 Eckles Ave., St. Paul, MN 55108, United States*

[c] *IT Innovation Center, General Motors, Warren, MI 48092, USA*

[d] *Jiangsu Provincial Academy of Environmental Science, Nanjing, Jiangsu 210036, China*

[e] *School of Energy Science and Engineering, Harbin Institute of Technology, Harbin, China*

Abstract: Computational fluid dynamics (CFD) simulation of combustible solid waste (CSW) pyrolysis in a fluidized bed reactor with feeding rate of 5 kg/h was performed in this study. The multi-phase flow was simulated using the Euler-Euler method, and a multi-component and multi-step reaction method was adopted to describe the pyrolysis process of CSW. The user-defined functions (UDF) of heterogeneous reactions were programmed and coupled with the computational fluid dynamics (CFD) software. The simulation methods used were validated by comparing the simulated temperatures and product yields with the experimental results.

Simulation results also indicated that the flow regime in the fluidized bed turned into stabilized fluidization gradually over pyrolysis time, and the mass flow rates of pyrolytic products at the outlet fluctuated within the range of ±10%. Pyrolysis temperature had the greatest effect on the product yield, while initial bed height had the most significant influence on product fluctuations. The CFD simulation results presented in this study provide valuable reference for the design and optimization of biomass pyrolysis process.

Keyword: CFD simulation; combustible solid waste; pyrolysis; fluidized bed reactor

Particle Scale Modelling of Bubble Dynamics in a Single Jet Fluidized Beds

Conference: Fluidization

Year: 2019

Proceeding: Fluidization XVI

Group: General Paper Pool

Session: 7C: Modeling and Simulation

Time: Wednesday, May 29, 2019 - 3:07pm-3:19pm

Authors:

Zhou, Z., Monash University
Shrestha, S., Monash University

Fluidized beds are widely applied in many industrial processes, and have many inherent features such as good mixing and excellent heat and mass transfer which are directly related to the presence of bubbles. The objective of the present work is to study the bubble properties mainly bubble size, bubble size distribution and bubble shape (aspect ratio and shape factor) in a bubbling gas solid fluidized bed operated with a continuous central jet. The results show that generally, the continuous injection of a central air jet to the bed leads to the formation of series of bubbles which rises through the bed and then finally bursts at the top of the bed. Initially, after the introduction of a central air jet to the bed, the first bubble is formed at a certain height above the distributor. Then there are several successive bubbles formed one after another. The average bubble diameters increase with the increase of jet velocity. The bubble size distribution in the overall bed follow gamma distribution function. A mono-peak positively skewed distribution is found for all the jet velocities. As the jet velocity increases, the long tail end pointing towards the right expands. At the upper part of the bed, the distribution is more scattered signifying the presence of small and large bubbles. The distributions of bubble shape in terms of aspect ratio are also examined, showing a normal distribution. The bubble aspect ratio has a very high scattered distribution at lower part of the bed compared to that at the upper part. The distribution of bubble aspect ratio shifts towards the left and the mode increases as the height of the bed increases. The analysis of the bubble size and bubble shape provides useful information for understanding the gas-solid contact and mixing in gas fluidisation.

Simulations of the Hydrodynamics in a Novel Spouted Bed Structure with the Swirling Flow Effects

Conference: Fluidization

Year: 2019

Proceeding: Fluidization XVI

Group: General Paper Pool

Session: 7C: Modeling and Simulation

Time: Wednesday, May 29, 2019 - 2:55pm-3:07pm

Authors:

Wu, F. Sr., School of Chemical Engineering, Northwest University, Xi' an, China

Simulations of the hydrodynamics in a novel spouted bed structure with

the swirling flow effects

Feng Wu[1]*, Zhenyu Huang[1], Xiaoxun Ma[1], Wenjing Zhou[2]

1,School of Chemical Engineering, Northwest University, Xi' an, China

2,School of Chemical Engineering and Technology, Xiâ€™an Jiaotong University, Xiâ€™an, China

*Corresponding author:

Email: wufeng@nwu.edu.cn

Tel: +86-15309202861

Abstract: The swirling flow technology is adopted in the inlet of the spouted bed so as to strengthen the particles radial mixing between spout and annulus. The gas-solid flow in a cylindrical spouted bed with a swirling flow generator installed on nozzle is numerically investigated by two-fluid model (TFM) with kinetic theory for granular flow. Simulations and analyses have been conducted on two types of spouted beds: conventional spouted bed and spouted bed with integral swirling blade nozzle (ISBN). The simulation results show that: the particle volume fraction near the cone region of spouted bed can be reduced significantly by swirling flow of gas phase cause by the ISBN, and the dead zone at the bottom of the annulus of spouted bed can be eliminated effectively. The value of turbulent kinetic energy of gas phase can be obviously enhanced by swirling flow along the axial direction in spouted bed, especially in the spout region. Also, the value of radial velocity and granular temperature of particle phase in spouted bed can be promoted by swirling flow, which is beneficial to the radial mixing of particles and gas phase between central spout and annulus in spouted bed.

Key words: Spouted bed; Integral swirling blade nozzle; Effect of swirling flow; Enhancement of particle movement; Numerical simulation

Transient Gas/Solid Flow Characteristics in Gas-Solid Cyclone Reactor Based on Euler/Lagrange Approach

Conference: Fluidization

Year: 2019

Proceeding: Fluidization XVI

Group: General Paper Pool

Session: 7C: Modeling and Simulation

Time: Wednesday, May 29, 2019 - 2:19pm-2:31pm

Authors:

Li, A., China University of Petroleum (East China)
Zhu, L., China University of Petroleum (East China)
Wang, Z., China University of Petroleum (East China)
Liu, C., China University of Petroleum (East China)

Abstract
A downward gas-solid cyclone reactor was designed for fluid catalytic cracking. The way mass and heat transfer of reactants is changed due to the swirling flow field in the new type reactor. A comprehensive three-dimensional numerical model was developed to deeper interpret the transient gas/solid flow behaviors in gas-solid cyclone reactor (GSCR). The continuous phase was modeled with turbulent computational fluid dynamics model and the dispersed phase was modeled with discrete phase model. The simulated dynamic pressure drop was agreed reasonably well with experiment. In addition, the solid flow pattern, particle concentration and gas velocity and kinetic energy were described respectively using the model. Simulated results show that the distribution of particle concentration is asymmetrical on the 0°-180°section and heterogeneous in time series. Particle groups with low kinetic energy were observed in conical segment and separation region. The phase interface of the gas and solid on different cross-section is obtained. Meanwhile, it can be seen that the size of the particle group is constantly changing. Gas phase has a higher downward angle in annular region and separation region than in mixing region. Besides, the velocity in the annulus is larger. The radial velocity distribution showed that the air velocity influenced by particle accumulation is not significant as its turbulent pulsation. Tangential and axial kinetic energy are accounting for 96.69%-97.11% of total kinetic energy and thus dominant.

An Integrated Model Simultaneously Considering Physical and Chemical Adsorption for CO2 Capture in the Fluidized Bed

Conference: Fluidization

Year: 2019

Proceeding: Fluidization XVI

Group: General Paper Pool

Session: 7D: Modeling and Simulation

Time: Wednesday, May 29, 2019 - 2:43pm-2:55pm

Authors:

Hu, X., Xiangtan University
Zhou, Y., Xiangtan University
Lu, Y., Xiangtan University
Liu, L., Xiangtan University
Kang, P., Xiangtan University

As is well known, the fluidized bed adsorption technology is beneficial for reduction of the greenhouse gas emission. Various supported porous sorbents are widely applied for CO_2 captu due to their high specific surface and thermostability, where the adsorption process simultaneously includes physical and chemical adsorption. However, the physical adsorption, a key factor affecting CO_2 capture efficiency, is always neglected in most numerical studies, whi consequently leads to distinct deviations between the simulation results and experimental dat Therefore, an integrated model, with simultaneous physical and chemical adsorption into consideration, is proposed in this study to solve the above mentioned problems. Besides, with regard to the particle agglomeration and its effect on the CO_2 capture efficiency, which will decrease gas-solid contact area and thus weaken the heat- and mass-transfer capabilities, EMMS (energy-minimization multi-scale)-based drag model is combined with the integrated adsorption model to study characteristics of CO_2 adsorption process and then the simulation results are validated by experimental data. Meanwhile the effects of inlet gas velocity, initial CO_2 concentration and gas temperature on the adsorption reaction rate and CO_2 capture efficiency are discussed through the thermodynamic analysis. In conclusion, the integrated model simultaneously considering physical and chemical adsorption will provide an effective guidance for numerical study of CO_2 capture in terms of dominate mechanism. Predictably, the proposed model can also be extended to other gas pollutant emissions treatments, such as desulfurization and denitrification.

CFD Analysis of the Influence of Particle Characteristics on the Hydrodynamics of a Bubbling Fluidized Bed

Conference: Fluidization

Year: 2019

Proceeding: Fluidization XVI

Group: General Paper Pool

Session: 7D: Modeling and Simulation

Time: Wednesday, May 29, 2019 - 2:07pm-2:19pm

Authors:

Uzair, M. A., Politecnico di Bari
Camporeale, S. M., Polytechnic University of Bari
Torresi, M., Politecnico di Bari
Fornarelli, F., Politecnico di Bari

Fluidized beds find their applications in various industries due to higher heat and mass transfer rates. A relevant domain of the wide-range of application is biomass-gasification. However, the hydrodynamics of fluidized beds are very complex and are difficult to quantify. Barriers to obtain enough experimental data include very high cost of pilot-scale reactors and limitation of experimental techniques to non-invasive ones due to opaqueness of dense gas-solid regimes. While there have been much investigations into hydrodynamics of two-phase flows, there have been fewer studies on mixing/segregation phenomenon of three phase flows involving binary mixtures of biomass and sand in a fluidized bed, and even more few for heat transfer. Another problem is the scale-up of reactors. When moving to large-scale fluidized bed, a set of parameters e.g. hydrodynamics (particle and bubble size and distribution), heat transfer change drastically, which in turn have strong negative impact on the operation of the fluidized bed. Computational fluid dynamics provides an alternative way to analyze the gas-solid and solid-solid behavior inside the fluidized beds.

For this purpose, 2D and 3D eulerian-eulerian approaches were used to perform simulations on a laboratory-scale bubbling fluidized bed for glass beads bed and validated against the experimental results [1]. The CFD simulations were performed on ANSYS FLUENT 17.2 using finite volume approach for 50 s. Gidaspowâ€™s drag law suitable for dense systems was used to measure gas-solid momentum exchange coefficient. Influence of restitution coefficient *e* was investigated on the hydrodynamics by varying *e* from 1 (perfectly elastic collisions) to 0.95, 0.9 and 0.8. Also, the influence of sphericity (Ï• = 0.95, 0.9, 0.85) of particles on the hydrodynamics was investigated. Particle diameter d was replaced in the relation for particle-air exchange coefficient by effective diameter d_{eff} given as d = Ï•d_{eff}. Preliminary results from both 2D and 3D simulations at *e* = 0.9 and Ï• =1 agreed in terms of time-averaged pressure drop across the bed, bed expansion and void fraction. Therefore, only 2D simulations were considered for further studies.

At *e* = 1, there was no bubble formation. As *e* was reduced to 0.95, bubbles started to form with size slightly increasing for *e* = 0.9 and 0.85. Also, there were not much variations in granular temperature for *e* = 0.85, 0.9 and 0.95. So, for further investigations in to the hydrodynamics for bed expansion, void fraction and pressure drop, *e* = 0.9 was chosen.

Initial simulations considering ideal spherical particles and an initial void fraction of 0.373 showed lesser time-averaged bed expansion and higher pressure drop across the bed, possibly due to the irregular shapes of particles, which allows for more space in between them and hence higher initial volume fraction of air inside the bed. This leads to lower mass of the bed compared to ideally spherical particles for the same bed height. Therefore, bed comprising non-spherical particles have lesser pressure drop and require low minimum fluidization velocity to fluidize the bed. The results revealed the strong dependence of hydrodynamics on the particle shape. Numerical results of time averaged pressure drop and bed expansion matched the experimental results for Ï• = 0.9 and 0.85. Glass beads are particles of uniform density, but biomass is a non-spherical particle as well as porous. Porosity of biomass indicates the voids inside the particle, thereby leading to assumption of wrong initial conditions required for CFD modeling e.g., initial void fraction of the bed and the bed height.

In the final paper we expect to show results concerning heat transfer and mixing/segregation phenomenon in a binary mixture of biomass and sand considering the particle shape and porosity. The objective is to perform numerical study mimicking real conditions, in order to have a proper understanding of the hydrodynamics. Initial results reveal deviations in the numerical and experimental results for bed expansion, pressure drop and solids distribution at various heights from the distributor plate, which initially show some dependence on the bulk density of bed. Further investigations will be carried out to account for the differences in the simulations.

CFD Modelling of Industrial-Scale Fluidized Systems: A Review of Recent Advancements

Conference: Fluidization

Year: 2019

Proceeding: Fluidization XVI

Group: General Paper Pool

Session: 7D: Modeling and Simulation

Time: Wednesday, May 29, 2019 - 2:19pm-2:31pm

Authors:

Akhavan, A., CPFD Software
Parker, J. M., CPFD Software

Computational fluid dynamics (CFD) is an important engineering tool for the research, design, and optimization of fluidized systems in many industries, including refining, petrochemical, cement, and power. At this scale, fluidized beds and other fluid-particle systems present a significant modelling challenge for traditional CFD due to the large solids inventory, discrete particle properties, and the complex coupled particle-fluid dynamics, reaction chemistry, and heat transfer. With recent advancements in CFD modelling tools and computing power, however, fluidized bed models are more accurate, efficient, and accessible than what was possible just a few years ago. This presentation will discuss these recent advancements in CFD modelling, including graphics processing unit (GPU) acceleration, and demonstrate the application of these advancements to the modelling of industrially-significant fluidized systems.

CFD Simulation of Biomass Gasification Fluidized Bed with a Thermal State-Based EMMS Drag Model

Conference: Fluidization

Year: 2019

Proceeding: Fluidization XVI

Group: General Paper Pool

Session: 7D: Modeling and Simulation

Time: Wednesday, May 29, 2019 - 2:55pm-3:07pm

Authors:

Kang, P., Xiangtan University
Liu, L., Xiangtan University
Lu, Y., Xiangtan University
Hu, X., Xiangtan University
Zhou, Y., Xiangtan University

As a promising biomass gasification technology, fluidized bed gasification can effectively alleviate the environmental and energy crisis. However, due to the high complexity and coupling relationship between gas-solid flow and chemical reaction in fluidized bed gasifiers, it is a great challenge to design and scale up such fluidized bed gasifiers. Emerging as an efficient studying approach, detailed CFD modeling may help to respond and address this key issue. The EMMS (energy-minimization multi-scale) model, compared with conventional drag models, shows relatively high performance in predicting gas-solid flow characteristics in the fluidized bed. However, most EMMS models are studied under cold model states, hence there are some deviations when they are directly applied to the thermal conditions due to real-time variation of the gas and solids properties. Moreover, most char conversion models are validated with the experimental values without further analysis of the reaction mechanism. In this work, a thermal state-based EMMS drag model coupled with a shrinking core model is proposed to determine the hydrodynamics, heat and mass transfer. With regard to the variations of phases properties due to chemical reactions, the real-time local properties of the two phases are accessed to calculate the heterogeneous drag correlation factor, which can greatly improve the generality and accuracy of EMMS model for the thermal state systems. Meanwhile, the transformation mechanism of control step between chemical reaction and diffusion are fully clarified in char conversion. Eventually, the proposed model is validated with experimental data and theoretical analysis. Predictably, the proposed EMMS model can provide more detailed and reliable information for the fluidized bed gasifier design and scale up.

Direct Numerical Simulation of the Two-Phase Interaction of Non-Uniform Structure in Fluid-Solid Systems

Conference: Fluidization

Year: 2019

Proceeding: Fluidization XVI

Group: General Paper Pool

Session: 7D: Modeling and Simulation

Time: Wednesday, May 29, 2019 - 1:55pm-2:07pm

Authors:

Zhao, W., Harbin Institute of Technology
Liu, G., Harbin Institute of Technology
Lu, H., Harbin Institute of Technology
Yin, X., Colorado School of Mines
Wu, Y., Harbin Institute of Technology

In this work, a direct numerical simulation (DNS) of the fluid particle two-phase flow system is performed. The DNS results span a rather wide parameter space: mean-flow Reynolds numbers on the order of 1 and 10, mean solid volume fractions from 0.1 to 0.4, solid-to-fluid density ratios from 2 to 1000 and elastic and moderately inelastic (restitution coefficient of 0.9) conditions. It can be observed from the results that non-uniform structures (clusters) appear inside the system. And as the parameters change, the particles distribution will exhibit a variety of structures. This paper focuses on the interphase interaction of different structures, considering not only the Reynolds number and the local concentration of the particles, but also the effect of the local concentration gradient of the particles. Unlike the averaging method of the two-fluid model, detailed information of particles is used for the local concentration of particles. A description of the non-uniform structure in the system is obtained by analyzing the phase interactions under different particle distribution structures.

Keynote Talk: Meso-Scale Modeling of Flow Regime Transitions in Gas-Solid Fluidized Beds

Conference: Fluidization

Year: 2019

Proceeding: Fluidization XVI

Group: General Paper Pool

Session: 7D: Modeling and Simulation

Time: Wednesday, May 29, 2019 - 1:30pm-1:55pm

Authors:

Tian, Y., Institute of Process Engineering
Wang, W., Institute of Process Engineering, Chinese Academy of Sciences

Gas-solid fluidized bed is a typical nonlinear nonequilibrium, dissipative system, featuring meso-scale structures, e.g., bubbles and clusters. To capture the effects of these meso-scale structures, we proposed the structure-dependent multi-fluid model (SFM), thereby analyzed the fluidization behavior under conditions of extremum energy dissipation rate. Two flow regime transitions were investigated. In particular, we found that the criterion of minimum energy dissipation rate applies only to homogeneous, dilute flow states, whereas the energy minimization multi-scale (EMMS) stability condition well predicts the choking regime transition of circulating fluidized beds. Further, we proposed a new model for characterizing the length of clusters. By invoking a stability condition of maximum energy dissipation rate, the solution of SFM was found to well predict the minimum bubbling velocity and the voidage jump transition adhered to Geldart group A particles.

Kinetics Investigation on the Pyrolysis of Lignite

Conference: Fluidization

Year: 2019

Proceeding: Fluidization XVI

Group: General Paper Pool

Session: 7D: Modeling and Simulation

Time: Wednesday, May 29, 2019 - 2:31pm-2:43pm

Authors:

Lian, W., Taiyuan University of Technology

The pyrolysis has been well known as a clean technology of making full use of the lignite resource. Commercializing the relevant process partially based on the cheap Computation Fluid Dynamics (CFD) simulation became possible if the relatively accurate and general reaction rate expression could be defined. However, the distributed activation energy model (DAEM), which was widely used to predict the coal pyrolysis, needed multiple integration to evaluate the reaction rate, making it difficult to be coupled into the species transport equations in CFD. In order to avoid the numerical integration, the lignite pyrolysis was assumed to be comprised of multiple parallel first order reactions in this work, and the total reaction rate was written as the sum of general Arrhenius equations. The kinetics parameters for each assumed reaction, namely, pre-exponential factor, activation energy and mass fraction, were obtained by fitting the thermogravimetry (TG) experimental data. The results showed that the general Arrhenius equations can more accurately predict the pyrolysis of lignite than DAEM. And then, based on the understanding that the chemical bonds broken in the pyrolysis would be same while the relevant fraction would be varied with the different kinds of lignite, we tried to use the activation energy and pre-exponential factor from one kind of lignite to fit the other three kinds of lignite, and got an acceptable result, which could validate the above theoretical analysis to some extent. In addition, for investigating the effect of particles diameter and heat transfer coefficient on the pyrolysis behavior, the model to predict the pyrolysis in a coal particle was developed based on the above kinetics expression. The results showed that the fluidized bed was a better choice to commercialize the lignite pyrolysis and the particles size should be smaller than 3 mm. Also, compared with DAEM, the computational time can be reduced from 51.9 s to 3.8 s under the same conditions, making it possible to design and optimize the relevant pyrolysis reactor based on the cheap CFD analysis. This investigation provides the reference of simulating the pyrolysis by coupling the Arrhenius kinetics reaction expressions into the species transport equations in CFD.

Keywords: Lignite pyrolysis, Kinetics, CFD, DAEM, Fluidized bed

A Pressure-Drop Method for Real-Time Monitoring the Solid Flux in Circulating Gas-Solid Process

Conference: Fluidization

Year: 2019

Proceeding: Fluidization XVI

Group: General Paper Pool

Session: 8A: Fundamentals of Fluidization

Time: Wednesday, May 29, 2019 - 4:12pm-4:24pm

Authors:

Gao, S., China University of Petroleum-Beijing
Fan, Y., China University of Petroleum, Beijing
Lu, C., China University of Petroleum-Beijing

A pressure-drop method for real-time monitoring the solid flux in circulating gas-solid processes

Sihong Gao, Yiping Fan*, Chunxi Lu

State
Key Laboratory of Heavy Oil Processing, College of Chemical Engineering, China University of Petroleum, Beijing, 102249, China

*Corresponding author.

E-mail
address: fanyipin@yahoo.com (Y. Fan)

Solid
(particles or granules) circulation flux has
significant influences on the most gas-solid processes including the circulating fluidized bedand the moving granular bed. However,
there are still no reliable and simple approaches in published studies for monitoring the instantaneous variation of the solid circulation flux during

operation, especially under the industrial conditions. A real-time monitoring method is developed for timely reflecting the instantaneous variation of the circulation flux in a gas-solid process, by measuring the pressure drop of the riser. In this study, two pressure taps are set between a fixed distance at the wall of the riser. One tap is 0.5 m away from the top of the pre-lift section; the distance between two taps is 4 m. The pressure drop between the 4 m riser is recorded continuously at 1 HZ by a digital manometer under a wide operating range. The experimental data clearly shown that the pressure drop of the 4 m riser has a good linear variation with the solid circulation flux. A theoretical correlation between the solid circulation flux and the pressure drop is proposed. The parameters in this mathematical correlation equation are experimentally explored and induced for monitoring and predicting the solid circulation flux.

Keywords: Gas-solid process; Solid circulation flux; Pressure drop; Riser

Cyclones with and without an Eccentrically Positioned Vortex Finder

Conference: Fluidization

Year: 2019

Proceeding: Fluidization XVI

Group: General Paper Pool

Session: 8A: Fundamentals of Fluidization

Time: Wednesday, May 29, 2019 - 4:00pm-4:12pm

Authors:

Muschelknautz, U., MK Engineering
Cocco, R., Particulate Solid Research, Inc.
Freireich, B., Particulate Solid Research, Inc.
Reddy Karri, S. B., Particulate Solid Research, Inc
Knowlton, T. M., Particulate Solid Research, Inc.

Usually the gas outlet tube immersed into the cyclone separator, also called the vortex finder, is located concentrically, i.e. its axis coincides with the axis of the cyclone. However, it has been shown in numerous recirculating cyclones of CFB power plants that eccentric positioning of the vortex finder axis improves the cyclone efficiency without affecting the pressure drop [1,2]. By applying this measure the circulating material became finer and its quantity increased resulting in a considerable improvement of the boiler performance: The heat transfer in the combustion chamber and, if installed in the fluid bed cooler, improved the optimum combustion temperature to ~900°C and reduced the amount of unburnt filter ash. In addition, the optimization led to cost savings of operational costs.

Trefz [3] gave a qualitative explanation of the positive effect of eccentrically shifting the vortex finder axis on the cyclone efficiency based on flow measurements in cyclones. At the top of the cyclone barrel region, there is a secondary flow through the boundary layer of about 10% of the total flow carrying particles towards the center. The vortex finder forces this secondary flow around the tube down to the opening allowing to remove particles from that flow by centrifugal force. In cyclones with a tangential inlet and a concentric vortex finder, however, this separation is not optimal: The axis of the vortex flow is pushed away from the geometrical center of the cyclone by the incoming gas flow and as a consequence the tangential velocity of the flow around the vortex finder is not uniform: In the section of the vortex finder surface which is closer to the cyclone axis the tangential velocity stagnates and the secondary flow streams vertically downwards with practically no circumferential velocity. In this section the particle separation is poor. By shifting the vortex finder axis into the â€œeyeâ€ of the vortex a uniform flow around the vortex finder is achieved and its separation efficiency improves considerably.

Recently, numerical granular-fluid simulations of high loaded cyclones with a concentric and with an eccentric vortex finder have been performed by applying the CFD code Barracuda® of CPFD Software LLC. The CFD results confirm that eccentrically shifting the vortex finder results in a better collection efficiency without compromising the pressure drop. The calculations are compared with measurement data of commercial high loaded cyclones in CFB boilers and the origin of the improvement due to the eccentric shifting the vortex finder is studied.

References

[1] Muschelknautz U, Muschelknautz E (1999). Separation efficiency of recirculating cyclones in CFB combustions. VGB PowerTech 4/1999, 48-53.

[2] Ipsen C, Roschek D, Muschelknautz U (2014) Optimierung eines Zyklonabscheiders einer zirkulierenden Wirbelschichtfeuerung. VGB PowerTech 4/2014:75-79

[3] Trefz M (1992). Die verschiedenen Abscheidevorgänge im höher und hoch beladenen Gaszyklon unter besonderer Berücksichtigung der Sekundärströmung. VDI-Fortschr.-Ber., VDI-Verlag, Düsseldorf, Reihe 3, No. 295.

Fluidization Behavior of Elongated Particles - CFD-DEM Simulations and X-Ray Tomography Experiments

Conference: Fluidization

Year: 2019

Proceeding: Fluidization XVI

Group: General Paper Pool

Session: 8A: Fundamentals of Fluidization

Time: Wednesday, May 29, 2019 - 5:00pm-5:12pm

Authors:

Mema, I., Delft University of Technology
Verma, V., Delft University of Technology
Wagner, E. C., Delft University of Technology
Kleijn, C. R., Delft University of Technology
van Ommen, J. R., Delft University of Technology
Padding, J. T., Delft University of Technology

Gas-solid fluidized beds are widely used in process and energy industries. The performance of a fluidized bed is majorly governed by the formation of gas bubbles and their distribution, facilitating rapid solids mixing and high heat and mass transfer rates. On the other hand, very large bubbles (slugs) reduce gas-solid contact, limiting the particle mixing, mass and heat transfer characteristics in the gas-solid bed. Therefore, the study of bubbles in fluidized beds is a key factor for its scale-up. Even though bubble formation in gas-solid fluidized beds has been the subject of a large number of investigations, most of these investigations considered only particles that are close to spherical. In recent years, there is an increased interest to use fluidized beds for biomass processing. Biomass particles which are dried and milled, and possibly pelletized, are characterized by an elongated shape and behave considerably different from spheres when fluidized. So far, the literature is quite limited when it comes to bubble dynamics of fluidized beds with non-spherical particles.

In this work, we investigate bubble dynamics in 3D fluidized beds with Geldart D spherocylindrical particles using both experimental and numerical approach. For numerical simulations, we use an open source CFDEM code which uses coupled OpenFOAM computational fluid dynamics (CFD) solver to describe the fluid phase and LIGGGHTS, the discrete element method (DEM) to account for particle-particle interactions. This CFD-DEM code was adapted to take in to account spherocylindrical particles [1, 2]. Hydrodynamic interactions between fluid and spherocylindrical particles are modeled using correlations developed by Sanjeevi et al. [3]. Results from the CFD-DEM simulations are compared with experimental results obtained with X-ray tomography. This non-invasive experimental method gives us insight into properties of bubble formation in a 3D fluidized bed [4-5]. Comparison is made on the basis of bubble size and shape, bubble rise velocity, and bubbles and void fraction distributions.

References

[1] V. V. Mahajan, T. M. J. Nijssen, J. A. M. Kuipers, J. T. Padding, (2018), â€œNon-spherical particles in a pseudo-2D fluidised bed: Modelling studyâ€, *Chem. Eng. Science*, **192**, 1105-1123

[2] I. Mema, V. V. Mahajan, B. W. Fitzgerald, J. T. Padding, (2018), â€œEffect of lift force and hydrodynamic torque on fluidization of non-spherical particlesâ€, *Chem. Eng. Science,* In Press, doi: 10.1016/j.ces.2018.10.009

[3] S. K. P. Sanjeevi, J. A. M. Kuipers, J. T. Padding, (2018), â€œDrag, lift and torque correlations for non-spherical particles from Stokes limit to high Reynolds numbersâ€, *Int. J. Multiphase Flow*, **106**, 325-337

[4] F. Schillinger, T.J. Schildhauer, S. Maurer, E. Wagner, R.F. Mudde, J.R. van Ommen, (2018), â€œGeneration and evaluation of an artificial optical signal based on X-ray measurements for bubble characterization in fluidized beds with vertical internalsâ€. *Int. J. Multiphase Flow*, **107**, 16-32

[5] S. Maurer, D. Gschwend, E. C. Wagner, T. J. Schildhauer, J. R. Van Ommen, S. M. A. Biollaz, R. F. Mudde, (2016), â€œCorrelating bubble size and velocity distribution in bubbling fluidized bed based on X-ray tomographyâ€, *Chem. Eng. J.*, **298**, 17-25

Fluidized Beds with Internals: A Real-Time Magnetic Resonance Imaging Study of Gas Bubbles and Particle Motion

Conference: Fluidization

Year: 2019

Proceeding: Fluidization XVI

Group: General Paper Pool

Session: 8A: Fundamentals of Fluidization

Time: Wednesday, May 29, 2019 - 5:12pm-5:24pm

Authors:

Conzelmann, N., Empa, Swiss Federal Laboratories for Materials Science and Technology
Boyce, C. M., Columbia University
Penn, A., ETH Zurich
Pruessmann, K. P., ETH Zurich and University of Zurich
Müller, C. R., ETH Zurich

Fluidized beds are frequently equipped with internals in order to modify the fluidization dynamics within the bed, such as reducing the size of gas bubbles, or enforcing specific particle circulation patterns[1,2]. Traditionally, the design of internals is often based on rules of thumb and empirical correlations derived from optical measurements in pseudo two-dimensional (2D) systems, or by inserting probes into 3D systems. However, such measurements are either influenced by wall effects (pseudo 2D systems) or do interfere with the particle flow (intrusive probe measurements). More recently, sophisticated tomographic techniques have been applied to study fluidization dynamics in freely bubbling fluidized beds, such as X-ray tomography[3], electrical capacitance tomography[4] and magnetic resonance imaging (MRI)[5â€“7]. In order to design and scale-up fluidized bed reactors with internals reliably and efficiently, we need fundamental insight into fluidization dynamics in large, 3D systems with internals.

In this work we used a recently developed real-time magnetic resonance imaging methodology[7], to quantify the motion of particles and the size and distribution of gas bubbles within a cylindrical model system of diameter 190 mm with internals of different sizes. Instantaneous ($T_{aquisition}$ = 4.7 ms) snapshots of particle position and velocity of the particle phase reveal strongly reduced bubbling and reduced particle mobility in the wake region of the horizontal obstacles with circular cross section. Interestingly, also the region below the internals showed changes in their fluidization dynamics, such as an increased number of gas bubble formation.

The measurements presented here, provide valuable fundamental insight into the effects of internals on the fluidization dynamics in fluidized bed reactors. We anticipate that they might prove helpful to design reactors and to test numerical simulations.

[1] D. Kunii, O. Levenspiel, Fluidization Engineering, 2nd ed., Butterworth-Heinemann, 1991.

[2] J.S. Dennis, Properties of stationary (bubbling) fluidized beds relevant to combustion and gasification systems, in: Fluid. Bed Technol. Near-Zero Emiss. Combust. Gasif., 1st ed., Woodhead Publishing, Cambridge, 2013: pp. 77â€"146.

[3] R.F. Mudde, Bubbles in a fluidized bed: A fast X-ray scanner, AIChE J. 57 (2011) 2684â€"2690. doi:10.1002/aic.12469.

[4] T.C.C. Chandrasekera, Y. Li, D. Moody, M.A.A. Schnellmann, J.S.S. Dennis, D.J.J. Holland, Measurement of bubble sizes in fluidised beds using electrical capacitance tomography, Chem. Eng. Sci. 126 (2015) 679â€"687. doi:10.1016/j.ces.2015.01.011.

[5] C.R. MÃ¼ller, J.F. Davidson, J.S. Dennis, P.S. Fennell, L.F. Gladden, a N. Hayhurst, M.D. Mantle, A.C. Rees, A.J. Sederman, Real-time measurement of bubbling phenomena in a three-dimensional gas-fluidized bed using ultrafast magnetic resonance imaging., Phys. Rev. Lett. 96 (2006) 154504. doi:10.1103/PhysRevLett.96.154504.

[6] C.M. Boyce, N.P. Rice, J.F. Davidson, A.J. Sederman, J.S. Dennis, D.J. Holland, Magnetic resonance imaging of gas dynamics in the freeboard of fixed beds and bubbling fluidized beds, Chem. Eng. Sci. 147 (2016) 13â€"20. doi:10.1016/j.ces.2016.03.005.

[7] A. Penn, T. Tsuji, D.O. Brunner, C.M. Boyce, K.P. Pruessmann, C.R. MÃ¼ller, Real-time probing of granular dynamics with magnetic resonance, Sci. Adv. 3 (2017) e1701879. doi:10.1126/sciadv.1701879.

Mixing Behavior of Binary Mixtures of Biomass and Silica Sand in a Fluidized Bed

Conference: Fluidization

Year: 2019

Proceeding: Fluidization XVI

Group: General Paper Pool

Session: 8A: Fundamentals of Fluidization

Time: Wednesday, May 29, 2019 - 4:36pm-4:48pm

Authors:

Emiola-Sadiq, T., University of Sask.
Zhang, L., University of Saskatchewan
Dalai, A. K., University of Saskatchewan
Wang, J., University of Sask.

Mixing and segregation of binary mixtures involving biomass materials in a fluidized bed was experimentally investigated. Three different biomass materials were studied: willow sawdust, pelletized soy and oat hull materials. Each of these three materials was prepared as a binary mixture with silica sand at six compositions ranging from 5% - 30% with an increment of 5%. The experiment was carried out at different superficial gas velocities in the bed ranging from the initial minimum fluidization velocity Umf of each sample mixture up to 4Umf with an increment of 0.5 Umf. The extent of mixing for each of the biomass composition was investigated using a mixing index based on the standard deviations of the sample at different axial positions in the bed. The radial composition was also investigated by means of digital imaging analysis using a high resolution digital camera located at the top of the bed. Ten uniform axial and radial positions were used during the experimental process. The experiment revealed the differences in the extent of particle distribution for pelletized biomass materials vs non-densified materials as a result of differences in particle size, density and importantly, particle shape of the material utilized. The results showed an increase in the extent of mixing as Umf increased for both pellet materials while the mixing index for sawdust decreased beyond a loading of 20%. With an increase in biomass loading, an increase in mixing index was found for the two pellet materials. A similar trend was observed for sawdust at the lower loading level. However, the mixing index started to decrease at higher loading level beyond 20%. The results would further help understand hydrodynamics of binary and multicomponent mixtures involving biomass materials in fluidized beds.

Study on the Staggered Electrodes for 3D Electrical Capacitance Tomography

Conference: Fluidization

Year: 2019

Proceeding: Fluidization XVI

Group: General Paper Pool

Session: 8A: Fundamentals of Fluidization

Time: Wednesday, May 29, 2019 - 4:24pm-4:36pm

Authors:

Shen, J., University of Chinese Academy of Sciences
Ye, M., Dalian Institute of Chemical Physics, Chinese Academy of Sciences
Meng, S., Dalian Institute of Chemical Physics, Chinese Academy of Sciences
Yang, W., The University of Manchester
Liu, Z., Dalian Institute of Chemical Physics, Chinese Academy of Sciences

Electrical capacitance tomography (ECT) has received attention in fluidized bed measurement for many years as a safe, economic, and non-intrusive measurement technique[1-3]. ECT is based on the measurement of the variation of capacitance of sensors composed of excitation and detection electrodes, which are normally installed externally of the fluidized bed vessels[4]. Compared with 2D ECT, 3D ECT can provide information concerning flow pattern and bubbles directly from the volumetric images of fluidized beds[5].

Low image quality, however, possesses a serve challenge for 3D ECT because of the sensitivity of sensors on the one hand and the non-linear ill-posed nature of image reconstruction on the other hand. For 3D ECT, the traditional rectangular sensors have been commonly used, in which the measured signal of the capacitance between electrodes in the layers far from each other would be extremely week compared to that between electrodes in the adjacent layers, causing low signal to noise ratio (SNR).

In this work, we

studied the ECT sensors with staggered electrodes, in particular the shape of electrodes with regard to its effect on the image quality. Three different shapes of electrodes, including the diamond, circular and hexagon shape, were studied. We studied the quality of reconstructed images with both numerical simulations and experiments for static objects, and analyzed quantitatively the image errors and correlation coefficients. The results show that the precision of the sensors with staggered electrodes are higher than that with rectangle electrodes. In addition, image quality obtained by sensors with staggered electrodes is superior than that by rectangle electrodes. This provides a potential way to improve the image quality for 3D ECT. The application of the staggered electrodes for 3D ECT in fluidized bed measurement was also illustrated.

References

1. Xie CG, Huang SM, Hoyle, BS, Thorn R, Lenn C, Snowden D, Beck MS. Electrical capacitance tomography for flow image: system model for development of image-reconstruction algorithms and design of primary sensors. Iee Proc G Circuits Dev Syst. 1992;139:89-98.

2. Liu S, Yan R, Wang H, Jiang F, Dong X. Applications of Electrical Capacitance Tomography in Two Phase Flow Visualization. J Therm Sci.2004;13:179-186.

3. Wang DW, Xu MY, Marashdeh Q, Straiton B, Tong A, Fan LS. Electrical Capacitance Volume Tomography for Characterization of Gas-Solid Slugging Fluidization with Geldart Group D Particles under High Temperatures. Ind Eng Chem Res.2018;57:2687-2697.

4. Yang W. Design of electrical capacitance tomography sensors. Meas Sci Tech.2010;21:042001.

5. Wajman R, Banasiak R, Mazurkiewicz L, Dyakowski T, Sankowski D. Spatial imaging with 3D capacitance measurements. Meas Sci Tech. 2006;17:2113-2118.

Synchrotron Based X-Ray Imaging Technique for Understanding of Pharmaceutical Powder Granulation

Conference: Fluidization

Year: 2019

Proceeding: Fluidization XVI

Group: General Paper Pool

Session: 8A: Fundamentals of Fluidization

Time: Wednesday, May 29, 2019 - 4:48pm-5:00pm

Authors:

Li, C., University of Saskatchewan
Zhang, L., University of Saskatchewan
Zhu, N., Canadian Light Source Inc.
Gasilov, S., Canadian Light Source Inc.

Fluidized beds are widely used in many unit operations. In the pharmaceutical industry, fluidized beds have been intensively used for granulation, coating, and drying. Due to the complexity of the granulation process in fluidized beds, observing and measuring the granulation process are challenging with conventional experimental methods. In this study, synchrotron-based X ray imaging techniques were for the first time applied to capture the dynamic granulation process with a single drop impacting method in pharmaceutical powder beds. Five common pharmaceutical excipients and two active pharmaceutical ingredients (APIs) were used as static powder beds. The dynamic interaction between the liquid binder (water) and solid powders was observed from X-ray images collected. In addition, porosities and morphologies of the final granules were investigated by X-ray computed tomography (CT). Primary results show that pharmaceutical powder properties, including particle size and hydrophilicity, have a significant impact on water penetration, dynamic granule growth, and the final granular product. Moreover, the present results also demonstrated that X-ray imaging techniques could be employed as powerful tools for advancing understanding various powder handling processes including granulation, drying and coating.

Design and Experimental Investigation of a Novel Interconnected Fluidized Bed for Chemical Looping Conversion of Solid Fuel

Conference: Fluidization

Year: 2019

Proceeding: Fluidization XVI

Group: General Paper Pool

Session: 8B: Fluidized Bed Applications

Time: Wednesday, May 29, 2019 - 5:25pm-5:37pm

Authors:

Liu, Y., Qingdao University of Science and Technology
Song, M., Qingdao University of Science and Technology
Zhang, X., Qingdao University of Science and Technology
Gao, M., Qingdao University of Science and Technology
Hu, X., Ningxia University
Guo, Q., Ningxia University

Chemical looping offers a versatile platform to chemical conversion and energy utilization in a clean and efficient manner. Except the chemical looping combustion/gasification, novel processes based on the chemical-looping thought have sprung up. Chemical looping conversion of solid fuels attracts more attention, and significant progress has been made in last decades. One technical challenge of using solid fuels involves the reaction enhancement between oxygen carrier and solid fuel. Because both the oxygen carrier and solid fuel are solid particles, the intensified mixing between the two different solid particles in fuel reactor seems to be of importance.

Chemical looping reactors are widely accepted as interconnected fluidized beds. Herein, a novel interconnected fluidized bed for chemical looping conversion of solid fuel was designed. Firstly, aiming to intensify the reaction between solid fuel and oxygen carrier particles, a novel impinging stream feeding device at the top or bottom of the fuel reactor was designed. Both the hot oxygen carrier particles from air reactor and solid fuel particles carried by the high-speed gasification gas flow were impinged in the opposite direction. The impinging stream can ensure the excellent mixing of two type particles by crushing solid fuels into fine particles at the high temperature. Secondly, to reduce unburnt volatiles and gasification products escaping out of fuel reactor, the fuel reactor was designed as two-staged fluidized bed. First stage was selected as riser or downer, of which flow pattern is approximate to plug flow, for the pyrolysis of solid fuel. The other stage was selected as turbulent fluidized bed in order to fully convert the solid fuel char.

To verify the stable operation of the designed interconnected fluidized bed, cold model setup was erected. Flow characteristics of particles in the built cold model were investigated. Experimental results indicated that the designed chemical looping reactor possessed excellent flow performance. It also can provide a new way for solid fuel conversion by the gas-solid-solid reaction system.

Full-Loop CPFD Simulation on the Multi-Scale Gas-Solids Flow Characteristics of a Pressurized Circulating Fluidized Bed

Conference: Fluidization

Year: 2019

Proceeding: Fluidization XVI

Group: General Paper Pool

Session: 8B: Fluidized Bed Applications

Time: Wednesday, May 29, 2019 - 5:37pm-5:49pm

Authors:

Zhu, X., Institute of Engineering Thermophysics, Chinese Academy of Sciences
Dong, P., Institue of Engineering Thermophysics, Chinese Academy of Sciences
Tu, Q., Institute of Engineering Thermophysics, Chinese Academy of Sciences
Zhu, Z., University of Chinese Academy of Sciences
Wang, H., University of Chinese Academy of Sciences

Complete
understandings of the gas-solids flow hydrodynamic characteristics are necessary for the design and optimization of the pressurized circulating fluidized bed (PCFB). However, few researches on the PCFB hydrodynamic behaviors based on CFD simulation have been reported in the literature. In this research, three-dimensional full-loop computational particle fluid dynamic (CPFD) simulation combined with pressure measurement was presented to improve the understandings of the multi-scale hydrodynamic behaviors of an industrial-scale cold PCFB test rig.

To
investigate the drag models effects on the simulation accuracy, the conventional homogeneous drag model (i.e. Wen-Yu model) and two schemes of sub-grid drag models, namely, the empirical scaling model (i.e. 0.6*Wen-Yu model) and the EMMS drag model, are implemented in the CPFD simulation respectively. The gas-solids flow behaviors predicted by those three drag models were compared and evaluated by

the measurement results based on pressure transducers. In addition, a series of operating pressure conditions were investigated to address the pressure effects on the PCFB hydrodynamic behaviors. Detailed CPFD simulation results are given and analyzed.

The results show that EMMS drag model not only predicts the most accurate macro-scale flow behaviors of the pressure distribution, particles volume fraction distribution and circulation flux with the experiment results, but also performs well in capturing the meso-scale coalescence and breakage of the particles clusters than the other two drag models. By contrast, Wen-Yudrag model heavily over-predicts the particles expansion and circulation flux, and 0.6*Wen-Yu drag model alleviates the over-prediction to some degree. The hydrodynamic behaviors of PCFB are enhanced with the increase of operating pressure,by means of the more homogeneous particles distribution, higher particles velocity and larger circulation flux.

Indirect Heating of Bubbling Fluidized Beds By Means of Liquid Metal Heat Pipes

Conference: Fluidization

Year: 2019

Proceeding: Fluidization XVI

Group: General Paper Pool

Session: 8B: Fluidized Bed Applications

Time: Wednesday, May 29, 2019 - 4:25pm-4:37pm

Authors:

Karl, J., Friedrich-Alexander-University Erlangen-Nuremberg

Indirect heating of bubbling fluidized beds by means of liquid metal heat pipes enables highest heat fluxes into atmospheric or pressurized fluidized bed reactors. An application for such indirectly heated fluidized bed technologies is the Biomass Heatpipe Reformer. The technology was firstly presented during the Fluidization X conference in Bejing 2001.

Heat pipes are closed tubes partially filled with a working liquid. Evaporation and condensation inside the tube guarantee highest heat transfer coefficients at both ends of the tubes and provide extraordinary high heat fluxes along the tubes.

The heat pipes provide the heat-of-reaction for endothermic biomass gasification reactions from a bubbling fluidized bed combustion into a pressurized steam-blown fluidized bed reactor. The concept allows a complete separation of combustion chamber and gasification chamber and enables therefore the use of different fuels and pressures in combustor and gasifier. Temperatures of 900°C in the combustion chamber and above 800Â°C in the gasifier enable the production of a particularly hydrogen rich syngas. The syngas suites ideally for the synthesis of Substitute Natural Gas (SNG) from biomass. A total of three prototypes have been tested at Technical University of Munich between 2001 and 2007. The start-up agnion Inc. the commercialization of the technology in 2007. The presentation will present the experiences from agnionâ€™s demonstration plants in the range from 500 kW and 1.3 MW. It will in particular discuss heat transfer and its impact on the reactor layout and dimensioning and long-term experiences with the heat pipes. Experimental observations have shown the essential impact of the fuel particles mixing on the syngas composition and quality.

An outlook will summarize further technical applications and recent developments for the application of liquid metal heat pipes in fluidized beds such as indirectly heated carbonate looping processes and pyrolysis applications.

Keynote Talk: Assessment of Combustion and Gasification Behaviour in Pilot Scale for Additivated Biomass

Conference: Fluidization

Year: 2019

Proceeding: Fluidization XVI

Group: General Paper Pool

Session: 8B: Fluidized Bed Applications

Time: Wednesday, May 29, 2019 - 4:00pm-4:25pm

Authors:

Pfeifer, C., University of Natural Resources and Life Sciences
Agu, C. E., University of South-Eastern Norway
Moldestad, B. M. E., University of South-Eastern Norway

During the storage of biogenic materials for energy utilization, destruction and conversion processes by bacteria and fungi take place inevitably. Besides the wood degradation, self-ignition of wood chip piles occurs consistently. Additivation of eg. Dolomite to the feedstock can inhibit microbial desctruction processes. Wood chips mixed with 3wt% calcium hydroxide was prepared and used as feed stock for a 20kW (fuel input) bubbling fluidized bed pilot plant. The reactor consists of a 100mm (innder diameter) bubbling bed reactor with 2m height and both gasification as well as combustion mode can be applied. The fluidization behaviour at different reaction temperatures (750-900°C) for combustion as well as for gasification with air was analysed. As feedstock the original biomass as well as the wood chips with additive was used. The producer gas was measured and compared for each feedstock and process parameter. The measurements are set into reference to calculation of mass and energy balances of the conversion of feedstock to syngas respectively thermal energy. Summarizing, the combustion as well as the gasification behaviour was hardly influenced by the additive.

Multi-Stage Fluidized Bed for Methanol to Aromatics

Conference: Fluidization

Year: 2019

Proceeding: Fluidization XVI

Group: General Paper Pool

Session: 8B: Fluidized Bed Applications

Time: Wednesday, May 29, 2019 - 5:49pm-6:01pm

Authors:

Qian, W., Tsinghua University
Hou, Y., Tsinghua University
Chen, Z., Tsinghua University

The conversion of methanol to aromatics (MTA) was expected as an alternative route to meet the increasing demand of aromatics in China. It is of importance since the technical route is crucial to China, which lack oil but was rich in coal. MTA process involves a series of consecutive reactions according to the hydrocarbon pool dual cycle concept. But, all of these reactions are mutually interactive and fully coupled in a single reactor. Apartment from aromatics, MTA process could producelarge amounts of olefinic intermediates and paraffinic by-products, especially with the reduction of catalyst acidity in reaction-regeneration cycles. The further conversion of non-aromatics products could improve the total aromatics yield. However, their conversion characteristic varied and depended on reaction temperature and acidic sites over zeolite. The characteristics of secondary conversion of MTA products with time was crucial to design reaction process and was investigated in a wide temperature range. It was clarified that the possible path of secondary conversion process of intermediates products depended on temperature and acidity. Corresponding products from different processes were determined. Based on these results and considering the variation of the temperature and acidity of catalyst during the fluidized bed reaction-regeneration cycle, the multi-stage fluidized bed process was proposed for three-step conversion of methanol for high-yield aromatic production. In this concept, the fresh catalyst characterized by high temperature and strong acidity from regeneration reactor flowed into the second stage of fluidized bed for the conversion of olefins and alkanes at high temperature of 550 °C. The partial deactivation of catalyst from the second stage overflowed into the first stage of fluidized for MTA at 470°C. The spent catalyst with medium acidity from the first stage was lift into the third stage for deep conversion of olefins at 470 °C. The fresh and mold Zn/ZSM-5 was used in this study and packing with each stage for simulate the proposed concept. The experimental results suggested that the yield of aromatics reached as high as 66% and its increase was about 10% in comparison of the result in the first stage.The undesired propane decreased obviously as is hoped by controlling the hydrogen transfer reaction and enhancing the dehydrogenation of alkanes.

Performance Assessment of a Vertical Pneumatic Conveying System for High Temperature Solar Particle Receivers Using a New Correlation for Minimum Transport Boundary

Conference: Fluidization

Year: 2019

Proceeding: Fluidization XVI

Group: General Paper Pool

Session: 8B: Fluidized Bed Applications

Time: Wednesday, May 29, 2019 - 5:13pm-5:25pm

Authors:

Saw, W. L., The University of Adelaide
Guo, P., The University of Adelaide
Ashman, P. J., The University of Adelaide
Nathan, G. J., The University of Adelaide
Lim, S., CSIRO

Solar
particle receivers for concentrated solar thermal (CST) have been widely investigated due to the high efficiency of direct heat transfer. In addition, particles, such as mineral sands and ceramics have high melting temperature, which allows them to be used directly as heat transfer medium at above >800°C. However, to realize the application of particle receivers in tower-mount CST plants requires the transportation of particles to and from the tower. Pneumatic conveying systems are proven to be industrially robust, having been widely implemented for over 100 years, however, no assessment of the potential to employ pneumatic conveying system under relevant to solar particle receivers has been reported previously. The minimum transport boundary is an important information required for the design and operation of a vertical pneumatic conveying system to ensure both a continuous and steady-state transport with minimum parasitic losses. The information on the minimum transport boundary for the suitable

particles under CST operating conditions is not available. Beyond this minimum transport boundary, further increase in solids flow or decrease in gas velocity could result in accumulation of solids at the bottom, which is defined as Type A chocking. Correlations have been suggested in the literature to predict Type A chocking. However, most of these correlations are only valid for particular experimental conditions, e.g., particle properties and conveying pipe diameter. In particular, relatively big deviation (>40%) was found by using these correlations for Geldart group B particles.

In the present study, a correlation, as shown in Eq 1, is developed to predict Type A chocking velocity (U_{ck}) for the selected Geldart group A and B particles, which suitable under CST environment, with a reasonable accuracy within 30%. The indexes a, b and c are all assumed to be governed by the Archimedes number (Ar). The value of b can be calculated for each Ar by the dimensionless velocity ((U_{ck} -U_t)/U_t, U_t is terminal velocity) and ($G/\rho_p U_t$) for the selected particle properties (particle density (ρ_p) and particle size (d_p)) and pipe diameter (D). Similarly the correlation of c with Ar can calculated by the dimensionless velocity and (D/d_p) for each Ar. For each dimensionless velocity, the value can be calculated for index a using the obtained values of b and c. Thus, a correlation of a with Ar can be calcuated. The correlations for indexes a, b and c are shown in Eqs 2-4. Comparison of present calculated chocking velocities with experimental chocking velocities in literature is shown in Figure 1. Therefore, further assessment on the performance of vertical pneumatic conveying systems can be conducted using this correlation.

Uck-UtUt=a×(GρpUt)b×(Ddp)c

$$\frac{U_{ck}-U_t}{U_t} = a \times \left(\frac{G}{\rho_p U_t}\right)^b \times \left(\frac{D}{d_p}\right)^c \qquad (1)$$

a=46000×Ar-2.08 Ar<21412.5×Ar-0.55 Ar>214

$$a = \begin{cases} 46000 \times Ar^{-2.08} & \text{Ar} < 214 \\ 12.5 \times Ar^{-0.55} & \text{Ar} > 214 \end{cases} \quad (2)$$

b=0.06×lnAr+0.08

$$b = 0.06 \times \ln(Ar) + 0.08 \quad (3)$$

c=0.2733×lnAr-0.9844

$$c = 0.2733 \times \ln(Ar) - 0.9844 \quad (4)$$

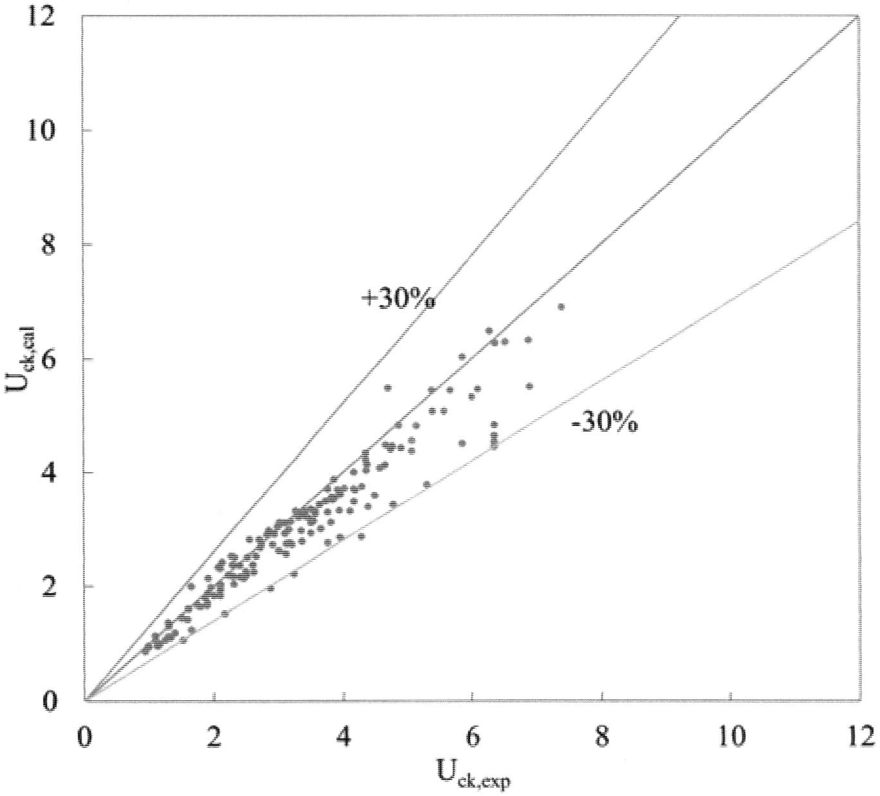

Figure 1 Comparison of the calculated chocking velocities with experimental chocking velocities reported from literature.

Performance of Large Scale Prepared Iron-Based Oxygen Carrier

Conference: Fluidization

Year: 2019

Proceeding: Fluidization XVI

Group: General Paper Pool

Session: 8B: Fluidized Bed Applications

Time: Wednesday, May 29, 2019 - 4:37pm-4:49pm

Authors:

Ren, T., Ningxia university
Guo, Q., Ningxia University
An, M., Ningxia University
Hu, X., Ningxia university

Performance of Large Scale Prepared iron-Based Oxygen Carrier

Ren tian, An Mei, Hu xiude, Guo Qingjie*

£¨State key laboratory of high-efficiency utilization of coal and green chemical engineering, Ningxia university, yinchuan 750021, China)

Abstract: To screen the low-cost and high-efficient oxygen carrier£¨OC£© that is suitable for coal chemical-looping gasification process, the natural hematite mixed cement, attapulgite (ATP) and bentonite were modified by mechanical mixing grinding ball method, and coal chemical-looping gasification experiments were carried out on the fluidized bed to investigate the redox and cyclic reactivity of modified oxygen carriers. Brunauer Emmett Teller (BET), scanning electron microscope (SEM) and X-ray diffraction (XRD) were used to characterize both fresh and used oxygen carrier particles. The results show that the reactivity order of iron-based composite oxygen carrier with different modifiers was as follow Fe@ATP>Fe@cemenct>Fe@bentonite at 950℃. The addition of cement effectively promotes the water-gas shift reaction, leading to the decrease of CO fraction and the increase of H_2 fraction in coal gasification products. The surface area and attrition resistance of Fe_2O_3/ATP OCs are improved significantly by using ATP as the support. Owing the synergy between ATP and Fe_2O_3, both are catalytic active, carbon conversion is enhanced significantly£®During the multiple cycles of experiments, the composition of the syngas remained stable and the sintering of the surface of the oxygen carriers resulted in a decrease in specific surface area and void volume. However, the particles kept a porous structure facilitating reactions between reactant gases and oxygen carrier particles Therefore, the compound iron oxygen carrier Fe@ATP is appropriate to be used for industrial production.

Key words: mechanical mixing grinding ball method; chemical-looping gasification; coal; oxygen carrier; hematite

Triboelectric Separation of Fine Coal By Using an Electric Field Fluidized Bed

Conference: Fluidization

Year: 2019

Proceeding: Fluidization XVI

Group: General Paper Pool

Session: 8B: Fluidized Bed Applications

Time: Wednesday, May 29, 2019 - 5:01pm-5:13pm

Authors:

Bai, X., China University of Mining and Technology
Wang, H., China University of Mining and Technology
Peng, Z., China University of Mining and Technology
Zhao, X., China University of Mining and Technology
Yang, J., China University of Mining and Technology
Zhang, G., China University of Mining and Technology

The occurrence of tribocharge is a universal phenomenon in gas-solid fluidization. After particle-particle or particle-wall contacts, coal particles obtain negative or and most mineral particles receive opposite polarity charge in fluidized bed. A novel electric field fluidized bed was set up to clean fine coal. The results show that the enrichment effect of fine coal in the fluidized bed is enhanced with the decease of bed height, and the separation performance increase with the increasing fluidization gas velocity and electric field. The sorting voltage is the most significant factor affecting the enrichment effect among the three. Under conditions of 60 kV voltage, 16.53 cm/s fluidization gas velocity and 2.5 mm static bed height, the clean coal with 9.84% ash content can be obtained after 30 minutes of operation, and the combustible recovery was 21.76%. It is confirmed that fine coal cleaning by the electric field fluidized bed is prospective, meanwhile further research work is needed to improve the recovery.

Verification of Coarse-Grained CFD-DEM in Multiple Flow Regimes

Conference: Fluidization

Year: 2019

Proceeding: Fluidization XVI

Group: General Paper Pool

Session: 8B: Fluidized Bed Applications

Time: Wednesday, May 29, 2019 - 4:49pm-5:01pm

Authors:

Lin, J., Zhejiang University
Luo, K., Zhejiang University
Wang, S., State Key Laboratory of Clean Energy Utilization
Hu, C., Zhejiang University
Fan, J., Zhejiang University

Abstract: The calculating capacity of traditional computational fluid dynamics-discrete element method (CFD-DEM) decreases with the increasing of tracking particles. The coarse-grained CFD-DEM can fewer the number of particles to be tracked and the collisions occur between particles by lumping several real particles into a coarse-grained parcel (CGP). The coarse-grained CFD-DEM was used to simulate the gas solid flow characteristics of different regime of fluidized beds (i.e., bubbling fluidized bed, spouting fluidized bed and circulating fluidized bed). The results of coarse-grained CFD-DEM were compared with the CFD-DEM simulation and experimental measurement. The characteristics of gas-solid flow were accurately predicted by coarse-grained method and compare well with traditional CFD-DEM and experiment data. Comprehensive assessment of coarse-grained CFD-DEM in predicting solid velocity is investigated in current work, while most of previous literatures force on the parameters of gas phase. What is worth mentioning is that coarse-grained CFD-DEM can observably improve the computational efficiency. The efficiency of coarse-grain method is about 5-8 times compares with the traditional CFD-DEM when the coarse-grained ratio is 2. Accordingly, the accuracy of coarse-grain method is validated in multiple flow regimes and the coarse-grained method can accelerate the calculation speed compared with the traditional CFD-DEM.

Bubbling Behaviour of Geldart Group B Particles in Pulsed Fluidization Mode: Experimental and Simulation Study

Conference: Fluidization

Year: 2019

Proceeding: Fluidization XVI

Group: General Paper Pool

Session: 8C: Modeling and Simulation

Time: Wednesday, May 29, 2019 - 5:37pm-5:49pm

Authors:

Akhavan, A., CPFD Software
Clark, R., CPFD Software
van Ommen, J. R., Delft University of Technology
Rhodes, M., Monash University
Coppens, M. O., University College London

Pulsed fluidization is a special mode of fluidization in which the flowrate of the fluidizing medium is intermittently changing. Pulsed fluidization has been proven to offer some unique advantages compared to steady-flow fluidization. These advantages include better mixing for cohesive particles [1], improved drying of wet particles in fluidized beds [2] and a more controlled bubble pattern in some bubbling fluidized beds [3].

Bubbling behaviour of group B particles in a two-dimensional column was studied experimentally. Solenoid valves were used to generate pulsation in the gas flow. Unlike the conventional concept of pulsed fluidization, total gas flow was not subjected to pulsation; a pulsing gas flow was imposed on a constant steady gas flow to the bed. In this paper, we are presenting the CFD simulation results of the system alongside the experimental findings. Barracuda VR was used for the CFD simulation. Glass beads with density of 2520 kg/m3 were used for particles. Air was used as the fluidizing medium and Wen-Yu was the drag model applied in most simulation cases. Effects of flow pulsation in general, and pulsation conditions such as frequency in particular, on bubbling characteristics were studied.

We found that pulsation in general leads to an increase in average bubble size, especially at lower pulsation frequencies; Up to 100% increase in average bubble size could be observed. The bubble size decreases with increasing pulsation frequency. This demonstrates that pulsation gives additional degrees of freedom to control fluidized-bed hydrodynamics.

[1] Hadi, B., van Ommen, J. R., & Coppens, M.-O. (2011). Enhanced particle mixing in pulsed fluidized beds and the effect of internals. Industrial & Engineering Chemistry Research, 51(4), 1713-1720.

[2] Akhavan, A., van Ommen, J. R., Nijenhuis, J., Wang, X. S., Coppens, M.-O., & Rhodes, M. J. (2008). Improved drying in a pulsation-assisted fluidized bed. Industrial & Engineering Chemistry Research, 48(1), 302-309.

[3] Wu, K., de Martín, L., & Coppens, M.-O. (2017). Pattern formation in pulsed gas-solid fluidized beds - the role of granular solid mechanics. Chemical Engineering Journal 329, 4-14.

CFD-DEM Simulation of Biomass Gasification in Fluidized Bed Reactors Under Thermally Thick Treatment

Conference: Fluidization

Year: 2019

Proceeding: Fluidization XVI

Group: General Paper Pool

Session: 8C: Modeling and Simulation

Time: Wednesday, May 29, 2019 - 4:49pm-5:01pm

Authors:

Wang, J., Zhejiang University
Ku, X., Zhejiang University
Lin, J., Zhejiang University

In the present paper, a thermally thick particle model is combined with the CFD-DEM model to simulate the biomass gasification behavior in fluidized bed reactors. The heat and mass transfer, homogeneous and heterogeneous reactions, pyrolysis, radiation, particle collisions, turbulence as well as the interactions of the particles and continuous phase are all taken into account. Effects of different operating parameters (e.g., reactor temperature, steam/biomass molar ratio, oxygen/carbon mass ratio, and injection position) on the biomass gasification behavior are systematically explored. The simulation results are analyzed in terms of product gas species, carbon conversion, steam decomposition, cold gas efficiency and H_2/CO ratio. Moreover, the big differences between thermally thin and thermally thick models are also highlighted.

Development of a Refined Grid Model for an Industrial Gas-Solid Flow System

Conference: Fluidization

Year: 2019

Proceeding: Fluidization XVI

Group: General Paper Pool

Session: 8C: Modeling and Simulation

Time: Wednesday, May 29, 2019 - 4:37pm-4:49pm

Authors:

Takabatake, K., The University of Tokyo
Sakai, M., The University of Tokyo

The DEM-CFD method is often employed in a simulation of a gas-solid flow. In the DEM-CFD method, local volume average technique is employed in the governing equations, and hence the CFD grid size must be sufficiently larger than particle size. In the DEM-CFD method, structure grids are often used in the calculation of a gas flow. The local volume average technique makes it difficult to create a narrow path or thin wall in a fluid system by structure grids, though these gas flows are often encountered in industrial systems. Therefore, the DEM-CFD has been applied to simple systems so far. In order to overcome the difficulty, refined grid model is developed, where fine grid is flexibly used in the DEM-CFD method. Several verification tests are performed to show adequacy of the refined grid. The DEM-CFD method introducing the refined grid model is applied to a fluidized bed inserted thin wall and a fluidized bed inserted a rotating screw. It is shown that not only adequacy of the gas flow is proven, but also qualitatively excellent behavior of solid phase can be simulated by the DEM-CFD method introducing the refined grid model.

This study was financially supported by JSPS KAKENHI (17H02825) and by Grant-in-Aid for JSPS Research Fellow from the Japan Society for the Promotion of Science (18J12267).

Effects of the Distributor and Baffle on the Hydrodynamic Performance of a Bubbling Fluidized Bed

Conference: Fluidization

Year: 2019

Proceeding: Fluidization XVI

Group: General Paper Pool

Session: 8C: Modeling and Simulation

Time: Wednesday, May 29, 2019 - 5:49pm-6:01pm

Authors:

Xing, X., University of Western Ontario

Effects of the distributor and baffle on the hydrodynamic performance of a bubbling fluidized bed

Xuelian Xing[a], Cedric Briens[*][a], Chao Zhang[b]

[a]*Institute for Chemicals and Fuels from Alternative Resources(ICFAR),*

The University of Western Ontario

1151 Richmond St, London, ON N6A 3K7

[b]*Department of Mechanical and Materials Engineering,*

The University of Western Ontario

1151 Richmond St, London, ON N6A 3K7

Abstract:

In this study, the multi-phase Eulerian-Eulerian two-fluid method (TFM) coupled with the kinetic theory of granular flow (KTGF) was used to investigate the hydrodynamics of particle flows (Geldart Group B) in a lab-scale bubbling fluidized bed. The focus of this study was to develop a technology to improve the bubble flow behavior inside the fluidized bed to reduce the formation of wet agglomerates when the liquid was sprayed into the fluidized bed. The effects of different gas distributors and baffles on the gas-particle mixing quality and bubble distribution for various superficial gas velocities were investigated. Different gas-solid drag models used in CFD simulation were tested and the results from those drag models were compared. The predicted voidage profiles under different operating conditions agree well with the experimental data.

Key words: Baffle, Bubbling fluidized bed, Hydrodynamic, Eulerian-Eulerian method, CFD simulation

Introduction

Bubbling fluidized bed has been widely applied in petroleum oil upgrading processes, owing to its inherent benefits, such as high rates of heat and mass transfer and quick mixing of solids. However, particle agglomeration occurs when the liquid is sprayed into a gas-solid fluidized bed. Undesired agglomerates will decrease heat and mass transfer rates, causing operating problems. Variation of the bubble flux across the section over fluidized bed plays an important role in helping the bed to reach an expected flow pattern and reduce agglomeration.

Various experimental methods have been used to determine the radial gas bubble distribution including digital image analysis, optical fiber probe, X-ray computed tomography and radioactive particle tracking[1 2]. However, each technique has its own drawbacks and limitation. Digital image analysis is limited to very dilute systems due to the opaque nature of gas-solid fluidization beds. Multiple bubbles are difficult to be detected by X-rays[3]. In this study, sturdy triboprobes were used.

Currently, the most commonly used approaches for simulating the gas-solid two-phase flow in a fluidized bed are Eulerian-Lagrange approach [also denoted as the Discrete element method (DEM)] and Eulerian-Eulerian (E-E) approach [or called as the Two-fluid model (TFM)]. In the TFM, both gas and particle phases are treated as interpenetrating continua, each phase has its own governing equations for momentum, continuity, and energy. The kinetic theory of granular flow (KTGF) is introduced to close the governing equations for the solid phase. In the DEM, the trajectory of each particle is tracked by Newtonâ€™s second law, which makes the DEM method more computational expensive compared to the TFM[4]. For these reasons, the DEM is not widely used for the simulations of two-phase flows in fluidized beds. In conclusion, the DEM method is more suitable for dilute flows, while the TFM) is more suitable for dense flows.

Therefore, in this present study, the hydrodynamics of particle flows in a bubbling fluidized bed was studied by a 2-D transient two-fluid model using a commercial CFD package (ANSYS Fluent 18.2). Afterwards, the effects of the gas distributor and baffles characteristics (shape and position) on the gas distribution in the bubbling fluidized bed were investigated.

METHODOLOGY

The experiments were carried out in a fluidized bed using a gas distributor with a 45Ëš slope. There were 20 tuyeres distributed evenly on the gas distributor, and the gas flow rate supplied to each tuyere was individually controlled to control the lateral distribution of the gas. Three gas distribution schemes were used, each with 10 active tuyeres: the â€œWestern caseâ€ used tuyeres on the higher side of the sloped distributor, the â€œEastern caseâ€ used tuyeres on the lower side and the â€œFlat caseâ€ used tuyeres uniformly distributed over the entire distributor. Air was used as the fluidizing gas with a superficial gas velocity ranging from 0.1 to 1 m/s. Silica sand with a Sauter-mean diameter of 190 Î¼m was used. Baffles inside the fluidized bed were added later to study and compare the bubble distribution profile with those without baffles. The baffle used in this research has a 45Â° angle, an 18 cm by 18 cm triangular geometry and occupies 36 % of the cross-sectional area of the bed.

The simulations were carried out for 30 seconds for each case, but only the results from the last 10 seconds were used to obtain the time-averaged voidage profile and gas velocity distribution, since stable results were obtained after 20 seconds.

Results and discussion

For the bubbling fluidized bed without the baffle, the sensitivity tests for different mesh sizes were performed. Different grid sizes in the gas distributor and fluidization sections were compared. Four different grid sizes (0.0005 m, 0.001 m, 0.002 m, 0.004 m) for distributor section and four different grid sizes (0.001 m, 0.002 m, 0.004 m, 0.006 m) for the fluidization section were calculated. All the simulations performed at a superficial gas velocity 0.2 m/s with the initial bed height 1.6 m, with further simulation runs planned at higher gas velocities. The time-averaged mean gas velocity and solid volume fraction at different bed heights were checked. It was found that, when the number of nodes is over 83392, the simulation results are independent on the mesh size. For this reason, the mesh with 83392 nodes was used. The effect of different drag models (Gidaspow, Gibilaro, Syamlal-Obrien, Wen-Yu, and Syamlal-Obrien-para models) on the simulations results was studied. It was found that the model from Syamlal-Obrien-para gave a better agreement with the experimental data.

The results indicated that the gas distributor and baffle significantly affected the gas volume fraction and velocity distribution in the fluidized bed. The geometry of the gas distributor played an important role in the solid volume fraction and gas velocity distribution. Furthermore, the presence of baffles is an important factor affecting the direction of the gas bubble flows.

Reference

1. Dubrawski, K. *et al.* Traveling column for comparison of invasive and non-invasive fl uidization voidage measurement techniques.**235**, 203â€“220 (2013).
2. Tebianian, S. *et al.* Investigation of particle velocity in FCC gas- fl uidized beds based on different measurement techniques. *Chem. Eng. Sci.* **127**, 310â€“322 (2015).

3. van Ommen, J. R. & Mudde, R. F. Measuring the Gas-Solids Distribution in Fluidized Beds -- A Review. *Int. J. Chem. React. Eng.* **6,** (2008).

4. ANSYS Fluent Theory Guide. (2017).

[*] Corresponding author.

Email: cbriens@uwo.ca (C. Briens)

Eulerian-Eulerian Modeling and Validation of Heat Transfer in Bubbling and Spouted Fluidized Beds

Conference: Fluidization

Year: 2019

Proceeding: Fluidization XVI

Group: General Paper Pool

Session: 8C: Modeling and Simulation

Time: Wednesday, May 29, 2019 - 4:25pm-4:37pm

Authors:

Porter, L., TU Bergakademie Freiberg
Richter, A., TU Bergakademie Freiberg
Kriebitzsch, S., TU Bergakademie Freiberg
Meyer, B., TU Bergakademie Freiberg

In this work an Eulerian-Eulerian model for studying the heat transfer in bubbling and spouted fluidized beds is introduced. The numerical model is based on data obtained in a previous experimental work conducted at TU Eindhoven. In the experimental work, the cooling of different types of particles was studied by use of a lab-scale pseudo-2D fluidized-bed test rig. The fluidized bed was operated in both bubbling and spouted fluidization regimes and investigated by a combination of infrared thermometry and PIV/DIA data acquisition. The experiment considered particles of three different size classes, namely Geldart A/B, B, and D particles. In each test run, the particles were first preheated to a defined temperature before being introduced to the fluidized-bed test rig.
The present work discusses carefully the capability of the numerical model to reproduce the experimental results for different types of fluidization and different kinds of particles, which is an important step towards the reliable modeling of reactive fluidized-bed systems.

Evolution of the Force Chain Topology with Increasing Particle Non-Sphericity in 2D and 3D Systems

Conference: Fluidization

Year: 2019

Proceeding: Fluidization XVI

Group: General Paper Pool

Session: 8C: Modeling and Simulation

Time: Wednesday, May 29, 2019 - 5:01pm-5:13pm

Authors:

Conzelmann, N., Empa, Swiss Federal Laboratories for Materials Science and Technology
Mueller, C., ETH
Poulikakos, L., Empa
Clemens, F., Empa, Swiss Federal Laboratories for Materials Science and Technology

That irregular distribution of inter-particle forces in compressed or sheared granular matter has become known as force chains. The irregular distribution influences the macroscopic behavior of granular matter and beyond that has been shown to lead to eddy like structures in slow granular flow and to influence the width of shear bands in granular flows.[1,2] Despite these implications the development of force chains has been primarily studied for systems of spherical particles, employing both experiments as well as simulations.[3â€"5] Do to the opacity of granular matter the experimental investigations have been focused on two dimensional systems.[6]

Particles occurring in natural and industrial processes are rarely perfect spheres or limited to two dimensional systems. In this work we use a discrete element method (DEM) to investigate the evolution of force chains for particle shapes varying from spheres to non-spherical and even concave shapes such as crosses and jacks. We further vary the system dimensions to represent 2D as well as 3D systems. The results provide insight into the inter-particle force distribution of real systems and allows for a better understanding of the macroscopic behavior of granular matter.

[1] P. Schall, M. van Hecke, Shear Bands in Matter with Granularity, Annu. Rev. Fluid Mech. 42 (2010) 67â€"88. doi:10.1146/annurev-fluid-121108-145544.

[2] F. Radjai, S. Roux, Turbulentlike Fluctuations in Quasistatic Flow of Granular Media, Phys. Rev. Lett. 89 (2002) 064302. doi:10.1103/PhysRevLett.89.064302.

[3] T.S. Majmudar, R.P. Behringer, Contact force measurements and stress-induced anisotropy in granular materials, Nature. 435 (2005) 1079â€"1082. doi:10.1038/nature03805.

[4] R.C. Hurley, S.A. Hall, J.E. Andrade, J. Wright, Quantifying Interparticle Forces and Heterogeneity in 3D Granular Materials, Phys. Rev. Lett. 117 (2016) 098005. doi:10.1103/PhysRevLett.117.098005.

[5] S.G. Bardenhagen, J.U. Brackbill, D. Sulsky, Numerical study of stress distribution in sheared granular material in two dimensions, Phys. Rev. E. 62 (2000) 3882â€"3890. doi:10.1103/PhysRevE.62.3882.

[6] L. Papadopoulos, J.G. Puckett, K.E. Daniels, D.S. Bassett, Evolution of network architecture in a granular material under compression, Phys. Rev. E. 94 (2016) 032908. doi:10.1103/PhysRevE.94.032908.

Fluidization and Sedimentation Patterns in a Secondary Sedimentation Tank

Conference: Fluidization

Year: 2019

Proceeding: Fluidization XVI

Group: General Paper Pool

Session: 8C: Modeling and Simulation

Time: Wednesday, May 29, 2019 - 5:25pm-5:37pm

Authors:

Fan, L., Hefei University
Dong, Q., Hefei University
Xu, N., Hefei University

Secondary sedimentation tank is an important unit for wastewater treatment processing. It often proves to be the bottleneck of the whole activated sludge wastewater treatment process. It is mainly used for the separation of sludge and water to guarantee the quality of effluent and the steady operation of biological system. However, many secondary sedimentation tanks encounter different problems due to inappropriate design. Existing facilities have to be utilized efficiently and economically to attain higher treating performance. The mixed liquor entering the secondary sedimentation tank should be sufficiently clarified in order to produce an effluent of acceptable quality. The sludge in the tank should also be adequately thickened so that the desired solids level in the bioreactors can be maintained through sludge re-circulation. Should any of these functions fail, suspended solids will be carried over the effluent weirs and escape with the effluent. Besides the resulting poor effluent quality, excessive loss of suspend solids may affect the whole biological process correspondingly. Hydrodynamic characteristic is inevitable to achieve the above goals. This paper employs both mixture model and discrete phase model to explore the fluidization and sedimentation of activated sludge flows in a horizontal secondary sedimentation tank. The influences of sludge particle diameter, density, entrance velocity are investigated and cross checked with two models. The flow and distribution of solid phase in the tank were obtained by the mixture model. Meantime, the DPM (Discrete Particle Method) was employed to explore different effects on particle paths. The particle flow paths at different situations were analyzed. The behavior of settled particle was also explored. The results from Euler-Euler model and those from DPM complement each other. Both of them draw a more complete picture of the flow and solid sedimentation in the secondary sedimentation tank. All these results could contribute to the design and optimization of secondary sedimentation tanks. It will also help to the separation and sedimentation of sludge in the tank.

Keynote Talk: Evaluation of Coarse Graining Strategy and Degree in DEM-CFD Simulations of Cyclone Flow

Conference: Fluidization

Year: 2019

Proceeding: Fluidization XVI

Group: General Paper Pool

Session: 8C: Modeling and Simulation

Time: Wednesday, May 29, 2019 - 4:00pm-4:25pm

Authors:

Di Renzo, A., University of Calabria
Di Maio, F. P., University of Calabria
Babiker Ahmed, H. M., Unical

Circulating fluidized beds are used in a variety of energy, chemical and petrochemical applications. Recent developments based on this configuration have been proposed in gas-phase reactors for the polymer industry, like the *MZCR*reactor technology [1], which already reached the commercial stage for polypropylene and polyethylene. Simulations are being increasingly used to provide insight and further improve scale-up and design of advanced multiphase process units and reactors. However, due to their prohibitive computational cost for medium- to large-scale systems, numerical models based on the DEM-CFD approach cannot be used, with their enormous predictive potential remaining unusable. In the recent years, methods that scale particle size by lumping smaller particles into representative *parcels*have been proposed [2], in an attempt to reduce the number of elements to consider and possibly increase the time-step. The concept behind the approach is simple and the derivation relatively rudimentary but these methods have been shown to provide highly powerful means to overcome computational limitations. A lot remains to be understood and characterized, however, such as the extent to which the predictive capability of the model is preserved.

In the present work, simulations of the two-phase flow in complex gas-solid cyclones are investigated at different gas and solid loadings. In particular, the attention is focused on the effect that the coarse graining strategy and degree exert on the replicability of the results as compared to pure DEM-CFD simulations and experimental data. A standard high-efficiency cyclone design is considered for simplicity, although more complex geometries do not introduce particular limitations. The number of particles represented by a *parcel*and the scaling of properties is examined both in terms of the macroscopic performance indicators, like separation efficiency and gas pressure drop, and the local distribution of the gas and solid flows inside the cyclone, including vortex formation, solids sliding, number of turns and particle entrainment.

References

[1] Covezzi, M., Mei, G. (2001). The multizone circulating reactor technology. *Chemical Engineering Science*,**56**, 4059â€“4067.

[2] Hilton, J.E., Cleary, P.W. (2014) Comparison of non-cohesive resolved and coarse grain DEM models for gas flow through particle beds. *Applied Mathematical Modelling*, **38**, 4197-4214

Mesoscale Simulation of Bubble Behavior in Bubbling Fluidized Beds

Conference: Fluidization

Year: 2019

Proceeding: Fluidization XVI

Group: General Paper Pool

Session: 8C: Modeling and Simulation

Time: Wednesday, May 29, 2019 - 5:13pm-5:25pm

Authors:

He, K., Chongqing University
Zheng, Z., Chongqing University
Chen, Z., Fuzhou University

Bubbles are typical mesoscale structures in gas-solid bubbling beds. The behavior of bubbles play a key role in gas-solid mass and momentum transfer. To accurately describe the bubble structure and its dynamic changes caused by complex gas-solid interactions, it is worth exploring the mesoscale model and conducting simulation studies. The lattice gas cellular automata (LGCA) model is a numerical simulation method from a mesoscopic view. The core of the LGCA model is to describe the complex macroscopic phenomena based on the microscopic or mesoscopic self-organized evolution according to specific model evolution rules. It features the simplicity in algorithms and the capability of dealing with complex boundaries.

Based on our previously proposed LGCA discrete mesoscale model for gas-solid two-phase flow (Zheng et al., 2011), the model evolution rules were revised for forces including particle gravity and bubble surface tension, and gas-solid interactions. In this paper, by analyzing the characteristics of the bubbles in fluidized beds, specific set of evolution rules were further defined corresponding to the local hydrodynamic properties, i.e. bed height, pressure, voidage, and particle velocities. With this set, a certain set of evolution rules was activated when the corresponding local hydrodynamic state is identified.

The model is then validated by simulating the formation, growth and collapsing of bubbles in the two-dimensional bubbling bed. The comparison between the simulation and the experimental characteristics of the corresponding cold model shows that the proposed self-organizing evolution mechanism of this LGCA model is capable of describing the macroscopic bubble behavior in the gas-solid bubbling fluidized beds.

A Mathematical Model of a Drum-Type Boiler with Thermal Stress Calculation in the Circulating Fluidized Bed Combustor

Conference: Fluidization

Year: 2019

Proceeding: Fluidization XVI

Group: General Paper Pool

Session: 8D: Modeling and Simulation

Time: Wednesday, May 29, 2019 - 5:13pm-5:25pm

Authors:

Ngo, X., Kunsan National University
Song, B., Kunsan National University
Shun, D., Korea Institute of Energy Research
Park, J., Korea Institute of Energy Research
Park, J. H., Korea Institute of Energy Research
Lee, J. G., Korea Institute of Energy Research

A mathematical model has been developed to predict the dynamic behavior of a natural circulating drum-type boiler system. The basic conservation rules of mass, momentum, and energy were applied to the system. The six state variables of the model include the total volume of water, steam pressure, steam quality at the top of the riser, drum water level, downcomer-riser loop flow, and the volumetric fraction of steam in the water-steam mixture under the drum water level. This model can simulate the dynamic drum water level directly based on the mass balance of the drum while the other models in the literature computed it by off-line with the aid of empirical equations and assumptions. A sub-model was applied to calculate the temperature and thermal stress distribution both in the metal layers and in the insulation layers of the drum shell. The sub-model was derived from a finite differences method based on the discretization of the Fourierâ€™s heat conduction equation. The simulation results with the dynamic model showed a good agreement with the data in the literature.

CFD-DEM Simulation of Fluidization of Non-Spherical Particles

Conference: Fluidization

Year: 2019

Proceeding: Fluidization XVI

Group: General Paper Pool

Session: 8D: Modeling and Simulation

Time: Wednesday, May 29, 2019 - 4:37pm-4:49pm

Authors:

Zhao, Y., Zhejiang University
Ma, H., Zhejiang University
Liu, Z., Zhejiang University
Zhao, H., Zhejiang University

The fluidization of non-spherical particles is very common in some industrial fields such as biomass utilization. It is well known that CFD-DEM (Computational Fluid Dynamics - Discrete Element Method) is a powerful tool for studying the fluidization. However, few researchers have investigated the fluidization of non-spherical particles using CFD-DEM. In this study, a robust and efficient CFD-DEM algorithm is proposed for the fluidization of non-spherical particles, in which the non-spherical particles (including rod-like particles, disk-like particles, tablet particles, and non-spherical particles in other shapes) are described by super-ellipsoid model, multi-super-ellipsoid model, and multi-sphere model. The volume fraction of the CFD cell and the drag force between the non-spherical particle and the fluid are calculated by dividing the particle into more than 100 parts to realize the two-way coupling between the non-spherical particles and the fluid. To validate the prediction accuracy of the CFD-DEM model, the simulations of the fluidization are then compared with the corresponding experiments. The results demonstrate that the proposed CFD-DEM model has a rather high efficiency and accuracy for modeling the fluidization of non-spherical particles in fluidized bed.

DEM Numerical Investigation of Mixing Behaviors of Binary Mixtures Containing Non-Spherical Particles in a Fluidized Bed

Conference: Fluidization

Year: 2019

Proceeding: Fluidization XVI

Group: General Paper Pool

Session: 8D: Modeling and Simulation

Time: Wednesday, May 29, 2019 - 4:25pm-4:37pm

Authors:

Ren, A., Harbin Institute of Technology
He, Y., Harbin Institute of Technology
Wang, T., Harbin Institute of Technology

DEM
numerical investigation of mixing behaviors of binary mixtures containing
non-spherical particles in a fluidized bed

Anxing Ren[1], Tianyu
Wang, Yurong He[1,*]

[1] School of Energy Science and Engineering, Harbin Institute of Technology, 150001, China

rong@hit.edu.cn

As
a renewable and sustainable energy source, the conversion and utilization of biomass are becoming increasingly popular worldwide, which can help to alleviate the energy crisis and mitigate global warming. The thermochemical processes of biomass such as combustion, pyrolysis and gasification are considered to be the most widely used techniques for converting biomass into energy or fuel. The fluidized bed has a series of advantages such as intense

mixing, large contact surface between the phases, and efficient heat and mass transfer. Thus, the process of biomass conversion and utilization is usually carried out in a fluidized bed[1]. However, fluidization of biomass particles is a difficult task because the non-spherical biomass particles are generally large and have a low density. In the actual industrial processes, inert particles such as silica sand, alumina, and calcite are often added to the biomass to assist its fluidization[2]. Therefore, understanding mixing behaviors of binary mixtures in a fluidized bed has specific significance for the design and optimization of related industrial processes.

In recent years, the macroscopic fluidization behaviors of the binary mixtures containing non-spherical particles have been experimentally investigated by many researchers. Fotovat et al.[3] investigated the effects of superficial gas velocity and the mixture composition on the mixing/segregation behaviors by analyzing the circulatory motion of the active tracer. Shao et al.[4] investigated the mixing behaviors in binary-component and multi-components particulate systems and found that the density of non-spherical particles has a more significant influence on the mixing characteristics than particle size and shape. Boer et al.[5] developed a Digital Image Analysis technique to explore the particle distributions in the fluidized bed and obtained qualitative and quantitative experimental results which can be used for validation of numerical models concerning non-spherical particle mixing.

Although studies on the mixing behaviors of binary mixtures containing non-spherical particles in fluidized beds have attracted attention, most studies focus on macroscopic flow characteristics. It is still difficult to obtain detailed information in the flow field through experimental research, such as the microscopic motion characteristics of the particles. With the rapid development of computer science and technology, numerical simulation has become an effective method to study particle behaviors in fluidized beds.

In this work, the Computational Fluid Dynamic-Discrete Element Method (CFD-DEM) with consideration of rolling friction is applied to investigate the mixing behaviors of binary mixtures containing non-spherical particles in a fluidized bed. The non-spherical particles are constructed by multi-sphere model[6].

In the multi-sphere method, a single non-spherical particle is represented by a composition of overlapping spheres of arbitrary size and the contact force and torque acting on a non-spherical particle are calculated as the sum of contact forces and torques on its spherical elements. The Gidaspow drag model[7] is employed to describe the momentum exchange between spherical particles and gas phase. The Di Felice correlation[8] and Hölzer/Sommerfeld model[9] are applied for non-spherical particles. The mixing processes of several rod-shaped particles with different aspect ratios but the same volume and spherical particles in the fluidized bed are simulated.

Both macroscopic and microscopic flow characteristics in the fluidized bed are investigated. The simulation results demonstrate that the addition of spherical particles is advantageous for the mixing and fluidization of the rod-shaped particles. The superficial gas velocity has an obvious effect on particle flow behaviors and particle mixing characteristics. The pressure drop and particle distribution in the fluidized bed are discussed. In addition, the effect of particle aspect ratio, density and size on flow and mixing characteristics is also analyzed.

Acknowledgement

This work was financially supported by the National Natural Science Foundation of China (Grant No. 91534112).

References

[1] Cui H, Grace J R. Fluidization of biomass particles: A review of experimental multiphase flow aspects[J]. Chemical Engineering Science, 2007, 62(1-2): 45-55.

[2] Fotovat F, Chaouki J, Bergthorson J. The effect of biomass particles on the gas distribution and dilute phase characteristics of sand–biomass mixtures fluidized in the bubbling regime[J]. Chemical Engineering Science, 2013, 102: 129-138.

[3] Fotovat F, Chaouki J, Bergthorson J. Distribution of large biomass particles in a sand]biomass fluidized bed: Experiments and modeling[J].

AIChE Journal, 2014, 60(3): 869-880.

[4] Shao Y, Zhong W, Yu A. Mixing behavior
of binary and multi-component mixtures of particles in waste fluidized beds[J]. Powder Technology, 2016, 304: 73-80.

[5] Boer L, Buist K A, Deen N G, , Padding J
T, Kuipers J A M. Experimental study on orientation and de-mixing phenomena of elongated particles in gas-fluidized beds[J]. Powder Technology, 2018, 329: 332-344.

[6] Favier J F, Abbaspour-Fard M H, Kremmer
M. Modeling nonspherical particles using multisphere discrete elements[J]. Journal of Engineering Mechanics, 2001, 127(10): 971-977.

[7] Gidaspow D. Multiphase flow and
fluidization: continuum and kinetic theory descriptions. Academic press, 1994.

[8] Di Felice R. The voidage function for
fluid-particle interaction systems[J]. International Journal of Multiphase Flow, 1994, 20(1): 153-159.

[9] Hölzer A, Sommerfeld M. New simple
correlation formula for the drag coefficient of non-spherical particles[J]. Powder Technology, 2008, 184(3): 361-365.

Keynote Talk: Development and Application of Flexible Eurelian-Lagrangian Method for Industrial Gas-Solid Flow Systems

Conference: Fluidization

Year: 2019

Proceeding: Fluidization XVI

Group: General Paper Pool

Session: 8D: Modeling and Simulation

Time: Wednesday, May 29, 2019 - 4:00pm-4:25pm

Authors:

Sakai, M., The University of Tokyo

The DEM-CFD method is well known as an Eulerian-Lagrangian method and has been employed to perform numerical simulations in industrial gas-solid flow systems. Modeling of arbitrary shape boundary becomes indispensable when the DEM-CFD method is applied to the industrial systems. This is because not the shape of the industrial systems always becomes simple. On top of that, efficient calculation technique is necessary so as to perform the simulations of industrial gas-solid flows because enormous number of calculated particles are used. In order to create the arbitrary shape calculation domain in the DEM-CFD simulation, the authorâ€™s group has developed an approach combining the sign distance functions and the immersed boundary method. This approach can create the arbitrary shape wall even if staggered grids are used. Very recently, the authorâ€™s group has developed a new flexible model by combining a scaling law model which is referred to as the coarse graining DEM with the wall boundary model. The coarse graining DEM makes it possible to perform a large-scale DEM-CFD simulations, where smaller number of calculated particles is used than actual one. This is because the coarse grain particle represents group of the original particles, where total energy agrees between them. In order to prove adequacy of the flexible model, several verification and validation tests has been performed in gas-solid systems such as fluidized bed and powder die-filling. The test results indicate that the flexible model can reproduce granular flow even in a complex shape domain. Consequently, adequacy of the new flexible model is proven.

This study was financially supported by JSPS KAKENHI (17H02825) and JSPS KAKENHI (17KK0110).

Numerical Simulation of Catalytic Coal Hydrogasification in a Lab-Scale Pressurized Fluidized Bed

Conference: Fluidization

Year: 2019

Proceeding: Fluidization XVI

Group: General Paper Pool

Session: 8D: Modeling and Simulation

Time: Wednesday, May 29, 2019 - 5:01pm-5:13pm

Authors:

Xia, Z., East China University of Science and Technology
Guo, J., East China University of Science and Technology
Yan, S., State Key Laboratory of Coal Conversion, Institute of Coal Chemistry, Chinese Academy of Sciences, China
Chen, C., East China University of Science and Technology
Bi, J., ENN Group State Key Laboratory of Coal-based Low Carbon Energy

Catalytic coal hydrogasification has been tested in a 36mm diameter pressurized fluidized bed at 3 MPa and 1123 K, and a 77.3 wt.% CH_4 yield in 30 mins is achieved with the Co-Ca catalyst. To help in the design and scale-up of an industrial pressurized bed gasifier, a thorough understanding of the bubble dynamics is needed.

In this presentation, we conducted a TFM-KTGF simulation of catalytic hydrogasification in the lab-scale bed operated at $3.7u_{mf}$. The predicted bed expansion height is validated with the empirical correlation. The average bubble size is found to be much smaller than the Mori-Wen correlation. As a key parameter, the bubble size distribution (BSD) is detected and statistically analyzed using a novel post-processing MATLAB code. The BSDs at different heights (0.25H, 0.50H, 0.75H, 1.0H) are shown to have a similar bimodal distribution. Furthermore, the bubbles are divided into three groups: the small (1 mm-3 mm), middle (3 mm-6 mm), and large (6 mm-9 mm) sized bubbles. The number of small bubbles decreases along the bed height, but the number of large bubbles increases. It is noted that the number of the middle sized bubbles is almost unchanged along the bed height. This suggests that the bubble coalescence and breakup tend to approach a balance. A catalytic hydrogasification model is coupled with the gas-solid hydrodynamics, and distributions of temperature and CH_4 concentration were predicted and compared with the experimental results. Further work will be focused on the effects of bed height, bed diameter and particle size on the overall performance of the gasifier.

Size Distribution of Fluidized Nanoparticle Agglomeratesfrom Agglomeration and Fragmentation: A Population Balance Study

Conference: Fluidization

Year: 2019

Proceeding: Fluidization XVI

Group: General Paper Pool

Session: 8D: Modeling and Simulation

Time: Wednesday, May 29, 2019 - 4:49pm-5:01pm

Authors:

de Martín, L., Chalmers University of Technology
Grillo, F., ETH Zurich
Chidambaram, K., Delft University of Technology
van Ommen, J. R., Delft University of Technology

Gas fluidization is emerging as a promising means for functionalizing nanoparticles in a scalable manner [1,2]. Yet, the practical use of fluidization in nanoparticle technology is currently hampered by a lack of fundamental understanding of the mechanisms behind nanoparticle fluidization. It is well known that nanoparticles are not fluidized as individual particles but rather in the form of highly porous fractal agglomerates that can be as large as hundreds of microns. Nanoparticle fluidization is in fact determined by the properties of such agglomerates (i.e., size and density) rather than by those of the individual particles [3]. For this reason, nanoparticle-agglomerate or nanopowder fluidization [4] are considered to be more accurate terms. Such large agglomerates are thought to be the result of a dynamic equilibrium between agglomeration and fragmentation events resulting from collisions induced by the fluidization process. However, a detailed understanding of how such dynamic equilibrium determines the size distribution of the fluidized agglomerates is still missing. In particular, the interplay between the inter-particle forces acting at the nanoparticle-scale, the multi-level structure of the agglomerates, and the collision events has not yet been fully described.

In this contribution, we present a study of the size distribution of fluidized nanoparticle agglomerates based on a population balance model that incorporates at the same time (1) a multi-scale description of the agglomerates [5] and (2) a detailed account of the outcomes of collision events. In particular, our model builds on the works of Brilliantov et al. [6] on agglomeration and fragmentation and of Liu [7] on the kinetic theory of granular flow applied to fluidization.

By analysing data from literature, we find that the experimental size distribution of fluidized nanoparticle agglomerates at dynamic equilibrium is independent of the nanoparticle properties and operating conditions when the distribution is rescaled with respect to the average agglomerate size. This suggests a universal mechanism for breakage and agglomeration under standard operation conditions. The population balance proposed here captures this feature when the size of the fragments resulting from a collision event follows an exponential law. Our model also sheds light into why the agglomerate size distribution of some nanoparticles, such as TiO_2 P25 (d_p = 21 nm), is nearly independent of the type of surface, i.e. hydrophilic vs. hydrophobic, whereas the size distribution for other nanoparticles, e.g. Al_2O_3 AluC (d_p = 13 nm) and SiO_2 A130 (d_p = 16 nm), is highly sensitive to the type of surface. We find that this is due to two combined factors: the high density of TiO_2 nanoparticles, which results in a large granular energy, and the relatively large nanoparticle size for this material, which results in a smaller tensile strength of the agglomerates. The way these two variables affect the outcome of a collision event in the balance explains the observed experimental results. TiO_2 fluidization is expected to be improved by using nanoparticles with smaller particle size and a hydrophobic coating. Such insights can be used to optimize and intensify nanopowder fluidization.

[1] Salameh, S., Gomez-Hernandez, J., Goulas, A., Van Bui, H., & van Ommen, J. R. (2017). Advances in scalable gas-phase manufacturing and processing of nanostructured solids: A review. *Particuology, 30*, 15-39.

[2] Van Bui, H., Grillo, F., & van Ommen, J. R. (2017). Atomic and molecular layer deposition: off the beaten track. *Chemical Communications, 53*(1), 45-71.

[3] Yao, W., Guangsheng, G., Fei, W., & Jun, W. (2002). Fluidization and agglomerate structure of SiO_2 nanoparticles. *Powder Technology,124*(1-2), 152-159.

[4] van Ommen, J. R., Valverde, J. M., & Pfeffer, R. (2012). Fluidization of nanopowders: a review. *Journal of Nanoparticle Research, 14*(3), 737.

[5] de MartÃn, L., Bouwman, W. G., & van Ommen, J. R. (2014). Multidimensional nature of fluidized nanoparticle agglomerates.*Langmuir, 30*(42), 12696-12702.

[6] Brilliantov, N., Krapivsky, P. L., Bodrova, A., Spahn, F., Hayakawa, H., Stadnichuk, V., & Schmidt, J. (2015). Size distribution of particles in Saturnâ€™s rings from aggregation and fragmentation. *Proceedings of the National Academy of Sciences, 112*(31), 9536-9541.

[7] Liu, L. (2011). Kinetic theory of aggregation in granular flow. *AIChE Journal, 57*(12), 3331-3343.

Plenary Talk: A Multistage fluidized bed reactor: Stability analysis, Suppression of back-mixing and its application in heterogeneous catalysis

Conference: Fluidization

Year: 2019

Proceeding: Fluidization XVI

Group: General Paper Pool

Session: Plenary Session #3

Time: Thursday, May 30, 2019 - 8:30am-9:10am

Authors:

Wei, F., Tsinghua University

Fluidized bed reactors have been widely applied in physical and chemical industries due to its easy fluidity and considerable heat and mass transfer capability. For its fluidity, since solids bed has highly inelastic energy dissipation, mal-distribution always occurs in dilute and dense fluidization [1,2]. A criterion, based on direct Lyapunov method, for predicting instability of uniform gas solids distribution is obtained to design a robust fluidized bed reactor [3]. Additionally, fluidized bed reactors with considerable transfer capability also have some inherent drawbacks like a high degree of gas solids back-mixing. It becomes a barrier to the application of fluidized beds to heterogeneous catalytic process, especially with the intermediates as the desired products. Therefore, the strategy, in the advantage of phase separation through a nozzle, to suppress gas solids backing to obtain plug-flow performance with narrow residence time distribution is a requirement for advanced gas solids reactors [4]. A review of the fundamental research as well as industrial applications on the stability analysis in uniform fluidization and multi-stage fluidized beds to inhibit gas solids back-mixing in the Fluidization Laboratory of Tsinghua University (FLOTU) will be provided.

[1] Chenxi Zhang, Weizhong Qian, Fei Wei, et al. *AIChE J.* **2016**, 62: 4251.

[2] Chenxi Zhang, Weizhong Qian, Fei Wei. *Chem Eng Sci.* **2017**, 173: 187.

[3] Chenxi Zhang, Weizhong Qian, Fei Wei, et al. *Chem Eng Sci.* **2018**, 176: 515.

[4] Chenxi Zhang, Yao Wang, Fei Wei, et al. *Can J Chem Eng.* **2019**. (Issue highlights).

Plenary Talk: Hydrodynamics of High Temperature Gas-Solid Fluidized Beds

Conference: Fluidization

Year: 2019

Proceeding: Fluidization XVI

Group: General Paper Pool

Session: Plenary Session #3

Time: Thursday, May 30, 2019 - 9:10am-9:50am

Authors:

Chaouki, J., Ecole Polytechnique Montreal
Shabanian, J., Polytechnique Montreal
Farag, S., Polytechnique Montreal

Gas-solid fluidized beds have been widely employed in chemical industries, such as thermal conversion
of fuels, oxidation of hydrocarbons, the FCC process, ore processing, and gas-phase polymerization.
Hydrodynamic characteristics of these units directly influence the heat and mass transfer rates, reaction
progress, and the overall performance of fluidized bed reactors. As many of these units are operated under
elevated temperatures, it is essential to adequately predict and/or thoroughly characterize the hydrodynamics of high temperature fluidized beds for proper design and performance evaluation. In some
industrial sectors, e.g., mining and petroleum industries, as well as fuel and power fields, the compositions
of feedstock are rapidly becoming more complex owing to the rarefaction of conventional resources. The
application of reduced quality feedstocks leads the corresponding fluidized bed based processes to operate
in the presence of cohesive interparticle forces (IPFs). This is while available hydrodynamic correlations,
which were developed based on experimental data collected at near-ambient conditions and

discard the
contribution of IPFs, are typically adopted for the design and performance evaluation of these units under
elevated temperatures. This, however, results in dubious predictions, in particular when IPFs are present
in the bed.

Application of clean electricity with microwave heating (MWH) technique, which is a volumetric and
selective heating technique, would make a revolution in the chemical industry. For multiphase systems,
as the response of each component to microwaves is different owing to different dielectric and magnetic
characteristics of adopted materials, a very large temperature gradient can be adeptly created within the
system. For instance, in the case of gas-solid systems, the application of MWH can yield particles to have
temperatures very different from the gas temperature. This feature can assist in promoting the desired
reactions on the surface of particles, if they act as catalysts since most catalytic agents are highly effective
microwave-to-heat converters, while suppressing the secondary gas-phase reactions since the gas-phase
components have negligible interactions with the electromagnetic waves.

In my presentation, I will first review our recent research activities on hydrodynamic characterization of
gas-solid fluidized beds at high temperature in the virtual absence and presence of IPFs. Second, I will
present a simple approach that we recently developed for quantifying the magnitude of cohesive IPFs in
a high temperature gas-solid fluidized bed. Particle agglomeration originates when the magnitude of
cohesive IPFs reaches to a high level. Hence, I will then review the mechanisms of particle agglomeration
in a high temperature gas-solid fluidized bed and present a novel, simple, effective, and robust approach
that we recently developed for early detection of defluidization in a high temperature bubbling gas-solid
fluidized bed reactor. I will next compare the performance of this newly developed approach with other
leading approaches proposed in the literature. In a following part, I will present a defluidization

map that
we recently developed to assist in the solid fuel formulations to avoid defluidization during co-combustion
of coal with waste in a fluidized bed reactor. I will then present an empirical correlation that we introduced
for estimating defluidization boundaries, and the results of our detailed study in identifying the most
effective counteractive or pre-emptive measures to delay or prevent defluidization. In the last part of the
presentation, I will review principles of MWH mechanism and present our recent inventions on enhancing
selected gas-solid, liquid-solid, and liquid-liquid reactions by application of MWH mechanism.

Plenary Talk: Scale-up of Fluidized Bed Drying: Hydrodynamics, Mixing and Heat and Mass Transfer

Conference: Fluidization

Year: 2019

Proceeding: Fluidization XVI

Group: General Paper Pool

Session: Plenary Session #3

Time: Thursday, May 30, 2019 - 9:50am-10:30am

Authors:

Glasser, B., Rutgers University

Fluidized beds are used extensively in the pharmaceutical and chemical industries as a batch or continuous process for drying moist powders and granular solids. Fluidized bed dryers provide good mixing of solids and intensive heat and mass transfer between the solid and hot gas phases in the system. It has been reported that an average of 12% of all energy consumed is used for drying, and the cost of drying could reach up to 60%-70% of the total cost of investments. Therefore, optimal design and operation of the drying process is sought for cost-effective manufacturing and to ensure a robust process with good product quality. In the pharmaceutical industry, the drying process removes moisture or solvents to provide pharmaceutical materials (active pharmaceutical ingredients (APIs) and granulations) with stable properties. Insufficient drying can lead to microbial growth and polymorph transformations. Often the API is heat sensitive so the temperature during drying must be carefully controlled. Complications include lengthy drying times, nonuniform drying, agglomeration and attrition. In order to understand fluidized bed drying we have to consider flow of the gas and solids, mixing, heat transfer and mass transfer which makes the process difficult to scale-up.

In this talk, we will first present results for scale-up of batch fluidized bed drying for a pharmaceutical application. Parametric analysis of four fluidized bed dryer operating parametersâ€“ initial moisture content, material loading, heating air temperature and air flux â €“ and their effect on drying was studied. Two different dryer scales were used to dry a pharmaceutical excipient powder. Based on these results we are able to introduce a practical scaling rule which can predict the drying behavior in the large scale unit based on results from a bench scale unit. We then discuss using knowledge gained from the batch fluidized bed dryers to predict the behavior of a continuous fluidized bed dryer, where solids continuously flow into and out of the drying chamber. Several pharmaceutical manufacturers are moving towards continuous manufacturing due to its potential to improve agility, flexibility, and robustness. The fluidized bed unit operation can be readily implemented as a continuous process for drying. A model for predicting the effluent moisture content for the continuous dryer was developed by combining the drying kinetics and residence time distribution (RTD) of powder in the dryer. The drying kinetics were obtained from the batch fluidized bed drying process. The RTD was characterized by carrying out a tracer response test in the continuous fluidized bed. A tank-in-series model was used to describe the RTD curves. A maximum mixedness model (MMM) was adopted for characterizing the micromixing in the continuous fluidized bed. The results of the model were compared to experimental results and we discuss how the operating conditions affect the resulting effluent moisture content. Finally, we will discuss the application of these results to improving the operation of pharmaceutical fluidized bed drying.

Plenary Talk: How Pyroclastic Flows Outsmart Granular Friction During Volcanic Eruptions

Conference: Fluidization

Year: 2019

Proceeding: Fluidization XVI

Group: General Paper Pool

Session: Plenary Session #4

Time: Thursday, May 30, 2019 - 11:30am-12:10pm

Authors:

Lube, G., Massey University
Jones, J. R., University of Massey
Fullard, L., Massey University
Dufek, J., University of Oregon

Granular-fluid flow and gas-fluidization traditionally constitute key processes in the investigation and modelling of violent volcanic and non-volcanic mass flows on Earth and other planets. These include phenomena such as pyroclastic flows, and avalanches of rock and snow. Amongst these natural gas-particle flow phenomena, pyroclastic flows constitute the most lethal volcanic process known. Causing more than fifty percent of volcanic fatalities globally, these hot mixtures of volcanic particles and gas exhibit an astonishing fluidity. This allows them to transport thousands to millions of tonnes of volcanic material across the Earthâ€™s surface over tens to hundreds of kilometres, by-passing rough and tortuous terrain with flat and upsloping surfaces. Here we give an account for the natural flow characteristics and hazard impacts of these flows to highlight a long-standing research question: how do pyroclastic flows attain their characteristic mobility? The fluidity of pyroclastic flows is attributed to an internal process that effectively counters granular friction. Aspects of this enigmatic fluidity have been variably explained by vertical gas escape, high internal gas pore pressure, acoustic fluidization, dynamic fragmentation, among others. However, the violence of real-world flows has precluded direct measurements, so that none of these processes has been quantitatively validated.

In this presentation we show, through large-scale experiments and numerical multiphase modelling, that pyroclastic flows generate their own air lubrication. This forms a near-frictionless basal region. Air lubrication develops under high basal shear when air is locally forced downwards by reversed pressure gradients and displaces particles upward. We demonstrate that air lubrication is enhanced through a positive feedback mechanism, explaining how pyroclastic flows are able to propagate over slopes much shallower than the angle of repose of any natural granular material. This discovery necessitates a re-evaluation of hazard models that aim to predict the velocity, runout and spreading of pyroclastic density currents.

Plenary Talk: Raiding the toolbox to tackle practical issues in particle technology

Conference: Fluidization

Year: 2019

Proceeding: Fluidization XVI

Group: General Paper Pool

Session: Plenary Session #4

Time: Thursday, May 30, 2019 - 10:50am-11:30am

Authors:

Hrenya, C. M., University of Colorado at Boulder

Cohesive particles are an integral and unavoidable presence in numerous practical applications â€" pharmaceutical powders, specialty chemicals, catalysts for energy production, to name just a few. The prediction of cohesive-particle flows is tricky at best, culminating in dozens of empirical correlations that can vary by orders of magnitude when applied to the same system. More recently, physically-based modeling endeavors for such flows have ensued, though their application to practical, industrial systems remains in its infancy for a number of reasons: cohesive forces like van der Waals forces are ridiculously sensitive to surface roughness, measuring surface roughness involves sophisticated equipment and highly-trained personnel, predicting cohesive forces for a given surface roughness is complex, converting particle-level cohesion into the necessary macro-scale (continuum) quantities is non-trivial, etc.

We recently undertook a 5+ year journey with our industrial sponsor to develop a physically-based and industrially-feasible framework to predict cohesive-particle flows. To overcome some of the stickiest obstacles along the way, we raided the toolbox of experimental and computational methods. Our trials and tribulations and (thankfully) successes will be overviewed.

Keynote Talk: Development of a New Methodology for Measurements of Particle Stresses in Fluidised Beds

Authors:
Ocone, R., Heriot-Watt University
Maione, R., Heriot Watt University
Ait Ali Yahia, L., Heriot-Watt University
Ozel, A., Heriot-Watt University
Piepke, T., Ruhr-Universität Bochum
Pollak, S., Ruhr-Universität Bochum
Barrett, R., Heriot-Watt University

Gas-solid fluidised beds are commonly used in industrial chemical processes such as fluid catalytic cracking, coal gasification and granulation [1]. In fluidised beds, large fluctuations in the solid volume fraction and phase velocities at a wide range of length and time-scales are observed. Researchers have investigated these flow behaviours in fluidised beds through experimental studies for a few decades. Nevertheless, accurate measurements of the particle stresses in fluidised beds are still not available [2]. Colafigli et al., [3] developed a device that allowed measurements of the bulk viscosity in a homogeneous gas-fluidised bed using silica particles. The device measurement principle is based on the Coaxial-Cylinder-Viscometer for liquids (CCV). The CCV is a well-known device used to evaluate the rheological characteristics of fluids. The instrument consists of a rotating inner cylinder, the so-called bob, and a static outer cylinder or cup. The fluid is placed in the annular gap between the two cylinders. The apparent torque, which can be measured during the rotation of the inner cylinder, can be converted to the dynamic viscosity and the shear stress of the fluid [4]. The main objective of the present work is to develop a new instrument to measure particle stresses in fluidised beds, based on the same rheological measurement principles employed in a CCV for liquids.

In our work, the Freeman FT4 Powder Rheometer, operated in its aeration mode was used as a smaller version of the CCV, where the original 48 mm rotating impeller was replaced with a new cylindrical cell designed for the purpose of the present study. This cell, which is 70 mm in height, 36 mm in diameter and composed of 6 blades is 3D printed using plastic material. It is placed inside the FT4 cylindrical cup and is analogous to a Virtual Couette Cell (VCC) with a 7 mm gap [5]. The cell is then rotated at different rotational velocities, and the torque needed to move the aerated powder placed in the cup is evaluated. A new device called FBR (Fluidized-Bed-Rheometer- Ruhr-UniversitÃ¤t, Bochum) was sized and developed in parallel with this. It is based on the same measurement principle of the Coaxial-Cylinder-Viscometer. Like the CCV, the FBR is composed of a rotating inner cylinder measuring 450 mm in length and 80 mm in diameter, while the outer cylinder is larger in both respects having a 600 mm length and 109 mm diameter. Each cylinder is composed of a different material with carbon fibre forming the inner cylinder and glass forming the outer. A fluidised bed is generated in the 14.5 mm gap between the cylinders, and the torque needed to rotate the bob is recorded. In order to avoid particles sliding on the rotating cylinder, the latter is coated with particles as the fluidised ones (glass beads).

Both technologies (FBR and VCC) were operated using mono (500 mm diameter) and poly (400 â€“ 600 mm) dispersed spherical glass beads. The corresponding torque was recorded as the shear rate was raised from 1.5 to 32 s^{-1}. The air velocity was set to 0 $m.s^{-1}$ for the VCC and 1.6 $m.s^{-1}$ for the FBR. The registered torque was then converted to shear stress t using the relations given by Marchal et al., [5], this while assuming a cylindrical shear zone in the VCC and a homogeneous stress distribution along the device's height. The first results obtained with the VCC with no aeration showed that the shear stress is independent of the increasing shear rate. The corresponding viscosity h is also evaluated as the ratio of the shear stress to the shear rate, and the obtained results are in good agreement with the ones obtained by Marchal et al., [5], using the same particles. Furthermore, the results obtained with the air velocities used in this study showed that the glass beads shear stress increases with shear rate, showing a Bingham plastic behaviour ($t = t_0 +$ hgwhere t_0 is the yield stress); this result applies to both the VCC and the FBR. Further measurements, including different air velocities and different particles varying in size and shape, will be performed in order to confirm the preliminary measurements and observations.

References

[1] M. S. Ray, â€œChemical Engineering, Volume 2: Particle Technology and Separation Processes, 4th edn, by J.M. Coulson and J.F. Richardson. Pergamon Press, Oxford, UK. 1991. 968 pp. ISBN 0-08-037957-5,â€• *Dev. Chem. Eng. Miner. Process.*, vol. 1, no. 2â€“3, pp. 172â€“172

[2] R. Borghi and F. Anselmet, *Turbulent Multiphase Flows with Heat and Mass Transfer*. John Wiley & Sons, 2013

[3] A. Colafigli, L. Mazzei, P. Lettieri, and L. Gibilaro, â€œApparent viscosity measurements in a homogeneous gas-fluidized bed,â€• *Chem. Eng. Sci.*, vol. 64, no. 1, pp. 144â€“152, 2009

[4] D. S. Viswanath, T. Ghosh, D. H. L. Prasad, N. V. K. Dutt, and K. Y. Rani, *Viscosity of Liquids: Theory, Estimation, Experiment, and Data*. Springer Netherlands, 2007

[5] P. Marchal, N. Smirani, and L. Choplin, â€œRheology of dense-phase vibrated powders and molecular analogies,â€• *J. Rheol.*, vol. 53, no. 1, pp. 1â€“29, 2008

Thermal Behaviour of Compartmented Fluidized Beds Under Uneven Fluidization Conditions

Authors:
Solimene, R., Consiglio Nazionale delle Ricerche
Bevilacqua, C., Istituto di Ricerche sulla Combustione, Consiglio Nazionale delle Ricerche
Cammarota, F., Consiglio Nazionale delle Ricerche
Chirone, R., Consiglio Nazionale delle Ricerche
Salatino, P., Università degli Studi di Napoli Federico II
Tregambi, C., Università degli Studi di Napoli Federico II
Troiano, M., Università degli Studi di Napoli Federico II

Particle solar receivers in Concentrated Solar Power (CSP) applications are becoming more and more relevant to develop technologies able to be coupled with high-efficiency thermodynamic cycles for energy production and/or with systems of thermal and thermochemical energy storage. As matter of fact, granular materials can be, in principle, operated as heat transfer fluid (HTF) and as energy storage media at high temperature (>1000°C) overcoming the technological issues related to the adoption of other HTF like molten salts. Among particle receivers, dense gas-solid fluidized suspensions have been recently proposed thanks to their excellent thermal properties, namely bed-to-wall heat transfer coefficient (several hundreds of W/m^2K) and effective thermal diffusivities (0.001-0.1m^2/s) associated with convective transfer due to bubble-induced and gross bed solids circulation. Both these features may be optimized by proper selection of fluidized solids type and size and fluidization regime. Non-conventional design and operation of fluidized beds based on uneven or unsteady (pulsed) fluidization may further enhance their thermal performances for CSP and thermal energy storage applications. Dense gas-solid fluidized beds have the potential to effectively accomplish three basic complementary tasks: a) collection of incident solar radiation; b) transfer of the incident power to heat exchange surfaces and henceforth to high-efficiency power cycles; c) thermal energy storage, aimed at equalizing the inherent time-variability of the incident radiation for stationary combined heat and power (CHP) generation. All these features have been exploited in a novel concept of solar receiver for CHP generation with inherent thermal energy storage. The concept is based on a novel design of the solar collector, based on a compartmented dense gas fluidized bed optimized so as to accomplish the three complementary tasks. The hydrodynamics of a near-2D dense gas-fluidized bed operated at ambient conditions and equipped with a compartmented distributor of fluidizing gas, consisting of two spargers, has been characterized to accomplish the effectiveness of compartmented fluidization of the bed without physical separations or internals inside the bed. Uneven fluidization was accomplished by partitioning of the fluidizing gas between the sections of the distributor: a larger gas superficial velocity (exceeding the incipient fluidization velocity) was established in the "reactive" section of the distributor, a smaller (or even 0) gas superficial velocity was established in the passive section of the distributor. Results confirm the expected inherent tendency of the bed to equalize uneven fluidization. Crossflow of fluidizing gas from the active to the passive sections of the bed, driven by lateral pressure gradients, promotes fluidization of the passive section and/or defluidization of the active section of the bed. Crossflow becomes increasingly important as the bed depth increases and this makes compartmented fluidization of deep beds more problematic. However, it was proven that proper fluidization maps can provide the basis for the design and operation of gas distributors to effectively achieve compartmented fluidization of the fluidized bed.

This work aims at investigating the thermal behaviour of compartmented fluidized beds under even and uneven fluidization conditions. In particular, the transversal thermal diffusivity was studied both by experiments carried out in the compartmented near-2D dense gas-fluidized bed and by CFD numerical simulations.

The experimental apparatus consists of a near-2D fluidized bed (2850x1860x200mm) equipped with an array of pressure and temperature taps at different locations in the bed. The fluidized bed can be considered as â€œnearly two-dimensionalâ€• because it is characterized by a thickness much smaller than the other dimensions, but at the same time large enough to prevent extensive wall effects for bubbles smaller than 120 mm. Accordingly, the test facility can be used to investigate bed hydrodynamical patterns along the width and the height of the bed as they would develop in full-scale 3D compartmented fluidized beds. The fluidization column is equipped with different diagnostic taps and inlet and outlet ports, necessary for air distribution system, temperature monitoring system, heating system and for discharge of bed material. The bed was equipped with two spargers acting as gas distributors, intentionally of different length so that it was possible to establish uneven fluidization of the bed with asymmetric patterns. Accordingly, the spargers, and the corresponding sections of the fluidized bed are called long and short. The bed material was fine silica sand with a mean Sauter diameter of 145 μm. Fine bed solids were chosen for CSP applications in order to obtain large heat transfer coefficient and effective thermal diffusivity even at small fluidization velocities. A temperature measurement system based on thermocouples was used to map thermal conditions inside the fluidized bed. The system consists of three movable probes which can be immersed directly in the bulk of the bed at different levels above the distributor. Additionally, four fixed probes are located at different locations at the lateral walls. The three movable probes are vertically inserted inside the bed from the top side of the apparatus. Two heating cartridges are located inside the short compartment and are regulated to keep the short compartment at a constant temperature. At steady state conditions the thermal behaviour of the fluidized bed under even and uneven fluidization conditions was evaluated by mapping the bed temperature at different vertical and horizontal positions. A simple model of the experiment was developed to estimate the transversal thermal diffusivity as a function of operating conditions investigated.

The experimental conditions investigated was also studied by CFD simulations using a Eulerian-Eulerian modelling approach. The main aim of the numerical simulations was to reproduce the experimental results to consolidate the proper choice of the frictional solid stress to properly capture the hydrodynamics of compartmented fluidized beds. In particular, the mixing length scales of the solid phase responsible of transversal thermal diffusivity under even and uneven fluidization conditions were specifically investigated to individuate the establishment of gross bed solids circulation.

Experimental Characterization and CFD Simulations of Gas Maldistribution Relevant to Fluidized Beds of Group A Particles

Authors:
Troiano, M., Università degli Studi di Napoli Federico II
Amblard, B., IFP Energies nouvelles
Solimene, R., Consiglio Nazionale delle Ricerche
Salatino, P., Università degli Studi di Napoli Federico II

An experimental and computational study of gas maldistribution in fluidized bed reactors is hereby presented. Gas maldistribution can be often experienced in industrial fluidized bed reactors, affecting the performance in terms of operation and potential damages of internals within the reactors.

The present study aimed at investigating the hydrodynamic behaviour of the fluidized bed in presence of a gas maldistribution and the capability of the CFD code Barracuda™ to predict the effects of the maldistribution of gas in a fluidized bed under well controlled conditions. For this purpose, experiments were carried out in a 20 cm diameter cold flow fluidized bed column connected to a single stage cyclone to capture entrained particles and to re-insert them into the fluidized bed by means of a O.D. 40 mm cyclone dipleg and an automatic L-valve. The column is filled with Group A particles (FCC catalyst, Ï• $_P$=1260 kg/m^3, D_{p50}=75 mm), with a bed height around 1.1 m. Two configurations of gas distribution have been investigated under turbulent fluidization regime (superficial gas velocity of 0.7 m/s). In the first operative configuration, called â€œeven fluidizationâ€• , all the gas is evenly fed at the bottom of the bed through a porous distributor. In the second operative configuration, the gas is partly fed to the bottom of the bed and partly injected through a nozzle located at the centre of the bed. This configuration is called â€œuneven fluidizationâ€• and it is aimed at purposely creating a gas maldistribution. Hydrodynamic characterization of the fluidized bed was carried out under even and uneven fluidization conditions by mapping solids concentration along radial and axial directions.

Local solid volume fraction profiles were measured by using optical probes at different bed heights and along the two radial directions of the column, normal to each other. Two types of optical probes are used: Type I, much larger than the particle diameter, and Type II, with a cross-sectional dimension closer to the particle diameter. Type I probes are one-channel fibre probes consisting of an 8 mm ID metallic rod. They are capable of detecting all the particles in the measuring volume and are commonly used to measure solids concentrations. Type II probes consist of two independent channels, separated by a calibrated distance. They can detect small groups of particles or single particles and are generally employed to determine particle velocities. In both configurations, channels consist of thousands of optical fibres arranged parallel to each other that can emit and reflect light. The arrangement is such that one layer consists of light emitting fibres and the next one consists of light receiving fibres. Both types have been used in the present experimental campaign, to measure the solid concentration. The optical probes were connected to a particle analyser consisting of photoelectric converter and amplifying circuits, signal pre-processing circuits, high-speed A/D interface card and acquisition software. The probe was inserted into the fluidized bed, perpendicular to the flow. When catalyst particles in the emulsion phase pass in front of the probe, a high percentage of the emitted light is reflected and transferred to the photocell, which converts it into a high voltage signal. In contrast, when a gas void passes, relatively little light is reflected to the probe and the photocell responds by giving a low voltage signal. The light reflection signals detected on each fibre was recorded with an acquisition frequency of 1 kHz. The sampling time was 60 s, chosen after trials with different sampling times (30 s, 60 s, 120 s), to minimize the sampling time without compromising reproducibility of average concentration profiles during the turbulent fluidization conditions. The concentrations of solid particles were measured in the axial direction (optical probes were inserted in the bed at 0.05, 0.1 and 0.15 m above the jet) and in the two main radial directions (11 measurement points for each direction). Solid concentration profiles have been compared with the average solid concentration obtained by means of pressure probes.

The experimental results have been then compared with CFD simulations in order to assess if the CFD code Barracuda™ is able to catch the effect of the gas injection configuration on the bed solid volume fraction. The gas/particles hydrodynamic modeling in Barracuda™ is based on the Multi-phase Particle in Cell method (MP-PIC). The gas phase is treated as a continuum in a Eulerian framework solving the averaged Navier-Stokes equations. Particles are treated with a hybrid Eulerian-Lagragian approach where equations are derived from the Liouville equation which governs the transport in an Eulerian framework of a particle distribution function f(x,u,m,t) where x is the spatial location, u and m are respectively the velocity and the mass of the particle and t is the time. The particle-phase equations are solved by discretizing the particle distribution function f(x,u,m,t) into computational parcels, each of which represents a certain number of real particles of identical size, velocity and position. For the resolution, computational parcels properties are interpolated onto the Eulerian grid to solve the solid-phase equations, once the equations are solved on the grid, the Eulerian grid properties such as gas velocities, gas pressure gradients and solids stress gradients are interpolated back to the parcel in order to update its position and velocity using a Lagrangian approach. Collisions between particles are modeled through the use of a particle stress function. The advantage of the MP-PIC method is that the coupling term between the gas and solid phases takes into account a distribution of particles with different sizes and velocities while a classic Eulerian approach for the solid phase only takes into account an averaged particle diameter and an averaged solid velocity for the interphase coupling term. In this paper, the particleâ€“fluid drag force was expressed using the Gidaspow drag function which combines the Wen-Yu model and the Ergun model. Then, the turbulence of the gas phase is modeled through a Large Eddy Simulation (LES) approach where the large-scale motion due to the large eddies are computed directly from the gas flow equation while the small-scale eddies motion is captured through the LES Smagorinsky subgrid scale model. Concerning the interactions with the wall, for the gas phase a no slip boundary condition is applied while for the solid phase, the velocity vectors of the particles after bouncing on a wall are computed through normal and tangential momentum retention (or restitution) coefficients. In the current simulation, the default retention coefficients of the software were used: 0.3 for the normal retention coefficient which means that the particles keep 30% of its normal velocity after bouncing and 0.99 for the tangential retention coefficient which means that the particles keep 99% of its tangential velocity after bouncing. The effect of drag correlation, inlet gas boundary conditions and jet injections have been investigated. The radial profile of solid concentration at three distances from the jet injection have been calculated with and without the presence of the central gas jet.

The experimental results allowed to compare the solid concentration profiles using the two different optical probes (one-channel and two-channel probes). The experimental profiles highlight a very good qualitative and quantitative agreement for the tests in the even fluidization configuration when using the two kinds of probes. Regarding the tests in the uneven fluidization configuration, both the one-channel and two-channel probes are able to catch the presence of the gas stream. However, results obtained with the two probes highlight some differences, probably related to the different measurement volume of the two kinds of probes. When comparing experimental and CFD results, it is noteworthy that the choice of drag correlation and boundary conditions strongly influences the agreement between the experimental and CFD results in terms of radial and axial profiles of solid concentration.

Comparative Study on Different Types of Liquid-Solid Fluidization Systems - Upflow, Downflow and Countercurrent Flow

Authors:
Nan, T., University of Western Ontario
Zhu, J., The University of Western Ontario
Comparative study on different types of liquid-solid fluidization systems - upflow, downflow and countercurrent flow

Tian Nan, Jesse Zhu*

Department of Chemical and Biochemical Engineering, University of Western Ontario, London, Ontario, Canada

This experimental study focusses on different types of liquid-solid fluidization systems categorized based on the flow directions of liquid and solids and relative liquid solid density. In general, heavy density solids tends to settle at the bottom, thus upflow of liquid is often required. On the contrary, downflow liquid is needed to fluidized light density solids from the top. In upflow liquid-solid system, both streams are flowing upwards with heavier density particles. While in downflow systems, liquid and solids are flowing downwards using light density particle. Countercurrent liquid-solid flow is also achieved when heavy particles are falling down while liquid is going up. In addition, the flow directions of liquid and solids can be reversed with light density particle which tends to rise with moderate downflow of liquid. The hydrodynamics under all these flow conditions with corresponding density of particles are experimentally studied and compared systematically. Solids holdup are measured at various superficial liquid velocities and solids flowrates. A flow regime map is generated to conclude different liquid-solid fluidization systems.

Applications of Tribo-Electric Probes in Fluidized Beds

Authors:
Li, Y., Western University / Institute for Chemicals and Fuels from Alternative Resources (ICFAR)
Jahanmiri, M., Icfar
Sanchez Careaga, F., Western University / Institute for Chemicals and Fuels from Alternative Resources (ICFAR)
Briens, C., Western University / Institute for Chemicals and Fuels from Alternative Resources (ICFAR)
Berruti, F., Western University / Institute for Chemicals and Fuels from Alternative Resources (ICFAR)
McMillan, J., Syncrude Canada Ltd

When fluidized bed particles carried by a bubble wake hit a metal probe, they strip electrons from its surface, inducing an electrical current that can be monitored. The main advantage of tribo-probes, when compared to optical fibers or capacitance probes, is that they can be used at high temperature and in turbulent fluidized beds, where high gas velocities are combined with high solids concentrations. By using multiple tribo-probes at various bed locations, one can gather local and global fluidization information simultaneously. This study shows how tribo-probes can be used to characterize the hydrodynamics of a fluidized bed.

This study was conducted at room temperature in a column with a rectangular cross-section (0.5 m x 0.1 m x 2.28 m) filled with silica sand (group B powder). Compressed air provided the fluidization gas. The gas distributor consisted of 20 tuyeres whose flow could be individually controlled: by opening 10 valves at different locations, bubbles could be concentrated to the western side, eastern side or center of the bed. Two baffle configurations modified the bed hydrodynamics: in the first configuration, a baffle with an open-ended triangular shape with an internal angle of 45 ° of the same width as the bed (0.18 m x 0.18 m x 0.1 m) was placed on one side of the bed, while in the second configuration two smaller baffles (0.125 m x0.125 m x 0.1 m) with the same geometry were attached on opposite sides of the bed. Gas and gas-liquid horizontal jets were formed with scaled-down versions of commercial nozzles. Experiments were conducted with a superficial gas velocity ranging from 0.1 to 1.0 m/s.

Each tribo-probe was a simple metal rod with a diameter of 6.35 mm that was connected to the electrical ground. The probe diameter was selected to avoid deformation at high gas velocities. A Teflon fitting isolated each probe from the grounded bed wall. An array of tribo-probes (5 rows × 9 lateral probe locations) inserted into the bed to half of the bed width (0.05 m) provided information over the whole bed. A new system was developed to measure the instantaneous current from each probe to the ground accurately and reliability, with a reasonable cost. Transient-voltage-suppression (TVS) diodes protect the current measurement system from damage caused by high-voltage discharges from the fluidized bed. Since conventional optical fiber or capacitance probes could not be used at high gas velocities, a radioactive non-invasive transmission method determined the local bed voidage to verify the lateral bubble gas distributions obtained from the tribo-probes. A radioactive particle was located outside the column, on one side of the bed and a radiation detector was placed just outside the opposite bed wall.

Tribo-probes were successfully applied to obtain the following information:

- The gas bubbles never spanned the distance between the two opposite bed walls. This was verified by using cross-correlation of signals from tribo-probes on opposite walls of the bed.
- The local bubble velocity and its evolution with bed height, gas distribution and gas velocity. This was obtained from cross-correlation of signals from probes located at different heights.
- The lateral distribution of the bubble gas, its evolution with bed height, superficial gas velocity, initial gas distribution, and the use of baffles. Results were validated with the radiation transmission method.

Three gas distributor configurations were studied. By opening 10 nozzles evenly distributed over the bottom of the fluidized bed, an even gas distribution is formed. When opening 10 nozzles close to the western side bed wall, in the eastern case, the bubbles concentrate on the eastern side bed wall.

A single baffle extending 18 cm from one wall (L=18 cm) greatly concentrates the bubbles to the baffle tip region, whichever gas distributor configuration is employed. However, when applying smaller symmetrical baffles (L=12.5 cm), the impact of the baffle on gas bubble profile dissipates with the increase of distance to baffle tip. The pure gas jet would have a minor impact on lower level bubble profiles.

Cross-correlation coefficient between two rows of tribo-probes provides the gas bubble velocity. The results show that the bubble velocity would increase with the increasing distance to the gas distributor, which applies to all 3 cases. The highest gas bubble velocity could reach around 3 m/s, which is validated with a standard correlation (Hilligardt & Werther et al., 1986).

- The penetration of horizontal gas and gas-liquid jets. The results obtained with the tribo-probes for the gas-liquid jets were validated with a separate method using a cold liquid and thermocouples. The results also agreed well with predictions from a standard correlation (Ariyapadi et al. 2005).

Interpenetrating Mixture Theory for Fluidized Beds:Thermodynamics and Turbulence

Authors:
Petty, C. A., Michigan State University
Benard, A., Michigan State University

Turbulent multiphase flows are encountered ubiquitously in the process industry (Soo, 1989; Gidaspow, 1994; Crow et al., 1998; Kolev, 2002; Kleinstreuer, 2003; Brenner, 2005; Ishii and Hibbiki, 2006). Examples include flows within hydrocyclone separators and fluidized bed reactors (Cocco et al., 2014; Yate and Lettieri, 2016). The potential for simulating these flows has primarily occurred because of advancements in computational hardware and software. However, longstanding deficiencies in turbulence closure models limit the practical utility of computational methods as an enabling technology (Davidson, 2013; Pope, 2000; Drew and Passman, 1999; Marchisio and Fox, 2013; Yeoh et al., 2014).

This presentation will extend a recently developed Reynolds stress closure for single phase fluids (Koppula et al., 2009, 2011, 2013) to a Reynolds stress closure for interpenetrating multiphase fluids (Manninen et al., 1996). The normalized Reynolds stress is a non-negative operator; therefore, it is essential that all of the eigenvalues of this operator must be non-negative for all turbulent flows in rotating and non-rotating frames-of-reference. This fundamental mathematical property cannot be compromised if turbulence modeling is to attain its full potential in predicting flows in complex geometries. Although current CFD technology can reproduce benchmark flows, the ability to predict low-order statistical properties beyond a calibrating flow is not possible. This weakness of turbulence modeling can be traced to the closure hypothesis that velocity fluctuations are objective vector fields. This assumption is embedded in the sub-grid closures associated with large-eddy simulations; the pressure/strain rate closures associated with the second-order moment equation for the Reynolds stress; and, the eddy viscosity models associated with the closure of the Reynolds averaged Navier-Stokes equation. The ad hoc assumption that the Reynolds stress is an objective operator similar to the Cauchy stress in continuum mechanics is not supported by direct numerical simulations of the Navier-Stokes equation and fundamental physical principles of thermodynamics and turbulence. Continuum scale hydrodynamic fluctuations, unlike molecular scale fluctuations, are not objective vector fields. Research at Michigan State University has identified a class of algebraic closure models for the normalized Reynolds stress that are realizable for all turbulent flows in rotating and non-rotating frames. This discovery has much potential to transform current CFD technology from an interpolating tool to a predictive tool.

References

Brennen, C. E., 2005, Fundamentals of Multiphase Flow, Cambridge University Press.

Cocco, R., Kari, S. B. R., Knowlton, T., 2014, â€œIntroduction to Fluidization: Back to Basicsâ€• , CEP, AIChE

Crow, C.T., C. M. Sommerfeld, and Y. Tsuji, 1998, Multiphase Flows with Droplets and Particles, CRC Press.

Davidson, P.A., 2013, Turbulence in Rotating Stratified and Electrically Conducting Fluids, Cambridge University Press.

Drew, D.A., and S. L. Passman, 1999, Theory of Multicomponent Fluids, Applied Math Series, Volume 135, Springer.

Gidaspow, D., 1994, Multiphase Flows and Fluidization: Continuum and Kinetic Theory Descriptions, Academic Press.

Ishii, M., and T. Hibiki, 2006, Thermo-Fluid Dynamics of Two- Phase Flow, Springer.

Kleinstreuer, C., 2003, Two-Phase Flow: Theory and Practice, Taylor & Francis.

Kolev, N.I., 2002, Multiphase Flow Dynamics, Volume I: Fundamentals; Volume II: Mechanical and Thermal Interactions, Springer.

Koppula, K. S., A. BÃ©nard, and C. A. Petty, 2009, â€œRealizable Algebraic Reynolds Stress Closureâ€• , Chem. Eng. Sci., 64, 4611-4624.

Koppula, K. S., A. BeÃ©nard, and C. A. Petty, 2011, â€œTurbulent Energy Redistribution in Spanwise Rotating Channel Flowsâ€• , Ind. Eng. Chem. Res., 50 (15), 8905-8916.

Koppula, K.S., S. Muthu, A. BÃ©nard, and C. A. Petty, 2013, â€œThe URAPS closure for the normalized Reynolds stressâ€• , Physica Scripta, T155.

Manninen, M., V. Taivassalo, and S. Kallio, 1996, On the Mixture Model for Multiphase Flow, Espoo, Technical Research Centre of Finland, VTT Publication 288.

Soo, S.L., 1989, Particulates and Continuum: Multiphase Fluid Dynamics, Hemisphere.

Marchisio, D.L., and R.O. Fox, 2013, Computational Models for Polydispered Particulate and Multiphase Systems, Cambridge University Press.

Pope, S. B., 2000, Turbulent Flows, Cambridge University Press.

Yates, J. G., and P. Lettieri, 2016, Fluidized-Bed Reactors: Processes and Operating Conditions, Particle Technology Series, Vol. 26, Springer

Yeoh, G. H., C. P. Cheung, and J. Tu, 2014, Multiphase Flow Analysis Using Population Balance Modeling: Bubbles, Drops and Particles, Butterworth-Heinemann.

Keynote Talk: An Innovative Hydrogen and Power Coproduction System Based on Integrated Exergy Recuperative Biomass Gasification and SOFC By Using an Internal Circulating Fluidized Bed

Authors:
Tsutsumi, A., The University of Tokyo
Ishizuka, M., The University of Tokyo
Tsutsumi, K., Exergy Power Systems, Inc.
Ishizuka, T., Hokkaido Electric Power Co., Inc.

A novel hydrogen and power coproduction system based on the exergy recuperative biomass gasification integrated with solid oxide fuel cell was proposed. In this system micro-tubular SOFC stacks are immersed in the internal circulating fluidized bed gasifier. The exothermic heat from SOFC is recuperated and utilized for endothermic heat of biomass steam gasification. In addition, providing the recirculation system of unreacted hydrogen separated from SOFC anode exhaust, gas turbine can be eliminated. Since the exergy loss of SOFC is much smaller than that of gas turbine because of no combustion, the net power generation efficiency of the combination of gasification and SOFC power generation system can be considerably improved. The internal circulating fluidized bed was composed of a moving bed high-temperature pyrolyzer with the function of tar cracking, a bubbling bed char gasifier equipped with micro-tubular SOFC stacks, and a fluidized bed combustor of remained char. A 20 kW-class pilot plant was constructed and tested. In the present study, the conceptual design of a integrated exergy recuperative biomass gasification and SOFC system for hydrogen and power coproduction and the performance of exergy recuperative biomass gasification are described.

NO Emissions Under Pulverized Char Mild Combustion in O2/CO2 Preheated By a Circulating Fluidized Bed: Effect of H2o Addition

Authors:
Zhu, S., Institute of Engineering Thermophysics, Chinese Academy of Sciences
Lyu, Q., Institute of Engineering Thermophysics, Chinese Academy of Sciences
Zhu, J., Institute of Engineering Thermophysics, Chinese Academy of Sciences
Zhang, Y., Institute of Engineering Thermophysics, Chinese Academy of Sciences
Pan, F., Institute of Engineering Thermophysics, Chinese Academy of Sciences
Liu, W., Institute of Engineering Thermophysics, Chinese Academy of Sciences

MILD (moderate or intense low-oxygen dilution) combustion of pulverized char in O_2/CO_2 was achieved by preheating fuel in a circulating fluidized bed (CFB). The purpose of this research was to study effects of H_2O addition on NO emissions under MILD combustion of pulverized char. First, the preheating characteristics were evaluated, including the conversion ratios of each fuel component and the composition of the coal gas. To calculate the nitrogen balance in the preheating process, the gas in the loop seal was changed to O_2/N_2. The results indicate that the concentrations of H_2 and CH_4 increased with the increasing H_2O concentration in the primary gas, and the conversion ratio of nitrogen was above 0.40. The heating value of high-temperature coal gas also increased. Next, the effects of H_2O addition on NO emissions were investigated. Increasing H_2O concentrations in the primary gas could reduce NO emissions. And the combustion efficiency was above 97.0%. These results demonstrate that the oxy-fuel preheating combustion technology with H_2O addition could achieve clean, efficient combustion of pulverized char.

Development of Gas-Solid Reactor for Activating Iron-Based Fischer-Tropsch Catalysts

Authors:
Cheng, C., National Institute of Clean-And-Low-Carbon Energy
Bu, Y., National Institute of Clean-and-Low-Carbon Energy
Men, Z., National Institute of Clean-and-Low-Carbon Energy
Du, B., National Institute of Clean-And-Low-Carbon Energy
Tong, R., National Institute of Clean-And-Low-Carbon Energy
Feng, L., National Institute of Clean-And-Low-Carbon Energy

In industries, Fischer-Tropsch technology uses syngas by coal gasification for producing clean oil through some iron-based catalysts, which should be firstly activated from its oxidation state to the reduced state before application. Currently, the reactor for activation of catalysts belongs to some kind of slurry bed with a gas-liquid-solid fluidization system. Due to the existence of liquid between gas and catalysts, the activation rate is affected directly by the liquid resistance. Furthermore, compared with gas-liquid-solid slurry bed, gas-solid fluidized reactor has advantages in aspects of shorting activation time, lowering investments, reducing operating costs, and online make-up, etc. Therefore, the idea of gas-solid fluidization reactor was presented by us for the activation of catalysts. The study of the fluidization behaviors for iron-based Fischer-Tropsch catalysts was carried out on a cold fluidization reactor by using air as fluidization medium, including particle characteristics, the degassing capacity and the distribution of catalysts concentration. The experimental results of fluidization showed that iron-based Fischer-Tropsch catalysts belonged to Geldart A and had an easier ability of fluidization than FCC catalysts. On the basis of fluidization results, a pilot gas-solid reactor for the catalysts activation was built to investigate the affecting factors, such as pressure, superficial velocity of syngas, CO concentration, and so on. The activation experiments on the pilot unit indicated that the greater the pressure the better the activation, and the turbulent state was much better than the bubbling state. The comparison of the activation effect between gas-solid fluidization reactor and gas-liquid-solid slurry bed showed that gas-solid fluidized activated catalyst almost has the same Fischer-Tropsch catalyst performance. The study helps to achieve the process package and promote the commercialization of gas-solid reactor for catalysts activation.

Performance of Anaerobic Fluidized Bed Microbial Fuel Cell with Different Porous Anodes

Authors:
Xiuli, Z., Qingdao University of Science and Technology
Liu, Y., Qingdao University of Science and Technology
Chunhu, L., Ocean University of China
Guo, Q., Ningxia University

Anode materials were used to construct the microbial fuel cell (MFCs), and the characteristics of the anodes are important for successful applied performance of MFCs. Using the cyclic voltammetry (CV) method, the experiments showed that 5 wt% multi-walled carbon nanotubes (MWNTs) were optimal for the PANI/MWNT film anodes prepared using 24 polymerization cycles. The maximum output voltage of the PANI/MWNT film anodes reached 967.7 mV with a power density of 286.63 $mW \cdot m^{-2}$. Stable output voltages of 860 mV, 850 mV, and 870 mV were achieved when the anaerobic fluidized bed microbial fuel cell (AFBMFC) anodes consisted of carbon cloth with carbon black on one side, copper foam and carbon brushes, respectively. Pretreatment of anodes before starting the AFBMFC by immersion in the stirred bacterial fluid can significantly shorten AFBMFC startup time. After continuously running the AFBMFC, the anode surfaces generate active microbial catalytic material.

Partial Regeneration of Spent SAPO-34 Catalyst in Methanol to Olefins Process Via Steam Gasification

Authors:
Zhou, J., Dalian Institute of Chemical Physics, Chinese Academy of Sciences
Zhang, J., Dalian Institute of Chemical Physics, Chinese Academy of Sciences
Zhang, T., Dalian Institute of Chemical Physics, Chinese Academy of Sciences
Ye, M., Dalian Institute of Chemical Physics, Chinese Academy of Sciences
Liu, Z., Dalian Institute of Chemical Physics, Chinese Academy of Sciences

The SAPO-34-catalyzed MTO reaction, working as a non-petrochemical route for producing the lower olefins, has been successfully industrialized in China.[1] This technology adopts a fluidized bed reactor-regenerator configuration, in which the continuous catalysts circulation leads inevitably to the age distribution of catalysts in the reactor and concomitantly uneven deposition of carbonaceous species onto individual catalyst particle. This non-uniform coking behavior of catalysts would average, and, to be more specific, lower down the overall ethene selectivity. Meanwhile, a common feature of the rather lower initial ethene selectivity raises an issue that the optimal operation window is too narrow to maximize light olefins for industrial MTO operation. In this work we studied the partial regeneration of the spent SAPO-34 catalyst via steam gasification, and showed that this method could not only reactivate the spent SAPO-34 catalyst, but also prompt the ethene selectivity and thus the lower olefins selectivity when compared to the partial regeneration of spent catalyst by traditional air combustion.

Firstly, the spent SAPO-34 catalyst was partially regenerated by steam gasification and air combustion, and then the MTO performance was evaluated under the same operation conditions in a fluidized bed reactor at atmospheric pressure. Figure 1a shows the catalytic performance over both the partially regenerated catalysts with nearly the same coke content (4 wt%). The selectivity of ethene over the SAPO-34-AC-30 min sample was much greater than for fresh SAPO-34 sample but lower than for SAPO-34-SG-120 min sample. That is, residue coke has a significant positive impact on ethene formation but depends on the regeneration gas. In view of this result, partial regeneration method is an efficient way for controlling the activity-selectivity-deactivation. We also gave the mechanistic understanding about the different catalytic performance. With several characterizations results, e.g., TGA, FTIR, GC-MS, indicated that both partial regeneration samples exhibited similar textual properties, while the SAPO-34-SG-120 min sample possessed lower acidity than the SAPO-34-AC-30 min sample. That is to say, the difference between two partially regenerated SAPO-34 catalysts in the catalytic performance in MTO reaction should be irrelevant to their textual properties. It should be noted that the low acid density favors the selectivity to propene and higher olefins.[2] Thus, the difference between the acidities of both partially regenerated samples does not also provide a fully satisfactory explanation for the difference in products distribution. Therefore, the higher selectivity to ethene over the SAPO-34-SG-120 min sample in comparison with the SAPO-34-AC-30 min points to different reaction mechanisms, as a result of the different property of the residual coke. As shown in Figure 1b, for SAPO-34-SG-120 min sample, a broad distribution of species ranging from methylated benzenes to methylated naphthalenes. Typically, the oxygenated compounds are the dominant components for SAPO-34-AC-30 min sample, with followed naphthalene and benzene. Methylated benzene and methylated naphthalene acting as active hydrocarbon pool species are more selective to the formation of ethene than oxygenated compounds, which has been reported in many literatures.[3-5]

Due to the effect of residual coke on the reaction and deactivation, control of the nature of the residual coke, also the coke content is an interesting way of improving the MTO process. By doing so, the ethene selectivity control will be no longer highly dependent on the age distribution of catalysts, which will beneficially extend the optimal operation window for industrial MTO operation. Additionally, steam gasification can beneficially transform invaluable coke to high-valued syngas (H_2 and CO) and simultaneously reduce greenhouse gas (CO_2) emissions.

Reference

1. Tian, P.; Wei, Y. X.; Ye, M.; Liu, Z. M. Methanol to olefins (MTO): from fundamentals to commercialization. ACS Catal. **2015**, 5, 1922−1938.

2. Batholomew, C. H.; Argyle. M. D. Advances in Catalyst Deactivation and Regeneration. Catalysts. **2015**, 5, 949-954.

3. Liu, Z.; Dong, X.; Liu, X.; Han, Y. Oxygen-containing coke species in zeolite-catalyzed conversion of methanol to hydrocarbons. *Cata Sci Technol* **2016,** *6*, 8157-8165.

4. Wang, C.; Xu, J.; Qi, G.; Gong, Y.; Wang, W.; Gao, P.; Wang, Q.; Feng, N.; Liu, X.; Deng, F. Methylbenzene hydrocarbon pool in methanol-to-olefins conversion over zeolite H-ZSM-5. *J. Catal.* **2015,** *332*, 127-137.

5. Song, W.; Fu, H.; Haw, J. F. Supramolecular origins of product selectivity for methanol-to-olefin catalysis on HSAPO-34. *J. Am.*

Pressurized Gas Switching Combustion in a Pre-Pilot Scale Reactor Cluster

Authors:
Zaabout, A., SINTEF Industry
Ugwu, A., Norwegian University of Science and Technology
Donat, F., Laboratory of Energy Science and Engineering, ETH Zürich
Albertsen, K., Euro SupportAdvanced Materials B.V
Cloete, S., SINTEF Industry
Müller, C. R., ETH Zurich
Amini, S., SINTEF Industry

CO_2 capture and storage (CCS) is increasingly seen as a vital cost-effective tool for bridging to a sustainable low carbon economy, especially given the importance of limiting global warming to 1.5 °C, as highlighted in the recently released IPCC special report [1]. CCS implementation is, however, still a far-reaching goal, requiring the development of high efficiency CO_2 capture processes, since energy penalty is the primary economic and environmental concern hampering current CCS technology.

Gas Switching Combustion, GSC, is an emerging technology for power production with integrated CO_2 capture, that incorporates the high energy efficiency of chemical looping combustion (CLC) and the design simplicity and practicality of a standalone dense fluidized bed [2]. This configuration avoids external solids circulation taking place in conventional chemical looping that uses interconnected fluidized beds, thereby avoiding many of the operational challenges, especially at high pressures. In the GSC concept, gas feeds are alternated into a bed of oxygen carrier, initially placed in a single reactor, to complete the redox cycle: the oxygen carrier is first reduced by a gaseous fuel in the fuel stage then oxidized by the oxygen from the air feed in the air stage. In this way, contact between air and fuel is avoided, resulting in inherent CO_2 and N_2 separation.

The GSC concept was demonstrated in a single reactor using different oxygen carriers (Ni-based, ilmenite and $CaMnO_{3\text{-}}$î′-based) and fuels (CO, H_2 and CH_4), under atmospheric and pressurized conditions [2-4]. In these studies, the GSC has shown very attractive features, such as ease of operation and control under pressurized conditions, in addition to sufficiently high CO_2 purity and capture efficiency.

Because of the dynamic nature of the gas switching concept, GSC requires a cluster of reactors operating in a coordinated manner to ensure a steady supply of hot gasses to a downstream power cycle. In this study, the standalone GSC reactor operated in previous studies [2-4], is scale up from 5 to 10 cm ID, then a cluster of three such reactors is constructed. The system is designed to operate at pressures up to 20 bar and temperatures up to 1100 Â°C. The $CaMnO_{3\text{-}}$î′-based oxygen carrier previously tested in the GSC reactor, was selected for this first of a kind GSC cluster testing, given its good performance with both syngas and methane, in addition to its exothermic reduction reaction that minimizes temperature variation across the GSC cycle.

[1] "IPCC, Global Warming of 1.5 Â°C (2018). Intergovernmental Panel on Climate Change."

[2] A. Zaabout, S. Cloete, S. T. Johansen, M. v. S. Annaland, F. Gallucci, and S. Amini, "Experimental Demonstration of a Novel Gas Switching Combustion Reactor for Power Production with Integrated CO2 Capture," *Industrial & Engineering Chemistry Research,* vol. 52, no. 39, pp. 14241-14250, Oct 2 2013.

[3] A. Zaabout, S. Cloete, and S. Amini, "Autothermal operation of a pressurized Gas Switching Combustion with ilmenite ore," *International Journal of Greenhouse Gas Control,* vol. 63, pp. 175-183, 2017/08/01/ 2017.

[4] A. Zaabout, S. Cloete, J. R. Tolchard, and S. Amini, "A pressurized Gas Switching Combustion reactor: Autothermal operation with a CaMnO3â^'Î´-based oxygen carrier," *Chemical Engineering Research and Design,* vol. 137, pp. 20-32, 2018/09/01/ 2018.

Gas-Solids Fluidized Bed Strippers - a Review

Authors:
Issangya, A., Particulate Solid Research, Inc.
Karri, R., Particulate Solid Research, Inc.
Cocco, R., Particulate Solid Research, Inc.
Knowlton, T. M., Particulate Solid Research, Inc.
Freireich, B., Particulate Solid Research, Inc.

Fluidized bed strippers are essentially flowing fluidized beds where an up-flowing gas removes (strips out) product entrained in a down-flowing stream of the emulsion phase (denser solid phase). The stripper has to facilitate the transfer of the gas in the interstices of the emulsion phase and that adsorbed on the particles into the bubbles of the stripping gas. The counter-current contacting between the down-flowing emulsion phase stream and the up-flowing bubbles stream is enhanced by the use of an assembly of horizontal or inclined baffles. In unbaffled fluidized beds, the bubbles movement is unrestricted so, as they rise, interact, coalesce and grow in size they tend to move laterally toward the center. This would result in too low of a mass transfer to be an effective stripper.

Fluidized bed strippers play a major role in hydrocarbon processing with respect to fluid catalytic cracking (FCC) and fluid coking. In the FCC process, the catalyst collected by the reactor cyclones contains significant amounts of the product hydrocarbon vapors. Failure to remove these vapors from the catalyst results in a loss of a valuable product and excessively high temperatures in the regenerator. Thus, entrained and adsorbed hydrocarbon vapors are typically removed from the catalyst in a fluidized bed stripper using steam. In a fluid coker stripper, steam is used to remove entrained and adsorbed hydrocarbons from the coke particles descending from the fluid coker reactor thus minimizing carry-under of valuable hydrocarbon product. For both unit operations, a variety of proprietary and standard baffles are used with these fluidized bed strippers. The most common types are disk and donut trays, horizontal sheds and structured packings. Horizontal grating trays, sometimes referred to as subway gratings, are also used but to a much lesser extent. This paper reviews studies available in the open literature on fluidized bed strippers as applied to the FCC and fluid coker unit operations. The review has shown that experimental work has dealt with stripping efficiencies, stripper flooding and other flow dynamics issues. Fundamental studies on mass transfer in fluidized bed strippers are few and the application of computational fluid dynamic techniques to model fluid bed strippers appears to be a fairly recent addition. Overall, fluid catalytic cracking strippers, in particular disk and donut strippers, have received the most attention, undoubtedly, due to their much wider industrial usage.

Keynote Talk: Long-Time Simulation of MTO Reactors with EMMS-Based Discrete Particle Method

Authors:
Xu, J., State Key Laboratory of Multiphase Complex Systems, Institute of Process Engineering, Chinese Academy of Sciences
Ge, W., Institute of Process Engineering, Chinese Academy of Sciences
Liu, X., Institute of Process Engineering, Chinese Academy of Sciences

The methanol to olefins (MTO) is an important method is to produce light olefins. However, the hydrodynamic behaviors and the reacting properties cause some difficulties in designing and optimizing the MTO reactor. In MTO reaction, the coke will deposit on the catalysts and the coke content will greatly affect the reacting behavior and the final products. However, the process of coke deposition is very slow and it usually takes hours of physical time for the catalysts to reach the optimal level of coke content. So, to accurately model the whole process of MTO reaction in a reactor, both the accuracy and efficiency of the simulating method are required. The discrete particle method (DPM) is adopted to accurately model the catalysts at the particle scale. A coarse-grained particle method is developed based on the energy-minimization multiscale (EMMS) model to reduce the computational cost of the particles. Furthermore, the computation of particles is accelerated with the graphics process unit (GPU), and the computational ability could be extended to hundreds of millions of computational particles. The simulation of 8 hours of physical time of a MTO reactor is achieved. The final productions are well accordance with the experimental results. The whole process of the coke deposition and the details of the movements of catalytic particles are obtained, which provides more information for scaling-up MTO reactors.

Predicting the Residence Time Distribution in Large-Scale Continuous Apparatuses Using Recurrence CFD

Authors:
Kieckhefen, P., Hamburg University of Technology
Höfert, M., BASF SE
Pietsch, S., Hamburg University of Technology
Heinrich, S., Hamburg University of Technology

In the design of continuous processes, the residence time distribution is a critical criterion. It quantifies both mixing and exposure of the material to the process at hand, thus correlating with product quality and homogeneity. The state of the art is the experimental measurement which is both costly and not suitable for rapid design iterations. Steady-state CFD simulations allow for the numerically cheap calculation of residence time distributions, but neglect inherently transient, but recurrent, mixing patterns. If the patterns occur on a sub-grid length scale, they can either be modelled, as frequently done in RANS simulations or be resolved in transient simulations that are computationally demanding. When the time scale of these mixing patterns is much faster than the time scale of residence times, their resolution by transient simulation may prove too numerically expensive to yield a true residence time distribution. If the length scale of the relevant, recurring mixing pattern is resolved by the grid and by orders of magnitude faster than the mean residence time, neither a transient simulation nor modelling in analogy to RANS are feasible.

Recurrence CFD (rCFD) offers an opportunity to time-extrapolate global recurrent mixing patterns in a physically consistent manner while achieving speedups of up to three orders of magnitude. This enables the calculation of true residence time distributions after capturing the system dynamics from a fully-resolved simulation.

In this study a pilot-scale spouted bed too large to simulate residence time distributions directly is considered, showcasing the utility of the method. CFD-DEM simulations were sampled and time-extrapolated using rCFD to derive residence time distributions. These were compared to experimentally obtained distributions. The influence of additional stabilizing draft plates was assessed to gauge their value in shaping the residence time distribution.

Eulerian Charge Model for Gas-Solid Flows with Bi-Disperse Particles

Authors:
Ceresiat, L., Heriot Watt University
Ozel, A., Heriot-Watt University
Kolehmainen, J., Princeton University

ranular materials acquire electrostatic charges after coming into frictional contact with themselves or with other materials. This phenomenon is called â€˜triboelectric chargingâ€™ or â€˜contact electrificationâ€™ [1]. It can be built in daily based experiences such as a plastic comb through hair and also naturally observed in sand storms and ash plumes of volcanic eruptions. Furthermore, triboelectric charging has significant implications in gas-solid flow systems such as spark discharge in pneumatic conveying of particles or wall sheeting in fluidized beds [2]. In these processes, the hydrodynamics is significantly altered by charge build-up and vice-versa. It is crucial to have better understanding of interplay between triboelectric charging and hydrodynamics to avoid overall inefficiencies of processes, hazard and safety issues for industrial practitioners.

In our recent papers, we studied gas-solid flows with tribocharging through CFD (Computational Fluid Dynamics) â€“ DEM (Discrete Element Method) approach in vibrated and fluidized beds [3,4]. In the CFD-DEM approach, the flow solver was coupled with a finite-volume based particle-particle and particle-mesh type Poisson solver for the electric field [5] and Laurentie et al. model [6] for charge transfer. We showed that the predicted mean charge was in good agreement with experimental data at different humidity levels for the vibrated bed and different fluidization behaviours seen at low and high humidity levels were well captured for the fluidization case. As CFD-DEM simulations are limited to relatively small systems, we have developed a kinetic-theory based transport equation for mean charge coupled with a two-fluid model [7] to investigate interplay between hydrodynamics and tribocharging in large-scale flow systems.

In this study, as a continuation work of Eulerian modelling of charge transfer, we have derived the transport equation for charge transfer of bi-disperse particles. The transport equation has been derived from the Boltzmann equation assuming Maxwellian distributions for particle velocities and charges. The charge collision rule accounts for the electric field as the collision kernel of the Boltzmann equation is integrated. The charge transport model predictions are then assessed through comparison with hard-sphere Eulerian-Lagrangian simulations.

References

[1] Jones, T.B., King, J.L. and Yablonsky, J.F., 1991. *Powder handling and electrostatics: understanding and preventing hazards*. CRC Press.

[2] Hendrickson, G., 2006. Electrostatics and gas phase fluidized bed polymerization reactor wall sheeting. *Chemical Engineering Science*, *61*(4).

[3] Kolehmainen, J., Sippola, P., Raitanen, O., Ozel, A., Boyce, C.M., Saarenrinne, P. and Sundaresan, S., 2017. Effect of humidity on triboelectric charging in a vertically vibrated granular bed: experiments and modeling. *Chemical Engineering Science*, *173*.

[4] Sippola, P., Kolehmainen, J., Ozel, A., Liu, X., Saarenrinne, P. and Sundaresan, S., 2018. Experimental and numerical study of wall layer development in a tribocharged fluidized bed. *Journal of Fluid Mechanics*, *849*.

[5] Kolehmainen, J., Ozel, A., Boyce, C.M. and Sundaresan, S., 2016. A hybrid approach to computing electrostatic forces in fluidized beds of charged particles. *AIChE Journal*, *62*(7).

[6] Laurentie, J.C., TraorÃ©, P. and Dascalescu, L., 2013. Discrete element modeling of triboelectric charging of insulating materials in vibrated granular beds. *Journal of Electrostatics*, *71*(6).

[7] Kolehmainen, J., Ozel, A. and Sundaresan, S., 2018. Eulerian modelling of gasâ€“solid flows with triboelectric charging. *Journal of Fluid Mechanics*, *848*.

3D Full-Loop Simulation of the Effect of Air Distribution on Gas-Solid Flow Uniformity in a Pilot-Scale CFB Apparatus

Authors:
Yang, Z., Institute of Engineering Thermophysics, Chinese Academy of Sciences
Song, G., Institute of Engineering Thermophysics, Chinese Academy of Sciences
Na, Y., University of Chinese Academy of Sciences
Yang, S., University of Chinese Academy of Sciences
Xiao, Y., Institute of Engineering Thermophysics, Chinese Academy of Sciences

The primary air distribution in the wind box of a circulating fluidized bed (CFB) boiler is one of the important factors that influence the bed material fluidization and coal combustion efficiency. To investigate the air distribution on gas-solid flow characteristics and to improve the gas-solid flow uniformity in a CFB, three simulation cases were implemented based on a three-dimensional (3D) full-loop pilot-scale CFB apparatus. The furnace width is 1.137 m and the depth is 0.392m. In the simulation, the air distribution plate was replaced by a baffle that allowed air to pass through and prohibited the passage of particles. The air distribution plate was divided into two parts: a central region and a surrounding region. The resistance of the central region was set as 2.5 kPa, while the resistances of the surrounding region were set as 2.5, 2, and 1.5 kPa, respectively, in three simulation cases. The simulation results showed that when the resistances of the air distribution plate in the central and the surrounding region were the same, the air velocity near the air distribution plate in the wind box exhibited a saddle profile in width direction and an inverted U-type profile in depth direction. The particle volume fraction near the air distribution plate in the furnace showed a W-type shape profile in width direction and U-type profile in depth direction. With the decrease of surrounding region resistance of the air distribution plate, the average air velocity of the central region of the wind box near the air distribution plate decreased, while that of the surrounding region increased. Correspondingly, the particle volume fraction in the central region of the furnace increased, while that in the surrounding region decreased. The uniformity of particle volume fraction was the best when the resistance of surrounding region was 1.5 kPa.

Theoretical Analysis of the Particle Collision Model for Multi-Phase-Particle-in-Cell (MP-PIC) Calculation of Dense Particle Flows with Application to Simple Shear Flow and Bi-Disperse Particle Settling

Authors:
Simonin, O., Université de Toulouse, CNRS-Toulouse
Fede, P., Université Paul Sabatier

The purpose of this communication is to present a theoretical analysis of collision model for multi-phase-particle-in-cell (MP-PIC) method by comparison of the first order particle velocity moment equations (number density, mass, momentum and second-order velocity correlation) derived from the corresponding PDF equation with the ones derived in the frame of kinetic theory of granular media.

The solid phase prediction in the frame of the MP-PIC method (Andrew and Oâ€™Rourke, 1996) is based on the Lagrangian computation of stochastic numerical particles (parcels) in order to solve a kinetic (or Liouville) equation of the particle distribution function , if particles have a distribution of velocity ( ) and mass ( ). One of the main advantage of such an approach, compared to more standard Eulerian particle moment approach derived in the frame of kinetic theory (Gidaspow, 1986) is that stochastic Lagrangian approaches are meshless and lead to very few numerical diffusion effect. In particular, they allow to predict (for an affordable cost) the small particle flow structures (clusters) which are known to play a crucial role in the macroscopic hydrodynamic behavior of fluidized beds (Agrawal et al., 2001).

In dilute flows, for solid volume fraction less than 1 %, particle-particle collision effect can be accounted for by using Monte Carlo (DSMC) method (see for example, Fede et al. ; 2015). For larger solid volume fraction, the DSMC method is no longer valid and, according to Andrew and Oâ€™Rourke (1996), particle interaction effect may be accounted for using an additional contribution in the particle acceleration term of the kinetic equation which represents the averaged effect of the surrounding particles. This term was originally written to account only for the collisional stress in the momentum equation (Andrew and Oâ€™Rourke, 1996) neglecting the dissipation and redistribution effect due to inter-particle collision which plays a crucial role, in non-homogeneous flows, on the particle fluctuating velocity variances and on the effective kinetic viscosity predictions (Boelle et al., 1995) and on the particle velocity in polydisperse flows (Gourdel et al., 1999). More recently, Oâ€™Rourke and Snider (2010) proposed to add a collision damping term which tends to equilibrate the numerical particle velocities to the local mass-averaged particle velocity with a collision damping time scale determined in order to give the correct decay rate of the fluctuating kinetic energy. Indeed, the expression of the collision damping term is written by comparison of the fluctuating kinetic energy equations derived from the corresponding MP-PICS kinetic equation with the ones derived in the frame of kinetic theory of granular media.

However, it is interesting to notice that the collision damping term added in the PDF equation leads to an averaged momentum and fluctuating kinetic energy transfer term between particle species in polydisperse solid mixture as shown by kinetic theory and Discrete Particle Simulation (DPS) results (Gourdel et al., 1999). Unfortunately, according to the proposed determination of the collision damping term, these momentum and energy transfer effects between particle species are equal to zero when the particle to particle collision are elastic while they should reach maximum values. Similarly, it can be shown that the redistribution effect between the separate components of the particle kinetic stress (or second-order velocity correlations) is not correctly accounted for by the proposed approach and is equal to zero for elastic particles, in contrast with kinetic theory of granular media and DPS results (Boelle et al., 1995 ; Parmentier and Simonin, 2012).

Detailed analysis of the collision terms appearing in the mean velocity and fluctuating kinetic energy, equation derived separately for each particle species in a bi-disperse mixture, and in the particle kinetic stress transport equations, in a mono-disperse granular flow, shows that the collision damping term cannot account for both dissipation and redistribution effect. Therefore, an additional collision term corresponding to a diffusion effect in the PDF equation, and to a Wiener process in the stochastic Lagrangian treatment of the numerical particles, is proposed. The explicit form is derived in order to satisfy both the dissipation and the redistribution effect for the particle kinetic stress tensor components according to kinetic theory of granular media. Then the corresponding collision transfer terms (averaged momentum and fluctuating kinetic energy) between particle species for a bi-solid mixture are derived and compared with the kinetic theory closure model (Gourdel et al., 1999).

References

Agrawal, K., Loezos, P.N., Syamlal, M., Sundaresan, S. 2001, â€œThe role of meso-scale structures in rapid gasâ€“solid flowsâ€• , J. Fluid Mech, Vol. 445, pp 151â€“185.

Andrews, M.J., Oâ€™Rourke, P.J., 1996, "The multiphase particle-in-cell (MP-PIC) method for dense particulate flows", Int. J. Multiphase Flow, Vol. 22, pp 379-402.

Boelle, A., Balzer, G., Simonin, O., 1995, "Second-Order Prediction of the Particle-Phase Stress Tensor of Inelastic Spheres in Simple Shear Dense Suspensions", in Gas-Particle Flows, ASME FED, Vol. 228, pp 9-18.

Gidaspow, D., 1994, Multiphase Flow and Fluidization Continuum and Kinetic Theory Descriptions, Academic Press, Boston, USA.

Gourdel, C., Simonin, O., Brunier, E., 1999, "Two-Maxwellian Equilibrium Distribution Function for the Modelling of a Binary Mixture of Particles", Circulating Fluidized Bed Technology VI, Proc. of the 6th Int. Conference on Circulating Fluidized Beds, J. Werther (Editor), DECHEMA, Frankfurt am Main (Germany), pp. 205-210.

Oâ€™Rourke, P.J., Snider, D.M., 2010, "An improved collision damping time for MP-PIC calculations of dense particle flow with applications to polydisperse sedimenting beds and colliding particle jets", Chem. Eng. Sci. 65, 6014â€“6028.

Parmentier, J.F., Simonin, O., 2012, â€œTransition models from the quenched to ignited states for flows of inertial particles suspended in a simple sheared viscous fluidâ€• , J. Fluid Mech., Vol. 711, pp 147-160.

Snider, D.M., Clark, S.M., Oâ€™Rourke, P.J., 2011, "Eulerian-Lagrangian method for three-dimensional thermal reacting flow with application to coal gasifiers", Chem. Eng. Sci., 66, 1285-1295.

CFD Modelling of Hydrodynamics in a Dual Circulating Fluidized Bed: Effect of Drag Models

Authors:
Li, S., the University of New South wales
Shen, Y., University of New South Wales

Chemical looping combustion (CLC) is an alternative to conventional oxy-combustion technologies, and a dual circulating fluidized bed (composed of two risers, two pot-seals, two cyclones and two downcomers) is one promising CLC process. In this work, a CFD model is developed to describe the complex behaviour in an interconnected dual fluidized bed based on the Eulerianâ€"Eulerian two-fluid model. In particular, the effect of various drag model on simulations results is investigated in aspects of pressure, particle distribution, and oxygen carrier recirculation rate. The drag laws used in this study are Gidaspow (1994), EMMS (1994), Rong et al. (2013), Tang et al. (2016). Both qualitative and quantitative results indicated that the drag model had a significant effect on internal flow behaviour. The results showed the EMMS drag model was found to achieve comparable predictions compare to the measurements in dual circulating fluidized bed, while other drag models can only predict certain fluidization regimes reliably. This work provides a guideline for choosing appropriate drag models for simulating oxygen carrier in dual circulating fluidized bed CLC system.

Experimental and Numerical Study of the Solid Circulation Rate in a Interconnecting Dual Fluidized Bed

Authors:
Lyu, K., Hamburg University of Technology

The solid circulation rate is extremely important in a connecting dual fluidized bed and/or looping system. In this paper, the experimental research has been done in a dual fluidized bed cold model and a computational particle fluid dynamic ï¼ˆCPFDï¼‰model has been setup and analyzed. It evaluates the dynamic fluidization characteristics of a binary system of olivine and char, mainly about the effects of fluid gas superficial velocity in the riser Q_1, return fluid superficial velocity Q_2 and loose fluid superficial velocity Q_3 in the loop seal on the binary materials circulation rate in the looping system. The influence of three different time steps and two different grids on the sensitivity of the computational model has been analyzed, respectively. The numerical model were successfully valid using the solid circulation rates in the experimental results. The well valid numerical model provides reference of theoretical and experimental for the hot circulating fluidized bed operation.

Investigating the Effects of Particle's Parameters and Operating Conditions on the Flow Hydrodynamics in a Wurster Type Fluidized Bed Based on DEM-CFD Simulation

Authors:
Hosseinipaeinkoulaei, S., Chinese Academy of Sciences
Wang, H., University of Chinese Academy of Sciences

Wurster type fluidized beds are widely used in the pharmaceutical industry in which incorporation of various sub-processes such as wetting, granulation, and circulation of particles for coating tablets and small particles. During the coating and granulation processes, numbers of parameters can influence the flow hydrodynamic behavior which plays an important role to control the coating quality. In this research, a DEM-CFD (Discrete Element Method, Computational Fluid Dynamics) coupled model is used to develop an understanding of the effects of particleâ€™s parameters including size, density, and moisture and operating conditions including temperature and velocity on the flow hydrodynamic behavior in a Wurster type fluidized bed coater. For this purpose, the drying and wetting conditions in a laboratory scale Wurster fluid bed coater are modeled and key process parameters including particle circulation, particle Cycle Time Distributions (CTD), Residence Time Distributions (RTD), the average solids concentration in the chamber, and the solid concentration in a cross-sectional area are given and analyzed. By designating appropriate operating and particleâ€™s parameters, the CTD and concentration in the chamber and cross-sectional area can be controlled. Based on the CFD simulation results, the optimum range of the operating condition and the particleâ€™s parameters for the Wurster fluidization process is discussed.

Evaluation of Particle Material Property Using Planetary Ball Mill

Authors:
Lau, R., Nanyang Technological University

This study aims to use planetary ball-milling as an evaluation tool of material properties and subsequent use to develop a model for the change in particle size distribution (PSD) during fluidization. It is believed that material properties such as hardness, density, brittleness, structure, etc play a crucial role in the particle attrition behavior. Unfortunately, little information on material properties is available, considering the wide variety of materials presence. As a result, planetary ball-milling is proposed as a fast assessment technique to identify the properties of different materials. Planetary ball milling devices are readily available in most laboratories and the reduction in PSD can resemble the particle attrition process during fluidization. A Rosin-Rammler (RR) distribution was used to describe the PSD for both fluidization and ball milling processes.

Experimental Study on the Hydrodynamic Characteristics and Flow Structures of Slugging Fluidized Bed with Electrical Capacitance Volume Tomography Measurement

Authors:
Wang, D., The Ohio State University
Pottimurthy, Y., The Ohio State University
Fan, L. S., The Ohio State University

Slugging fluidization, which mainly occurs in the bed with large height-to-diameter aspect ratio, represents one of the major gasâ€“solid fluidization regimes. There are two different kinds of axisymmetric slugs based on the specific particle size employed, i.e., round-nosed slugs and square-nosed slugs. Round-nosed slugs occur when fine particles are used as the fluidized particles. Square-nosed slugs occur when coarse particles are used in a small diameter bed, resulting in significant particle bridging and â€œlockingâ€• effects between particles and the bed wall. For both kinds of slugging fluidized bed, the friction between the solid particles and the wall plays a significant role to the pressure drop through the bed. Due to the limited availability of the experimental measurement methods, characterization of the slugging phenomena was mainly conducted through visualization by measuring the slug rise velocity, slug length, inter-slug spacing, and slug frequency. Pressure drop measurements were normally used to reconstruct the cross-sectional average gas-solids volume fraction profiles in the slugging fluidized bed. These methods normally canâ€™t capture the detailed characteristics of the slugging fluidized bed. For example, due to the difficulties of identifying the contribution portion of friction to the pressure drop of the bed under different conditions, the pressure-drop derived solids volume fraction in the solids-dense region is generally larger than the actual value. With recent advances in the sensor development and image construction techniques for the electrical capacitance volume tomography (ECVT) for 3-D multiphase flow imaging, it is now possible to use the ECVT to obtain the reconstructed 3-D image of multiphase flow phenomena in real time. In this study, three-dimensional ECVT sensors were applied to imaging the flow structures of gasâ€“solids slugging fluidized beds. The detailed 3-D images were obtained and analyzed to reconstruct the real time 3-D gas-solids volume fraction profile of the bed. The entire evolving process of the slugging fluidization from its formation to breakage was analyzed. The solids volume fraction profiles of the slugging bed were plotted under different gas velocity conditions. Particles with different sizes were used in the experiments to investigate the different hydrodynamic characteristics of round-nosed and square-nosed slugging beds.

Forces on an Immersed Horizontal Baffle Made of Multiple Inclined Slats in Fluidized Beds

Authors:

Liu, D., China University Of Petroleum ,Beijing

Zhang, S., China University of Petroleum,Beijing

Zhang, Y., China University of Petroleum,Beijing

Baffles made of multiple inclined slats are often used in fluidized bed reactors to improve gas-solids contact and optimize residence time. However, the baffles have to suffer continuous dynamic forces induced by vigorous bubble and particle motion in beds. The information on the dynamic forces acting on immersed baffles is very helpful to their reasonable strength design. In our previous study[1], forces on a single horizontal slat were investigated systematically, which is usually placed in the bed center in experiment. For the horizontal baffles made of multiple slats used in fluidized bed reactors, information on forces acting on slats at different lateral positions and the influence of major geometrical parameters on forces on slats are more interested. In this study, distribution of forces on different slats of an immersed horizontal baffle made of multiple inclined slats and the influence of slat pitch were measured in detail.

The experiment was conducted in a cold fluidized bed with a square cross-section of 300 mm×300 mm. The horizontal baffle made of multiple slats with an inclined angle of 45° were inserted in the bed at height of 500 mm from the bottom distributor. One of the inclined slats is test slat that adhered strain gauges on its surface. The bed material was coarse silica sand belonging to Group B particle. The force distribution of slats located at different lateral positions and effect of slat pitch were investigated by moving the position of the test slat and changing the number of inclined slats in the experiment.

The distribution of forces on the different slats shows the forces on the slat located far from the oriented side of the baffle were much larger than that at other positions, which resulting from guide fuction of the baffle. The forces on the test slat located in bed center with increasing slat pitch indicates that forces on the slats made up of the baffle significantly decreased when the slat pitch was less than 71 mm, especially at high superficial gas velocity. However, the forces increased a little when the pitch was larger than 71 mm. The forces on the slat with a small distance of 35 mm was about 1/3 of forces on a single slat immersed in a freely fluidized bed (i.e. L_s=104 mm). This result demonstrates that decreasing the distance of inclined slats is an effective measure to reduce dynamic forces on the horizontal baffles immersed in fluidized beds.

[1] D. Liu, S. Zhang, R. Wang, Y. Zhang. Dynamic forces on a horizontal slat immersed in a fluidized bed of fine particles. *Chemical Engineering Research & Design*, 117 (2017): 604-613.

Adsorption of Volatile Organic Compounds with Multistage Fluidized Bed

Authors:

Song, W., Sino-Danish College, University of Chinese Academy of Sciences,

WANG, L., Sino-Danish College, University of Chinese Academy of Sciences,

DU, L., Institute of Process Engineering, Chinese Academy of Sciences

Zheng, J., Institute of Process Engineering, Chinese Academy of Sciences

Volatile Organic Compounds (VOC) are the most common air pollutants emitted from different industrial process. A multistage fluidized bed adsorption/desorption process coupled with catalytic oxidization was proposed for VOC emission control. The experimental and modeling study for VOC adsorption/desorption process in multistage fluidized bed was carried out. A 20000 m^3/h capacity system for VOC control was also constructed and successful demonstrated.

The influence of bed height on the VOC adsorption process in one stage fluidized bed is investigated with different gas velocity and solid flow rate. The increase of bed height results in the increase of gas phase efficiency and VOCs removal efficiency. When the bed height is 50 mm, the gas phase stage efficiency and VOCs removal efficiency reaches the highest value of 85% and 79% respectively.

The effects of gas velocity, gas/solid ratio and total stage number on the performance of multistage fluidized bed adsorption/desorption process have also been tested. The results show that the gas phase stage efficiency is decreasing with the stage from top to bottom. The stage efficiency and VOC removal efficiency are increasing with the decrease of gas-solid ratio. For the desorption process, the VOC concentration in the concentrated stream is more than ten times higher than that of adsorption stream and that is suitable for the following catalytic oxidation process.

The equilibrium model for multistage fluidized bed was established to describe the VOC adsorption/desorption process. The simulated result shows good agreement with the experimental results. The model is used in the design and operation of the industrial demonstration project.

Numerical Analysis of Electrostatic Force and Electric Potential Distribution in Charged Gas-Solid Fluidized Bed

Authors:

Tan, Z., Institute for Process Modelling and Optimization

Liang, C., Southeast University

Chen, X., Southeast University

Electrostatics occurs in the gas-solid fluidization processes, which influences the hydrodynamics of gas-solid flow. Numerical simulation has been applied to study the electrostatic phenomena in gas-solid flow in recent years, including Eulerian-Eulerian and Eulerian-Lagrangian approaches. However, few detailed research has investigated the influence mechanism of electrostatics on the hydrodynamics of gas-solid flow. In this work, a hybrid approach is applied to study the electrostatics in gas-solid fluidized bed which consists of two parts: (1) Computational Fluid Dynamics-Discrete Element Method (CFD-DEM). (2) The electrostatic field model. DEM is implemented to analyze the electrostatic force acting on the particles. The magnitude and direction of the electrostatic force is obtained by solving Coulombâ€™s law directly. The electrostatic field model is used to obtain the electric potential distribution. The model is described by the electric potential scalar equation which is derived by Maxwellâ€™s equations and the Bruggemanâ€™s equation.

Hydrodynamics of a Novel High-Speed MTO Loop Reactor: Profiles of Solids Concentration and Circulation Rate

Authors:

Wang, F., China University of Petroleum(Beijing)

The methanol-to-olefins (MTO) is an important reaction for producing basic petrochemicals from nonoil resources such as coal and natural gas. The reactor design plays a vital role for the hydrodynamics of gas-solid phase, which relate closely with the conversion and selection of reaction. A novel high-speed MTO loop reactor (turbulent or fast fluidized bed upward flow in the draft tube in parallel with bubbling or turbulent bed downward flow in the annulus) was developed aiming at improving the flow characteristics of gas-solid phase for MTO reaction. The profiles of solids concentration in the four regions (the bottom region, the draft tube region, the annulus region and the gas-solid separation region) and the solids circulation rate between the annulus region and the draft tube region were investigated in the present paper. They were measured by the optical fiber probe. The fiber probe was placed at different axial and radial positions in the four regions. An empirical relation of local and average solids concentration within the draft tube region was obtained based on the experimental results. Moreover, a model about the solids circulation rate was established on the basis of energy equilibrium principle.

Experimental Investigation of High-Temperature Fluidization Using Endoscopic Particle Image Velocimetry-Digital Image Analysis Technique (ePIV/DIA)

Authors:

Cruellas, A., Eindhoven University of Technology

Gallucci, F., Eindhoven University of Technology

Sint Annaland, M. V., Eindhoven university of Technology

ABSTRACT

Enhanced mass and heat transfer, resulting in a more homogeneous heat and mass distribution, makes the use of fluidized bed reactors an interesting alternative to conventional configurations such as packed bed reactors, especially for processes operated at elevated temperatures. However, detailed studies on the hydrodynamics of fluidized bed reactors at high temperatures are still scarce, and a deeper understanding of the influence of the temperature on the fluidization behaviour is required for the modelling and optimization of high-temperature fluidized bed reactors.

The main objective of this work is to investigate high-temperature fluidization experimentally, using the novel non-intrusive ePIV/DIA technique. This technique combines endoscopic particle image velocimetry with digital image analysis, with which detailed information on both the gas and solid phases of the fluidized bed can be obtained simultaneously with high spatial and temporal resolution at the required high operating temperatures.

Using a fluidized bed consisting of zirconia particles of 400-600 mm diameter (Geldart B type), ePIV/DIA experiments have been carried out up to a temperature of 500 °C, keeping the excess velocity constant using the measured minimum fluidization velocity at the respective temperature. From the experimental data, detailed information on the solids circulation patterns and bubble properties as function of the position in the bed at different temperatures have been obtained.

A very strong influence of the operating temperature has been observed. The effect of particle raining increases significantly at elevated temperatures, especially at 400-500 °C, affecting considerably the hydrodynamics of the system. The increase in the extent of particle raining at higher temperatures results in a decrease in the average bubble diameter, but also in an increase in the number of bubbles recorded per image. The larger number of smaller bubbles at higher temperatures improves the fluidization behaviour, since it enhances the gas-solid contact and reduces the gas by-pass through the bubbles. However, it has also been observed that at elevated temperatures the solid fluxes in the bed are smaller, most probably also related to the increase in particle raining reducing the upwards solids movement. It should be noted that decreased solid fluxes could result in reduced solids mixing, consequently lowering the quality of the fluidization.

The increased particle raining at elevated temperatures can be caused by the decrease in gas density resulting in a decrease in the hydrodynamic forces, or because of the increase in the importance of interparticle forces, as suggested by different authors in the literature. In this work, the relative importance of the hydrodynamic and interparticle forces have been analysed using the obtained detailed hydrodynamic data as a function of temperature.

ACKNOWLEDGEMENTS

This project has received funding from the European Unionâ€™s Horizon 2020 research and innovation programme under grant agreement No 679933 (MEMERE project).

The present publication reflects only the authorâ€™s views and the European Union is not liable for any use that may be made of the information contained therein.

Coal Separation Intensification and Bed Density Computational Model By Using a Binary Mixture Dense Media Gas Solid Fluidized Bed

Authors:

LIU, C., China University of Mining and Technology

Zhao, Y., China University of Mining and Technology

He, J., China University of Mining and Technology

Abstract: The binary dense media fluidized bed separator was used to realize the separation intensification of the coal. The mixture of the magnetite powder and the fine coal with the mean diameter of 187.5Î¼m and 750Î¼m for different fine coal weight proportion of 0%, 4%, 8%, 12%, 16%, 20% was used as dense media for fluidization and separation experiments. The computational model between the incipient fluidized velocity U_{mf} and the weight proportion of the fine coal $x_c$was achieved. The bed density computational model in the binary dense media gas-solid fluidized bed was derived based on the two-phase fluidization theory with experimental verification. The stability of bed density in the binary dense media gas-solid fluidized bed was also studied: the optimal gas velocities were almost the 1.3 or 1.4 times of the incipient fluidization velocities for different fine coal weight proportions with the corresponding bed densities of 2.21, 1.97, 1.85, 1.78, 1.66 and 1.54 g/cm^3, respectively. The satisfactory yield and ash content of clean coal and the optimum probable error E_p value of 0.065 or 0.07 g/cm^3 was achieved when the weight proportion of the fine coal less than 16%, which indicated the separation performance is notably intensified by using the binary dense media fluidized bed.

High Recovery of Rees through Acid Baking of Their Bearing Minerals in a Fluidized Bed Reactor

Authors:

Latifi, M., Ecole Polytechnique Montreal

Chaouki, J., Ecole Polytechnique Montreal

There is a fast growth in new applications of Rare Earth Elements (REE), especially, in energy, environment and high technology fields where durability, high efficiency and low carbon emissions are required. The REE are called â€œrareâ€ not because their abundance is low but because these elements are found in the deposits from which extraction is extremely difficult. Also, due to the fact that the REE have similar chemical properties, there was no known technique to produce these elements in individual forms until the twentieth century. **Due to serious difficulties in the REE extraction, the total process would include a complex network of unit operations**.

A REE production process from the mine to the final oxide form includes geology, mining, ore physical beneficiation, baking, hydrometallurgy and separation of individual elements. A concentrated ore is usually produced through a physical beneficiation process. Afterwards, it undergoes an acid baking process to crack the heavy molecules of the REE minerals as well as the associated gangues in order for easier extraction of the REE from undesired elements in a downstream hydrometallurgical process.

Being the most common process, baking with sulfuric acid at a temperature below its boiling point is carried out in rotary kilns; however, it suffers from serious challenges such as mass transfer limitations (i.e. causing high consumption of acid) and material handling due to stickiness of material inside the reactor.

Authors, have developed and commissioned a new fluidized bed reactor where a bed of concentrated REE ore particles (with average Sauter mean diameter of 100 Î¼m) is fluidized at a temperature higher than sulfuric acid boiling point (~ 350 °C) where acid is atomized into the bed and the vapor gets in contact with ore particles. Bastnaesite and monazite are the main REE bearing minerals of the ore. Upon optimization of the superficial gas velocity, acid consumption declines, the bed stays being fluidized (with negligible detrimental effect of particles agglomeration) whereas conversion of REE bearing minerals to leachable REE sulfates is above 95%.

Gas Solids Phase Separation in Multi-Stage Fluidized Beds

Authors:

Zhang, C., Tsinghua University

WANG, Y.

Qian, W., Tsinghua University

Wei, F., Tsinghua University

Back-mixing of gas and solids can be prevented effectively in both dense and dilute fluidized beds with baffles, resulting from the phase separation, i.e. the formation of â€˜gas cushionâ€™ in dense system or â€˜solids cushionâ€™ in dilute system, across baffles. In present work, the equilibrium Mach number (Ma_e) is introduced here as the indication of dense and dilute phase in fluidization. Then, gas solids flow across a baffle is treated as equilibrium flow of gas solids mixture through a nozzle. And the principle of gas solids separation through a nozzle can be reasonably explained by a diagram illustrating the effect of throat area on Ma_e. Furthermore, various applications with baffles in fluidization are discussed in light of our model.

Impact of Fluidized Bed Hydrodynamics on the Distribution of Liquid Sprayed into the Bed

Authors:

Li, Y., Western University / Institute for Chemicals and Fuels from Alternative Resources (ICFAR)

Briens, C., Western University / Institute for Chemicals and Fuels from Alternative Resources (ICFAR)

Berruti, F., Western University / Institute for Chemicals and Fuels from Alternative Resources (ICFAR)

Sanchez Careaga, F., Western University / Institute for Chemicals and Fuels from Alternative Resources (ICFAR)

McMillan, J., Syncrude Canada Ltd

The Fluid Coking Technology™ is one of the most advanced upgrading processes for heavy oil reserves. In Canada, for example, it is used to upgrade around 15-20 % of heavy oil production to a light, synthetic crude. In a Fluid Coker, preheated bitumen (350 â„ƒ) is sprayed through steam-gas atomization spray nozzles into a fluidized bed of hot coke particles (500-550 â„ƒ) to be thermally cracked to condensable vapors, non-condensable gases and solid coke. However, when spraying the bitumen into a bed of hot coke particles, a portion of the liquid is trapped in wet agglomerates formed with the solid coke particles. Coke particles flowing down from the spray region go through a stripper section before being conveyed to a burner vessel, where they are reheated. Wet agglomerates entrained with the particles flowing through the stripper are responsible for fouling of the stripper sheds, which causes major operating problems. It is, therefore, essential to minimize the formation of wet agglomerates.

This research focuses on reducing agglomerate formation and enhancing agglomerate breakup by modifying the bed hydrodynamics in the spray region. It investigates the impact of changes in the distribution of gas bubbles that can be achieved by either modifying the distribution of the fluidizing gas or adding internals such as baffles.

This study uses a safe system that has been found to simulate agglomerate formation and breakup in a Fluid Coking process: an aqueous solution of Gum Arabic is sprayed into a fluidized bed of sand particles maintained at 120 °C. Experiments were conducted in a column with a rectangular cross-section (1 m × 0.5 m × 2.28 m). Liquid was atomized into the bed using nitrogen and a scaled-down version of a typical commercial spray nozzle. Liquid was injected for about 11 s, the bed was dried and its particles were screened to recover the agglomerates, which were then sieved to get their size distribution. The recovered agglomerates were analyzed to determine their Gum Arabic concentration, from which the proportion of the injected liquid that was trapped in agglomerates could be determined. Experiments were conducted with superficial velocities ranging from 0.18 m/s to 1.0 m/s.

The flow patterns of the gas bubbles were determined with an array of triboprobes. Triboprobes consist of metal rods connected to the ground. When sand particles carried by the bubble wakes hit a probe, they strip electrons from its surface, inducing an electrical current that can be monitored. The main advantage of triboprobes, when compared to optical fibers or capacitance probes, is that they can be used at high temperature and with high gas velocities.

The gas distributor was equipped with 20 gas nozzles in two parallel rows, whose flow could be controlled individually to modify the gas bubble distribution. Both amplifier and resistance have been applied and compared for triboelectric method. By applying the resistance, the voltage (is recorded as the raw signal. And the Data acquisition is protected with transient-voltage-suppression (TVS) diodes from the damage of high-voltage from the fluidized bed. With this improvement, we greatly lower the cost of recording data from multi local positions contemporaneously.

By changing the gas distributor configuration, opening 10 nozzles each time, three extreme cases of gas distributions are obtained. By opening 10 nozzles near the back-bed wall, an even gas distribution is formed, which is flatter than the gas distribution from a conventional fluidized bed due to bubble coalescence. By opening 10 nozzles close to the western side bed wall, the bubbles would concentrate on the western bed side, which is called western case. Similarly, we can have the eastern case when the bubbles concentrate on eastern side bed wall. For these three extreme cases, the results have been verified with both amplifier and the resistance method.

The gas bubble distribution was also modified by inserting baffles in the fluidized bed. An open-ended right triangle shape with an internal angle of 45 Â° having the same width as the bed (0.18 m x 0.18 m x 0.1 m) is set as the basic baffle geometry. To simulate the frusto-conical baffle having a central aperture applied in industry, two lengths of O-shaped flux-tubes are put at the center of the basic geometry baffle separately; one is the same length as the bottom edge of the baffle, the other one is half length of the full-length flux-tube.

The results from triboelectric method show that all three geometries of baffle would redirect the bubbles to the baffle side. However, thereâ€™s no great impact on the bubble profiles from different baffle geometries, where the measuring level is right above the spray. When applying a basic geometry baffle, by changing the gas distributor configuration, for all the three cases, even, western, and eastern case, the bubbles are redirected to the baffle side, while keeping the even case is in-between of the western and eastern case at spray region.

The results from Gum Arabic method shows that by changing the gas distributor, the amount of water trapped in agglomerates can be reduced, with the best results obtained when the gas is evenly distributed at the spray level. This could be because this configuration both provides a high solids entrainment into the spray jet cavity, near the nozzle tip, and high rate of agglomerates break-up near the spray end.

The basic geometry baffle decreases the portion of trapped liquid for all gas velocities, gas distributor configurations, and vertical distance from the spray nozzle. However, the geometry of the flux-tube baffle needs to be optimized.

Fluidized Bed Combustion of Solid Lignin-Rich Residues from Bioethanol Production

Authors:

Solimene, R., Consiglio Nazionale delle Ricerche

Troiano, M., Università degli Studi di Napoli Federico II

Cammarota, A., Consiglio Nazionale delle Ricerche

Chirone, R., Consiglio Nazionale delle Ricerche

Salatino, P., Università degli Studi di Napoli Federico II

The behaviour of lignin-rich residues from second-generation bioethanol production plant has been investigated during fluidized bed combustion. With the depletion of fossil fuels and the issues related to the emissions of greenhouse gases, the utilization of alternative automotive fuels has attracted worldwide attention. Biofuels are unique among available alternative energy sources in their compatibility with existing liquid transport fuel. The global production and use of biofuels have increased from 18.2 billion liters in 2000 to 60.6 billion liters in 2007, with about 85% of this being bioethanol. The production of second-generation bioethanol, i.e. ethanol produced from lignocellulosic biomass and scrap of agricultural crops, will be of 155 billion liters by 2020. The composition of lignocellulosic biomass is typically: cellulose (35-45%), hemicellulose (25-30%) and lignin (25-30%). Differently from cellulose and hemicellulose, lignin is a polymer mainly consisting of not fermentable phenylpropanoids. Therefore, the solid residues after ethanol distillation and separation in the second-generation bioethanol production, are characterized by high lignin content. These lignin-rich residues can be used to produce chemicals as well as energy. Fluidized bed combustion/co-combustion can be considered as promising and viable options to recover thermal power from lignin-rich residues of a second-generation bioethanol production plant. However, the understanding of the mechanisms of devolatilization, conversion and attrition of particles of lignin-rich residue in fluidized beds is scarce and it deserves further investigations.

This work aims at investigating the combustion behaviour of a lignin-rich residue in lab-scale fluidized beds. In particular, devolatilization, char burnout and comminution phenomena of single fuel particles were studied. Devolatilization experiments aimed at characterizing the devolatilization rate, the average volatiles molecular weight and the devolatilization time of wet fuel particles while varying particle size and operating temperature based on the analysis of time-resolved pressure signals. Fragmentation and char burnout experiments aimed at evaluating the particles fragmentation tendency during the devolatilization (primary fragmentation) and char burnout (secondary or percolative fragmentation) steps. Elutriation by attrition experiments were carried out to estimate the carbon elutriation rate under devolatilization and combustion conditions.

A stainless-steel bubbling fluidized bed reactor 17 mm ID and 1 m high was used for the devolatilization experiments. The reactor consists of a lower section, an upper section and an inter-flange of connection between the two logs. The inter-flange length contains most of the bed material. A thermocouple is inserted in the fluidized bed at about 0.01 m above the high-pressure-drop porous gas distributor plate. A stainless-steel tube is laterally welded to the column for rapid discharge of bed particles. The lower section is used as wind box. Technical nitrogen from cylinders was used as fluidizing gas. The fluidizing gas-flow rate was established by setting the upstream pressure of a critical orifice located along the gas feeding line. The orifice (0.187 mm dia.) was operated in critical flow conditions. The upstream relative pressure varied between 1 and 3.5 bar. The reactor was operated at a pressure slightly larger than atmospheric pressure by means of a calibrated flow restriction at the exhaust consisting of an orifice with an adjustable cross section. The reactor was heated by means of a cylindrical oven (1 kW) 0.4 m high equipped with a PID temperature control system. A high-precision (accuracy better than 0.06 mbar) piezo-resistive electronic pressure transducer with small-time response (less than 1 ms) and 100 mbar full-scale was used to measure the time-resolved gas pressure in the reactor. The pressure signal was acquired by means of a data acquisition unit consisting of an A/D converter and a personal computer.

Fragmentation, elutriation and combustion experiments were carried out in a stainless steel atmospheric bubbling fluidized bed combustor 41 mm ID and 1 m operated at 850°C. Three different configurations of the reactor were used. In the first configuration, used for particle fragmentation experiments, the top section of the fluidization column was left open to the atmosphere. A stainless steel circular basket was inserted from the top to retrieve fragmented and un-fragmented particles from the bed. In the second configuration, used for fines elutriation rate experiments, a two-exit head was fitted to the top flange of the fluidization column. Flue gases were conveyed alternately to two removable filters made of sintered porous alumina. Batches of material were into the bed via a hopper connected sideways to the upper part of the freeboard. A paramagnetic analyzer and two NDIR analyzers were used for online measurement of O_2, CO, and CO_2 concentrations, respectively, in the exhaust gases. The third configuration, used for single particle combustion experiments, also consisted in leaving open the top section of the fluidization column. A stainless-steel probe was inserted from the top of the column to convey a fraction of the exit gases directly to the gas analyzers. Data from the analyzers were logged and further processed on a PC.

Experiments were carried out using a lignin-rich residue coming from a second-generation bioethanol production plant (LHV (as received) =5645 kJ/kg, moisture 60.7 wt%, volatile matter 25%, fixed carbon 8.1%, ash 6.1%). The lignin content was obtained by Klason method. For devolatilization experiments, fuel particles with a diameter from 3 mm to 8 mm were used, while for fragmentation/combustion/elutriation experiments fuel particles sized between 10 and 17 mm were used. Gases used in the experiments were air and nitrogen of technical grade.

Regarding devolatilization experiments, pressure was continuously monitored in the fluidized-bed reactor during the experiments. The selection of the steady-state overpressure in the reactor, dictated by the flow rate of fluidizing gas and by the calibrated flow restriction at the exhaust, determines the time evolution of the phenomena. An overpressure of about 10 mbar was chosen for the tests at 750°C and 850°C, and an overpressure of 20 mbar at 650°C. The bed of sand (400-600 mm, 13 g) was fluidized with nitrogen to rule out the influence of volatile matter and char combustion on the recorded pressure profiles. The upstream pressure at the critical orifice was kept at a constant value of about 1, 1.65 and 2.3 bar at 650, 750 and 850°C, respectively. The mass of each batch of fuel particles was fixed at 100 mg for the sizes 3-5 mm, 135, 160 and 200 mg for 6 mm, 7 mm and 8 mm-sized particles. The tests consisted of the injection of a batch of fuel particles in the fluidized bed followed by complete devolatilization. The pressure signal was continuously logged during the experiments at a sampling rate of 1,000 Hz. Pressure time series were analyzed to determine the volatile emission rate as a function of time. The time for 95% devolatilization degree and the mean molecular weight of the emitted volatile compounds were computed from the raw data. Primary fragmentation experiments were performed using the basket-equipped configuration. Single fuel particles were injected into the bed from the top of the column. After about 3 min, required to completely devolatilize the fuel, the resulting char was retrieved by means of the basket to investigate the number and size of the produced fragments. For secondary fragmentation experiments, single fuel particles were first pyrolyzed under minimum fluidization condition, then, nitrogenâ€“oxygen mixtures (4.5% of O_2 volume concentration) were used as inlet gas. Experiments were carried out at a superficial gas velocity of 0.4 and 0.8 m/s. Char particles were injected into the bed from the top of the column. The char particles were retrieved from the bed at definite time intervals and the number and size of the fragments were recorded. The experimental run was stopped when the char combustion was complete as detected by the time-resolved profiles of concentration of gaseous species measured at the exhaust. Regarding the attrition rate experiments, elutriated fines were collected by means of the two-exit head by letting the flue gas flow alternately through sequences of filters for definite time intervals. Fines collected in the filters were further analyzed to determine their fixed carbon content. This procedure enabled time-resolved measurement of carbon elutriation rates. Furthermore, combustion experiments were carried out at 850°C by injecting single fuel char particles with a gas superficial velocity of 0.4 and 0.8 m/s. Continuous analysis of the exhaust gases enabled measurement of the carbon combustion rate as a function of time until complete burnout was reached.

Results highlighted that devolatilization rate and times were relatively long and increased with decreasing temperatures. Fuel particles did not undergo primary fragmentation, whereas secondary and not percolative fragmentation occurred during the late stage of char burnout. Attrition by surface wear was limited. However, results indicated that carbon loss by attrition increased with the superficial gas velocity and it was more pronounced under inert conditions than under combustion environment.

Directly Irradiated Fluidized Bed Reactor for Thermochemical Energy Storage and CO2/H2O Splitting

Authors:

Tregambi, C., Università degli Studi di Napoli Federico II

Bevilacqua, C., Istituto di Ricerche sulla Combustione, Consiglio Nazionale delle Ricerche

Di Benedetto, A., Università di Napoli Federico II

Solimene, R., Consiglio Nazionale delle Ricerche

Salatino, P., Università degli Studi di Napoli Federico II

Landi, G., Istituto di Ricerche sulla Combustione, Consiglio Nazionale delle Ricerche

Montagnaro, F., Università degli Studi di Napoli Federico II

Extensive research and developments activities are in progress to exploit the huge amount of solar energy falling on Earth. Concentrating Solar Power (CSP) is a fast-growing renewable technology in which the solar energy, upon concentration, is focused onto a receiver whence it is converted to electricity or industrial process heat. One of the great advantages of CSP is the relatively easy integration with thermal energy storage (TES) systems. TES systems allow to decouple the two steps of solar energy absorption and exploitation, enabling dispatchability of solar energy. In this way, it is possible to overcome the intrinsic intermittent behaviour of the solar source, providing onâ€‘demand power delivery even during the night or in not favourable weather conditions. TES systems can rely on sensible heat storage, latent heat storage and thermochemical energy storage (TCES). In TCES the concentrated solar energy is used to perform reversible chemical reactions characterized by high latent heat. During the endothermic step, the solar energy is collected, and the reaction products are separated and stored. Then, when energy is required, the products are put back into contact to release the stored energy. TCES is a quite complex technology, but rewards with high density of energy storage and improved stability over long time-scales, as the incident energy is stored in the noble form of chemical bonds. An ambitious strategy for TCES aims to convert the harmless and inexpensive H_2O and CO_2 molecules into H_2 and CO in twoâ€‘step reaction cycles by using metal oxides systems, which act as catalysts at relatively high temperatures (1000â€“1500 Â°C). In a first endothermic step, the metal oxide is thermally reduced at high temperature, with the consequent release of O_2. The reduced metal oxide is then oxidized back with H_2O/CO_2 to eventually produce H_2/CO and close the cycle. In order to make this process economically feasible and compete with current available technologies, two challenges need to be addressed. A first one â€“ material oriented â€“ concerns the development of a metal oxide owning high reversibility and cyclability, fast reaction kinetics and a not too high temperature for the reduction step. The second one â€“ reactor oriented â€“ aims at developing a feasible and reliable reactor configuration to maximize the conversion of solar energy into solar fuels. Among the investigated materials, cerium oxide is one of the most interesting one, as it is characterized by a high reversibility and fast reâ€'oxidation kinetics. However, its conversion degree for each reaction cycle is low, and very high reaction temperatures (~ 1500 Â°C) are required to fulfil the endothermic reduction step. Since last years, mixed perovskites materials are widely investigated for H_2/CO production as they feature significant redox properties, structure stability and a wide range of formulations.

This work deals with the development of a directly irradiated labâ€'scale fluidized bed reactor targeted at maximizing the collection of solar energy, withstanding the highâ€'concentrated flux typical of highâ€'temperature CSP applications (> 1 MW/m^2) and ensuring a uniform temperature distribution of the reactive material. The reactor is made of two concentric circular columns: the internal one (OD=12 mm, ID=10 mm, length=80 mm) is referred as riser, the external one (OD=21 mm, ID=16 mm, length=70 mm) as annulus. The upper part of the annulus column is connected to a conical section (internal angle=30Â°, height=120 mm), which represents the freeboard of the reactor and hosts, at its extremity, a transparent quartz window to let the solar radiation enter keeping seal-tight operation of the reactor. During the reactor operation, a gas stream is fed into the internal column to induce the rise of a dense particle suspension, typically under fast fluidization conditions. At the outlet of the riser, the dense particle suspension directly interacts with the solar concentrated radiation (simulated by mean of a shortâ€'arc Xe lamp) and increase its temperature. In the freeboard zone, due to the cross-section increase, the gas velocity is progressively reduced, and the solid particles fall down on the conical section and descend along it to eventually convey into the annulus column. Here, the particles move by gravity towards the bottom, where they re-enter the riser by mean of four small holes drilled in the riser column. During their motion along the annulus section, the â€œhotâ€• particles transfer their heat to the riser column, thus preheating the rising dense suspension prior to its interaction with the concentrated solar radiation. This feature allows an autothermal operation of the reactor, enabling the possibility of reaching higher reaction temperatures. Gas exit is provided at the top of the conical section, just below the transparent window, by two small tubes (OD=6 mm, ID= 4 mm). The total inventory of the reactor is in the order of 15â€“25 g. As mentioned above, the solar concentrated radiation is simulated by a shortâ€'arc Xe lamp of 7 kW_{el} coupled with an elliptical reflector. The use of Xe lamps to simulate the solar spectrum is well established in the literature since the differences with the solar spectrum are quite small, especially in the visible spectral range. At full power, a peak flux of 2 MW/m^2 is obtained in the focal point, while the total thermal power supplied to the system is of about 1.8 kW_{th}. Temperature measurements are made by two K-type thermocouples: one at the bottom of the riser (T_{down}), the other at the top of the annulus (T_{up}). The former measure the temperature of the â€œcoldâ€• material before its rise, the latter measure the temperature of the â€œhotâ€• material before it descends along the annulus. It is worth noting that T_{up} does not represent the higher temperature reached by the material, as the highest temperature is probably experienced by the particles during their upward flight in the conical freeboard section. Unfortunately, this temperature cannot be currently measured in an affordable way. Exhaust gas are sent to a mass spectrometer to analyze the concentration of the relevant species, namely Ar, CO, CO_2, N_2, O_2 for the performed experiments.

The experimental campaign on the fluidized bed reactor consisted in: i) thermal and hydrodynamic characterization under inert conditions; ii) characterization of a thermochemical energy storage process; iii) characterization of a CO_2 splitting process.

The thermal and hydrodynamic characterization of the fluidized bed reactor was performed by means of different inert granular material, characterized by different density, colour and particle dimensions. Silica sand, silicon carbide and bauxite with a mean Sauter diameter in the 150â€“300 mm were used as bed materials. Steady state temperature of 900â€“1100 Â°C were reached in the system depending on material and fluidization conditions. Some of the materials exhibited a tendency to agglomerate and sinter during the reactor operation and were then rejected for the subsequent tests. This phenomenon was probably emphasized by the presence of small quantities of impurities which promoted the processes of agglomeration and sintering.

Concerning the thermochemical energy storage process, carbonation and calcination of a Caâ€‘based sorbent was achieved. More into detail, limestone particles were used as bed materials, and several iterated calcination and carbonation reaction cycles were successfully performed.

Finally, the CO_2 splitting process was performed using solid particles of a perovskite material made of a mixture of La and Sr as â€œAâ€□ cation and Fe as â€œBâ€□ cation. To perform these tests, the system was initially operated using an inert material and heated up to 1000 Â°C using air as fluidizing gas. Then, the inlet gas was switched to pure N_2 (99.999%$_v$) and, through a pneumatic system, a small amount (1â€“2 g) of perovskite was fed to the system by one of the two outlet tubes. Exhaust gas was continuously analyzed to follow the O_2 evolution. Once the material was completely reduced (no further O_2 release), the inlet gas was switched to pure CO_2 (99.999%$_v$) so to re-oxidize the perovskite material and produce CO. During this step, the exhaust gas was continuously analysed to monitor the CO concentration. Finally, when no more CO was produced, the inlet gas stream was switched back to pure N_2 to reduce the material again and start a new reaction cycle. The oxidation/reduction cycles were then performed isothermally at a temperature of about 1000 °C. Several iterated reaction cycles were performed and, for each cycle, the amount of O_2 and CO produced were evaluated. A first estimate of thermal/chemical efficiency for the proposed reactor configuration was performed.

CPFD Model Optimization for Simulation of Segregation in a Fluidized Bed of Particles with Different Sizes and Densities

Authors:

Kokourine, A., Hatch

Adham, K., Hatch Ltd.

Many industrial fluidized beds comprise of mixtures of particles with different sizes and densities. In order to achieve uniform fluidization and avoid segregation, the particle feed sizes can be optimized versus the fluidizing velocity. Optimization for different operating scenarios can be achieved by CPFD modeling, if the modeling parameters are fine-tuned based on a proper set of experimental data. Here, reported test data trends of particle segregation are compared with multiple sets of CPFD models, to show that pertinent modeling can reasonably predict the actual behavior of solids mixture, under a range of velocities.

Impact of Column Geometry and Internals on Gas and Particle Flows in a Fluidized Bed with Downward Solids Circulation

Authors:

Cochet, Y., Western University

Briens, C., Western University / Institute for Chemicals and Fuels from Alternative Resources (ICFAR)

Berruti, F., Western University / Institute for Chemicals and Fuels from Alternative Resources (ICFAR)

McMillan, J., Syncrude Canada Ltd

Fluid Coking^{TM} is a process employed by the petroleum industry for thermal conversion of heavy hydrocarbon molecules, such as bitumen from oil-sands, into distillate products. It is particularly important in Canada as synthetic crudes produced in Fluid Cokers represent about 15% to 20% of total oil production. In Fluid Cokers, heavy oil is upgraded into synthetic crude oil through a thermal, non-catalytic cracking process. Industrial Fluid Cokers are composed of two vessels, a fluidized reactor and a burner: in the reactor, heavy oils are sprayed on hot coke particles where it cracks to solid coke and vapors that leave the reactor from the top. Coke particles flow down from the spray region through a stripping section where hydrocarbon vapors are displaced by steam. Cold coke is then conveyed from the bottom of the reactor to the burner where it is reheated by combustion before reinjection in the reactor. A variation of the Fluid Coking technology, FlexiCoking, uses a third vessel to gasify coke.

A major operability issue in industrial Fluid Cokers is the fouling of stripper sheds in the lower reactor section. This fouling is attributed to wet particles and agglomerates that move too quickly from the liquid spray zone to the stripper section, thus carrying unreacted liquid to the stripper section. To minimize the presence of unreacted liquid in the stripper section, Fluid Cokers are currently operated at temperatures that are significantly higher than the optimum temperature for the conversion of bitumen to the desired liquid products, leading to overcracking and a lower yield of valuable products. Therefore, the objective of this study is to determine how reactor geometry and operating conditions can be modified to minimize the presence of unreacted liquid in the stripper section. Different aspects of Coker design and operation are studied, such as column geometry, number and location of spray nozzle banks; and number and location of baffles.

The study is conducted in a cold recirculated fluidized bed, with a circular cross-section. The bed of coke particles is fluidized with air and operated at room temperature. The hydrodynamics of Fluid Cokers are greatly influenced by the hydrocarbon vapors evolving from the lateral injections of heavy oil feedstock, at various heights. In this study, the vapors are simulated with air injection, distributed over 5 banks of 8 nozzles each. The design of the bed was chosen to match key parameters of industrial Fluid Cokers, such as the heights of the 5 nozzle banks, the diameter/bed height ratio, and the solids recirculation flowrate. Since industrial units are often tapered, to moderate the increase in superficial gas velocity caused by the feedstock injections, two different, typical taper shapes were studied.

Ring baffles have been proposed to reduce stripper shed fouling, by minimizing the bypassing of wet particles from the injection zone to the stripper region. The impact of ring baffles on the Residence Time Distribution (RTD) of wet agglomerates and particles was, thus, studied with different column geometries and feedstock distribution schemes. Studies were also conducted to evaluate the potential fouling of ring baffles.

The main experimental tool of this study is non-intrusive Radioactive Particle Tracking. A radioactive tracer is encapsulated to achieve the required size and density. The unit is surrounded by 12 scintillation detectors at known locations. For each experiment, the system tracks a single tracer during a long time (18 to 60 hours). These long recording times allow statistically significant analysis of the data to extract information about agglomerates or particles behavior. Geiger counters at the solids inlet and outlet also provide the solids residence time distribution for the whole bed. Radioactive Particle Tracking measurements were analyzed to provide the following information:

- The overall residence time distribution (RTD) of the tracer. These results were compared to the results obtained by injecting dyed coke particles in the solids inlet and sampling particles from the solids outlet. Because, in this system, dyed particles sampled from the solids outlet may have gone more than once through the column, the dyed particles experiments cannot provide the true RTD. In this study, they were used to determine which radioactive tracer best simulates individual particles.
- The distribution of time-spans between the agglomerate formation zones (located in front of each lateral injection) and the stripper. This information is essential to develop strategies that will reduce stripper fouling in Fluid Cokers.
- The flow patterns of agglomerates and particles in the Coker model. This information helps design more effective baffles and nozzle positioning.
- The flow patterns of the gas bubbles. These results were compared to measurements performed with invasive triboelectric probes. This information provides information on vapor flows in cokers.

Experiments to date have shown that the addition of ring baffles, i.e. one baffle below each of the 4 top lateral injection banks, decreases the proportion of short residence time over the whole bed, from the solids inlet to the column to the solids outlet from the column to the recirculation line. For instance, the proportion of tracer going through the whole bed in less than 10 s would decrease from about 7 to 5 % with the addition of the baffles. However, the baffles increase the proportion of agglomerates that go from the zone where agglomerates are produced to the stripper in less than 10 s, in average over the 5 banks, from 14 % to 20 %. Further experiments will investigate how the number, location and geometry of the baffles can be modified to enhance Fluid Coker performance.

Preliminary results have shown that redirecting all the liquid that normally flows to the lowest nozzle banks to the other 4 nozzle banks reduces the proportion of agglomerates that go from the zone where agglomerates are produced to the stripper in less than 10 s. Further experiments will confirm these results for different baffle configurations for publication at the conference.

Future studies, to be presented at the conference, will investigate the effect of changing the column taper, when adjusting the feed distribution between the various nozzle banks. The impact of baffles with a different column taper will also be studied.

Additional work is ongoing to better understand the phenomena at play through the analysis of tracer trajectories and bubble flow patterns: for example, it seems that the right combination of baffles and injection banks promotes the formation of zones where wet agglomerates that would otherwise move too quickly are captured and held for a while before being released.

CFD-DEM Simulation of the Fluidization Behavior of Particles during Coating in a New Developed Spouted Bed Apparatus

Authors:

Grohn, P., Technische Universität Kaiserslautern

Weis, D., Technische Universität Kaiserslautern

Antonyuk, S., Technische Universität Kaiserslautern

For various products, e.g. in the pharmaceutical, chemical and food industry the coating of solid particles with different substances is an important production step. For this purpose, spouted beds or Wurster coaters are often used (Grace und Epstein 2011; Heinrich et al. 2015). To control and optimize these processes, it is essential to know and describe the particle kinematics and dynamics as well as the growth of the coating layer on the particles. An essential tool to obtain these data are numerical simulations of the multiphase flow with the computational fluid dynamics (CFD) coupled with the discrete element method (DEM). With CFD, the flow field of the gas in the process apparatus can be determined by the Eulerian-Lagrangian approach, which treats the fluid phase as a continuum. In DEM, the interactions are calculated for each single particle based on models describing physical properties of particles and their mechanical behavior during slow, fast and repeated stressing (Antonyuk et al. 2015; Breuninger et al. 2018; Salikov et al. 2015).

In this work, we studied the particle dynamics during a coating process in a self-developed spouted bed apparatus, consisting of a conical and a cylindrical part, which is operating at negative pressure. The apparatus consists of an annular gap through which the gas enters with a high velocity. The base plate can be adjusted by a fine threat in height and thus the gap height can be varied. With the help of a two-fluid nozzle the liquid was injected into the apparatus. As coating substance, water with coloring agent was used. The coating process was calculated during 3D CFD-DEM simulation in order to obtain additional information about the fluidization behavior of the particles and the influence of the injected liquid on the particles.

For the calculation of the particles contacts, a novel contact elastic-plastic model was implemented. In the model, the restitution coefficient depends on the load level and number of collisions due to contact hardening. The drying rate of the water droplets for the process parameters was determined by experimental investigations. For that, droplets were applied on a substrate with a known contact angle and the time to complete evaporation was measured under different humidities, temperatures and gas velocities. The size distribution of the droplets formed by the used nozzle were obtained by static light scattering. Furthermore, the wetting of the particles with the injected droplets was implemented in the simulation and determined. The additional attractive forces due to the existence of liquid layers after droplet deposition were taken into account based on the model of Tsuazawa et al. (Tsunazawa et al. 2016).

The results were used to determine the boundary conditions under which a successful fluidization and coating of particles in the self-developed spouted bed can be realized. With the help of the simulations the particle movement in this spouted bed was analyzed. Another focus was on the behavior of the particles during the coating process as well as the different contact scenarios (particle-particle or particle-wall contact). The numerical results were compared with the conducted experiments.

References

Antonyuk, Sergiy; Heinrich, Stefan; Gurikov, Pavel; Raman, Subrahmanyam; Smirnova, Irina (2015): Influence of coating and wetting on the mechanical behaviour of highly porous cylindrical aerogel particles. In: Powder Technology 285, S. 34â€“43. DOI: 10.1016/j.powtec.2015.05.004.

Breuninger, Paul; Weis, Dominik; Behrendt, Isabell; Grohn, Philipp; Krull, Fabian; Antonyuk, Sergiy (2018): CFDâ€“DEM simulation of fine particles in a spouted bed apparatus with a Wurster tube. In: Particuology. DOI: 10.1016/j.partic.2018.03.015.

Grace, John R.; Epstein, Norman (Hg.) (2011): Spouted and spout-fluid beds. Fundamentals and applications. Cambridge: Cambridge University Press.

Heinrich, Stefan; Dosta, Maksym; Antonyuk, Sergiy (2015): Multiscale Analysis of a Coating Process in a Wurster Fluidized Bed Apparatus. In: Mesoscale Modeling in Chemical Engineering Part I, Bd. 46: Elsevier (Advances in Chemical Engineering), S. 83â€“135.

Salikov, Vitalij; Heinrich, Stefan; Antonyuk, Sergiy; Sutkar, Vinayak S.; Deen, Niels G.; Kuipers, J.A.M. (2015): Investigations on the spouting stability in a prismatic spouted bed and apparatus optimization. In: Advanced Powder Technology 26 (3), S. 718â€“733. DOI: 10.1016/j.apt.2015.02.011.

Tsunazawa, Yuki; Fujihashi, Daiki; Fukui, Sho; Sakai, Mikio; Tokoro, Chiharu (2016): Contact force model including the liquid-bridge force for wet-particle simulation using the discrete element method. In: Advanced Powder Technology 27 (2), S. 652â€“660. DOI: 10.1016/j.apt.2016.02.021.

Simulation on the Cal System with Dual Fluidized Bed Reactor

Authors:

Li, Z., Zhejiang university

Wang, Q., Zhejiang university

Fan, H., Zhejiang university

Fang, M., Zhejiang university

Luo, Z., Zhejiang university

The high-temperature calcium-based carbon capture system uses a dual fluidized bed reactor as a carbonization furnace and a calcination furnace.Flue gas enters the carbonization furnace and reacts with CaO for decarburization treatment, and the generated $CaCO_3$ is sent to a calciner for high-temperature calcination decomposition. High concentration of CO_2,which is condensed and compressedâ€‹storaged and utilized. In this paper, the Aspen plus software is used to optimize internal parameterization simulation of carbon capture system, simulate the carbon capture system integrated with the 600MW coal-fired power plant system, and research carbon capture system waste heat utilization.

For the optimization of the internal parameterization simulation of carbon capture system, including temperature,Ca/C ratio,etc.The results show: Under the condition of the CO_2 capture rate is more than 90%, simulated the optimal temperature is 635â„ƒ,and carbon capture efficiency decreased as carbonization temperature increasing. The Ca/C molar ratio was 3.9, and change of carbon capture efficiency was not obvious with increase of Ca/C molar ratio.

Through the simulation of the carbon capture system integrated with the 600MW unit coal-fired power plant system, four units including coal combustion,carbon capture system, air separation,steam turbine and power generation,etc.The results showï¼šHeat release of carbon capture system integrated with coal-fired power plant system flue gas and CO_2 condensation is 601.203MW,air separation unit consumes energy of 171.64MW.

For waste heat utilization, the waste heat of the flue gas is used to heat cold air entering the coal-fired boiler and boiler condensed water for steam turbine power generation, and heat release of CO_2 condensation is all used to heat the $CaCO_3$ into calciner. The results show:the net efficiency of the power plant increases with the proportion of waste heat used to heat the cold gas, and CO_2 condensation can reduce the energy of air separation unit by 195.5 MW.

The significance of the research in this paper is to carry out the carbon capture system and integrated simulation between the systems, provide data help for the actual application of the system, supply a good theoretical foundation, save the time consumed in actual application of the system, and reduce possibility of system error.Subsequent research includes optimization of integration between systems optimization of utilization of waste heat and measures to improve system efficiency.

Experimentations and Simulations in a 4m Height Heated Ubfb Solar Receiver

Authors:

Ansart, R., Université de Toulouse, CNRS-Toulouse

Sabatier, F., Université de Toulouse, CNRS-Toulouse

Zhang, H., School of Life Science and Technology, Beijing University of Chemical Technology

Kong, W., School of Life Science and Technology, Beijing University of Chemical Technology

Baeyens, J., European Powder and Process Technology

Flamant, G., CNRS

Simonin, O., Université de Toulouse, CNRS-Toulouse

Concentrated Solar Power (CSP) plant efficiency is limited by the heat transfer fluid used and especially its operating range of temperature. Indeed, increasing the maximum operating temperature would allow to increase the efficiency of the power conversion cycle deployed.

A new alternative to conventional heat transfer fluids (HTF) is solid particles suspension. It features a good thermal capacity and allows for low-cost heat storage without the usual disadvantages of currently used HTF (operating temperature range, safety issues, storage, corrosion and maintenance costs [1]).

In order to reach heat transfer coefficient from the solar heated surfaces to the HTF in the range of existing industrial receivers, working with conventional fluids, an innovative receiver conception is to use an Upflow Bubbling Fluidized Bed (UBFB) [1][2].The Next-CSP (http://next-csp.eu) European project aims at improving the reliability and performance of CSP plants with this new technology, by demonstrating the concept feasibility with a $4MW_{th}$ working pilot, and upscale it to the industrial scale by designing a $150MW_{el}$ plant.

The critical design specification for the UBFB receiver is to maximize the heat transfer between the solar heated surfaces and the HTF, which is directly linked to the gas-solid hydrodynamics. In addition, the UBFB technology implies the use of Geldart group A particles [3].

Heat transfer coefficient was measured up to 400-1100 W/m^2K during on-sun experiments with small 1m height receiver tubes [4,5] while industrial scale receiver tubes could reach up to 4m height.

However, hydrodynamics in the UBFB strongly depends on the tube height [6]. Indeed, during the fluidization of Geldart group A particles within a tall small internal diameter tube, it has been established that the fluidized bed can be separated into three zones. In the bottom, the bubbles are freely bubbling and starts to coalesce. Then bubbles trend to go large enough to form wall slugs. Finally, wall slugs enlarge to become axi-symetrical slugs[7], which are known to strongly reduce the particle mixing and thus the heat transfer coefficient. In addition, operating temperature strongly impact on the starting height of slugging: operating at higher temperatures trend to shift slugging area higher, and reduce the wall slugging zone.

In order to optimize the design of an industrial UBFB solar receiver, slugging mechanism and especially the impact of temperature should be investigated and understood.

Experimental measurements of fluidization of Geldart group A powder (cristobalite of d_{sv}= 55µm) in a 3.5m height, 5cm internal diameter, UBFB receiver tube were conducted. The UBFB was maintained by external heating elements at a steady state temperature of 20°C, 475°C and 600°C. Then three-dimensional numerical simulations were carried out using the unstructured parallelized multiphase flow code, NEPTUNE_CFD, based on a Eulerian n-fluid modelling approach [8], including the solving of the coupling between hydrodynamics and heat transfer, imposed by a thermal flux at the wall. The convective/diffusive heat transfer between the gas and particles, and the particle-particle radiative transfer was accounted for [9]. Significant cells size had to be used to ensure mesh independent solution [10].

Numerical simulations were compared to experimental measurements of bed voidage, slugs size and frequency. Emphasis focused on the impact of high temperature on these parameters and numerical simulations allow to study for the trajectory and particles repartition within the receiver. Heat transfer between the tube walls and the particles will be investigated especially through wall boundaries conditions.

Acknowledgements

This project has received funding from the European Unionâ€™s Horizon 2020 research and innovation program under grant agreement No 727762, Next-CSP project and supported by the Beijing Advanced Innovation Center for Soft Matter Science and Engineering of the Beijing University of Chemical Technology. This work was granted access to the HPC resources of CALMIP under the allocation P1132 and CINES under the allocation gct6938 made by GENCI.

REFERENCES

[1] Benoit, H., Spreafico, L., Gauthier, D. & Flamant, G., 2016. Review of heat transfer fluids in tube receivers used in concentrating solar thermal systems: Properties and heat transfer coefficients. Renew. Sustain. Energy Rev. 55, 298â€“315.

[2] Zhang, H., Benoit, H., Gauthier, D., DegrÃ¨ve, J., Baeyens, J., PÃ©rez, LÃ³pez.I., Hemati, M. & Flamant, G., 2016. Particle circulation loops in solar energy capture and storage; gas-solid flow and heat transfer considerations. Appl. Energy 161, 206â€“224.

[3] Boissiere, B., Ansart, R., Gauthier, D., Flamant, G. & Hemati, M., 2015. Experimental hydrodynamic study of gas-particle dense suspension upward flow for application as new heat transfer and storage fluid, Can. J. Chem. Eng. 93, 317â€“330.

[4] Benoit, H., Lopez, P. I., Gauthier, D., Sans, J. L., & Flamant, G., 2015. On-sun demonstration of a 750Â°C heat transfer fluid for concentrating solar systems: Dense particle suspension in tube. Solar Energy, 118, 622â€"633.

[5] Flamant, G., Gauthier, D., Benoit, H., Sans, J.L., BoissiÃ¨re, B., Ansart, R. & M. Hemati, 2013. A new heat transfer fluid for concentrating solar systems : Particle flow in tube. Energy Procedia, Volume 49, 616-626.

[6] Sabatier, F., Ansart, R., Zhang, H., Baeyens, J. & Simonin, O., 2018. Experiments support simulations by the NEPTUNE_CFD code in an Upflow Bubbling Fluidized Bed reactor. Chemical Engineering Journal, submitted.

[7] Kong, W. B., Tan, T. W., Baeyens, J., Flamant, G. & Zhang, H. L., 2017. Bubbling and Slugging of Geldart Group A Powders in Small Diameter Columns. Ind. Eng. Chem. Res. 56, 4136â€"4144.

[8] Hamidouche, Z., Masi, E., Fede, P., Ansart, R., Neau, H., Hemati, M. & Simonin, O., 2018. Numerical Simulation of Multiphase Reactive Flows, in: Advances in Chemical Engineering, volume 52, 51â€"124.

[9] Benoit, H., Ansart, R., Neau, H. , Garcia TriÃ±anes, P., Flamant, G. & Simonin, O.,2018. 3D numerical simulation of upflow bubbling fluidized bed in opaque tube under high flux solar heating. AIChE Journal, Volume 64, No 11.

[10] Parmentier, J., Simonin, O. & Delsart, O., 2012. A functional subgrid drift velocity model for filtered drag prediction in dense fluidized bed, AIChE J. 58, 1084â€"1098.

Minimum Spouting Velocity for Fountain Confined Conical Spouted Beds of Fine Particles

Authors:

Tellabide, M., University of the Basque Country

Estiati, I., University of the Basque Country

Pablos, A. Sr., University of the Basque Country

Altzibar, H. Sr., University of the Basque Country

Aguado, R. Sr., University of the Basque Country

Olazar, M. Sr., University of the Basque Country

MINIMUM SPOUTING VELOCITY FOR FOUNTAIN CONFINED CONICAL SPOUTED BEDS OF FINE PARTICLES

Mikel Tellabide*, Idoia Estiati, Aitor Pablos, Haritz Altzibar, Roberto Aguado and Martin Olazar
University of the Basque Country; Dept. Chemical Engineering
PO Box 644, 48080 Bilbao, Spain
*T: +34-94-601-5994; F: 94-601-3500; E:mikel.tellabide@ehu.es

Nowadays, fluidized beds are commonly used in industrial operations when fixed are not suitable. Nevertheless, these contact methods perform poorly when particle size exceeds 1 mm. Mathur and Gishler (1) developed the spouted bed regime and the main difference with the fluidized bed lies in the cyclic movement of the particles. This regime has been successfully applied in operations with coarse particles, those of irregular texture, wide size distribution and sticky ones.

Despite the great versatility of this contact method, it has severe problems when scaling up, specifically those involving a key parameter, as is the ratio between the inlet diameter and particle diameter (2). In fact, the inlet diameter should be no more than 20-30 times larger than the average particle diameter in order to achieve stable spouting. Thus, values of this ratio greater than 30 lead to unstable regime due mainly to slugging formation. Therefore, a draft tube must be used to avoid this problem, and therefore attain stable operation. Nevertheless, operation with fine particles in a draft tube spouted bed leads to severe entrainment, which is a serious problem in most physical and chemical processes conducted in spouted beds.

A novel internal device called fountain confiner has been designed to allow operation with fine particles. In fact, this device avoids even the need for a draft tube and improves greatly solid circulation and gas-solid contact (3). This device confines the fountain, stabilizes the system and changes the trajectory of the gas leading to an additional contact between the gas and the solid in the fountain.

The main aim of this work is to study the hydrodynamics of the fountain confined spouted bed and propose a correlation for predicting the minimum spouting velocity in a wide range of operating conditions and particle properties. Runs have been carried out following an experimental design in order to allow a reliable data analysis. Once the factors of greater influence have been identified, correlations have been developed for the calculation of the minimum spouting velocity for fountain confined conical spouted beds with and without different types of draft tubes and particles of different size and density.

The results show that the most significant factors are static bed height, gas inlet diameter and particle properties. Minimum spouting velocity increases as the height of the static bed and particle size and density are increased and gas inlet diameter is decreased.

REFERENCES

1. B. Mathur, P.E. Gishler, A technique for contacting gases with coarse solid particles, AICHE J. 1 (1955) 157â€"164.
2. Olazar, M.J. San Jose, A.T. Aguayo, J.M. Arandes, J. Bilbao, Stable operation conditions for gas-solid contact regimes in conical spouted beds, Ind. Eng. Chem. Res. 31, (1992) 1784â€"1792.
3. Pablos, R. Aguado, M. Tellabide, H. Altzibar, F.B. Freire, J. Bilbao, M. Olazar, A new fountain confinement device for fluidizing fine and ultrafine sands in a conical spouted beds, Powder Technol. 328, (2018) 38-46.

Fully Resolved Simulation of Char Particle Combustion By Immersed Boundary-Lattice Boltzmann Method

Authors:

Jiang, M., Huazhong University of Science and Technology

Liu, Z., Huazhong University of Science and Technology

Yu, J., Huazhong University of Science and Technology

There are many challenges for fully resolved simulation of char particle combustion as it couples many physical and chemical process, such as the momentum, heat and mass transfer in inter-phases and the existed heterogeneous and homogeneous reactions. Limited relevant reports can be found to date, especially for the suspended particles. Immersed boundary-lattice Boltzmann method (IB-LBM) has been widely used for the simulation of cold or hot particle-laden flows. A new improved IB-LBM is developed for fully resolved simulation of char particle combustion. The fluid reactive flow is solved by LBM with multi-distribution function scheme. The boundary Dirichlet and Neumann conditions at the particle surface, such as velocity, temperature, and chemical species concentration, are well enforced by the improved immersed boundary method. The combustion reaction in the domain is described by two semi-global heterogeneous reactions and one homogeneous reaction. in addition, MPI parallel computing strategy is adopted to perform the accelerated and large-scale calculation. A benchmark validation is implemented to find a satisfactory agreement between the present simulation results and experimental data in previous literatures. The method is then used to investigate the combustion property of a single char particle. The effects of the temperature of coming flow, Reynolds number, oxygen concentration and environment of combustion (air and oxygen) on burning rate are detailed analyzed. Furthermore, the combustion of particle group with different number of char particles are simulated. And the effect of flow field disturbance on each chemical component as well as the effect of char particles number and spatial distribution on burning rate are discussed.

Modeling of Oxy Gasification of High Ash Coal in Fluidized Bed Gasifier

Authors:

Roy, S., Indian Institute of Technology Delhi

Roy, S., Indian Institute of Technology Delhi

Syngas derived from coal gasification has been a principal source for power and several chemicals [1]. Among various prevalent technologies for producing syngas from coal, fluidized bed gasification has been pioneered for applications involving high ash (20-40%). Based on the requirement of the syngas, the gasifying agent can be either oxygen or air. In applications such as power generation, the primary gasifying agent is a mixture of steam and air. Whereas, when the demand is for production of chemicals of high purity, such as methanol (and its derivatives like DME (dimethyl ether)), pure oxygen along with steam is employed for gasification. Even though the subsequent technology for conversion of syngas to methanol has been commercially developed worldwide, the real bottleneck lies in reliable gasification of high ash coal such that it ensures cheaper gas cleanup and its conversion to methanol without any detrimental effect on catalyst and reactor downstream. It has been observed that fluidized beds operated with high ash coal predominately suffer from low carbon conversion with intermittent bed agglomeration. In order to address such issues, it is thus required to develop an efficient yet simple modeling framework which should estimate the quality of syngas produced (specifically CO/H_2) along with non-isothermal bed behavior based on varying operating conditions.

With this context, the present study thus envisages on developing a mathematical model to predict the exit syngas composition, temperature distribution, and carbon conversion. The model considers eight significant reactions: four each of homogeneous and heterogeneous. To incorporate the effect of bed hydrodynamics, a two-phase description of fluidized beds is considered, wherein it is assumed that the coal (heterogeneous) reactions take place in the emulsion phase, whereas, the gaseous (homogenous) phase reactions occur in both the emulsion as well as the bubble phase [2,3]. It was further assumed that the devolatilization of the incoming fresh coal feed was instantaneous as compared to the actual char gasification reactions. Pure oxygen was used a gasifying medium with molar ratio with steam being maintained at 2. When the bed was run at atmospheric pressure and 900 °C, preliminary results indicate that with an increase in the height the molar concentrations of CO and H_2 increase in both bubbles as well as the emulsion phase. Once such an analysis is performed in a non-isothermal bed, it is further expected to yield temperature distributions of coal particles which can be indicative of their tendency to agglomerate. A detailed analysis of the model with the influence of pressure and inlet oxygen concentration will be discussed in the final presentation.

References

[1] L. Chen, S.Z. Yong, A.F. Ghoniem, Oxy-fuel combustion of pulverized coal: Characterization, fundamentals, stabilization and CFD modeling, Progress in Energy and Combustion Science. 38 (2012) 156â€"214. doi:10.1016/J.PECS.2011.09.003.

[2] M.L. de Souza-Santos, Comprehensive modelling and simulation of fluidized bed boilers and gasifiers, Fuel. 68 (1989) 1507â€"1521. doi:10.1016/0016-2361(89)90288-3.

[3] F. Chejne, E. Lopera, C.A. LondoÃ±o, Modelling and simulation of a coal gasification process in pressurized fluidized bed, Fuel. (2011). doi:10.1016/j.fuel.2010.06.042.

AUTHOR INDEX